Aircraft Noise

Aircraft Noise

Editors

Lothar Bertsch
Adrian Sescu

MDPI • Basel • Beijing • Wuhan • Barcelona • Belgrade • Manchester • Tokyo • Cluj • Tianjin

Editors

Lothar Bertsch
Aerodynamics and Flow Technology
German Aerospace Center (DLR)
Göttingen
Germany

Adrian Sescu
Aerospace Engineering
Mississippi State University
Jackson
United States

Editorial Office
MDPI
St. Alban-Anlage 66
4052 Basel, Switzerland

This is a reprint of articles from the Special Issue published online in the open access journal *Aerospace* (ISSN 2226-4310) (available at: www.mdpi.com/journal/aerospace/special_issues/aircraft_noise).

For citation purposes, cite each article independently as indicated on the article page online and as indicated below:

LastName, A.A.; LastName, B.B.; LastName, C.C. Article Title. *Journal Name* **Year**, *Volume Number*, Page Range.

ISBN 978-3-0365-4898-2 (Hbk)
ISBN 978-3-0365-4897-5 (PDF)

Cover image courtesy of Lothar Bertsch

Contents

About the Editors . . . . **vii**

Preface to "Aircraft Noise" . . . . **ix**

Jonas Meister, Stefan Schalcher, Jean-Marc Wunderli, David Jäger, Christoph Zellmann and Beat Schäffer
Comparison of the Aircraft Noise Calculation Programs sonAIR, FLULA2 and AEDT with Noise Measurements of Single Flights
Reprinted from: *Aerospace* **2021**, *8*, 388, doi:10.3390/aerospace8120388 . . . . **1**

Chunhui Wu and Stephane Redonnet
Prediction of Aircraft Noise Impact with Application to Hong Kong International Airport
Reprinted from: *Aerospace* **2021**, *8*, 264, doi:10.3390/aerospace8090264 . . . . **17**

Michel Nöding and Lothar Bertsch
Application of Noise Certification Regulations within Conceptual Aircraft Design
Reprinted from: *Aerospace* **2021**, *8*, 210, doi:10.3390/aerospace8080210 . . . . **57**

Umberto Iemma and Francesco Centracchio
Sound-Quality-Based Decision Making in Multiobjective Optimisation of Operations for Sustainable Airport Scenarios
Reprinted from: *Aerospace* **2022**, *9*, 310, doi:10.3390/aerospace9060310 . . . . **81**

Vincent Domogalla, Lothar Bertsch, Martin Plohr, Eike Stumpf and Zoltán S. Spakovszky
Low-Noise Design of Medium-Range Aircraft for Energy Efficient Aviation
Reprinted from: *Aerospace* **2021**, *9*, 3, doi:10.3390/aerospace9010003 . . . . **103**

Michel Nöding, Martin Schuermann, Lothar Bertsch, Marc Koch, Martin Plohr and Robert Jaron et al.
Simulation of Landing and Take-Off Noise for Supersonic Transport Aircraft at a Conceptual Design Fidelity Level
Reprinted from: *Aerospace* **2021**, *9*, 9, doi:10.3390/aerospace9010009 . . . . **133**

Junichi Akatsuka and Tatsuya Ishii
System Noise Assessment and Uncertainty Analysis of a Conceptual Supersonic Aircraft
Reprinted from: *Aerospace* **2022**, *9*, 212, doi:10.3390/aerospace9040212 . . . . **157**

Victor F. Kopiev, Ivan V. Belyaev, Andrey I. Dunaevsky, Andrey A. Poukhov and Igor L. Trofimovsky
On the Fundamental Possibility of a Supersonic Civil Aircraft to Comply with ICAO Noise Requirements Using Existing Technologies
Reprinted from: *Aerospace* **2022**, *9*, 187, doi:10.3390/aerospace9040187 . . . . **173**

Grazia Piccirillo, Nicole Viola, Roberta Fusaro and Luigi Federico
Guidelines for the LTO Noise Assessment of Future Civil Supersonic Aircraft in Conceptual Design
Reprinted from: *Aerospace* **2022**, *9*, 27, doi:10.3390/aerospace9010027 . . . . **187**

Christopher Teruna, Leandro Rego, Damiano Casalino, Daniele Ragni and Francesco Avallone
A Numerical Study on Aircraft Noise Mitigation Using Porous Stator Concepts
Reprinted from: *Aerospace* **2022**, *9*, 70, doi:10.3390/aerospace9020070 . . . . **213**

Joshua D. Blake, Adrian Sescu, David Thompson and Yuji Hattori
A Coupled LES-Synthetic Turbulence Method for Jet Noise Prediction
Reprinted from: *Aerospace* **2022**, *9*, 171, doi:10.3390/aerospace9030171 . . . . **241**

Colby Niles Horner, Adrian Sescu, Mohammed Afsar and Eric Collins
Numerical Investigation of a Rectangular Jet Exhausting over a Flat Plate with Periodic Surface Deformations at the Trailing Edge
Reprinted from: *Aerospace* **2021**, *8*, 314, doi:10.3390/aerospace8110314 . . . . **265**

About the Editors

Lothar Bertsch

Dr. Lothar Bertsch has worked for DLR since 2006 and leads two small teams of Post-Docs, PhD, and graduate students. He has 10+ years of experience in overall aircraft noise simulation and low-noise aircraft design and initiated the "Aircraft Noise Simulation Working Group"(ANSWr) between DLR, ONERA, and NASA in 2012. Furthermore, he organized several international workshops on aircraft noise and is the initiator and lead editor of a dedicated {CEAS Aeronautical Journal} Special Issue (2019). Since 2014, he has acted as a Visiting Assistant Professor at TU Delft's "Aircraft Noise and Climate Effects"section of Prof. Simons. At TU Braunschweig, he lectured for 5 years and from 2020 he has been a guest lecturer at RWTH Aachen in the field of aircraft noise.

Adrian Sescu

Adrian Sescu is currently an Airbus Helicopters Endowed Associate Professor and Graduate Coordinator in the Department of Aerospace Engineering, Mississippi State University (MSU), and affiliated with the High-Performance Computing Collaboratory at MSU. He holds a Ph.D. degree in Mechanical Engineering from the University of Toledo and MS and BS degrees in Aerospace Engineering from the Polytechnic University of Bucharest. Previously, he was a postdoctoral fellow at Johns Hopkins University, where he conducted research on turbulence and environmental flow. Dr. Sescu's research interests cover theoretical and computational aeroacoustics, transition in boundary layers, and turbulent flows, with over 100 journal articles and conference proceedings in these areas. He received the ASEE SFFP Award in 2013, 2014 and 2016, and spent three summers at AFRL/RQVA where he studied boundary layer receptivity. He has been a PI and co-PI on several projects funded by NASA, AFOSR, US Army, and FAA. Dr. Sescu is an Associate Fellow of the American Institute of Aeronautics and Astronautics and a member of ASME, APS, and Sigma Gamma Tau (Aerospace Engineering Honor Society).

Preface to "Aircraft Noise"

This reprint is about aircraft noise. Multidisciplinary topics are covered. One focus is supersonic transport aircraft and their noise generation during approach and take-off. The articles within this study present recent and ongoing research activities from research entities and universities from around the world. This book is beneficial for engineers and researchers that are working the field of aircraft noise.

Lothar Bertsch and Adrian Sescu
Editors

aerospace MDPI

Article

Comparison of the Aircraft Noise Calculation Programs sonAIR, FLULA2 and AEDT with Noise Measurements of Single Flights

Jonas Meister *, Stefan Schalcher, Jean-Marc Wunderli, David Jäger, Christoph Zellmann † and Beat Schäffer

Laboratory for Acoustics/Noise Control, Empa, Swiss Federal Laboratories for Materials Science and Technology, 8600 Dübendorf, Switzerland; stefan.schalcher@empa.ch (S.S.); jean-marc.wunderli@empa.ch (J.-M.W.); david_jaeger@gmx.ch (D.J.); christoph.zellmann@n-sphere.ch (C.Z.); beat.schaeffer@empa.ch (B.S.)

* Correspondence: jonas.meister@empa.ch; Tel.: +41-58-765-42-46

† Present address: n-Sphere AG, 8045 Zurich, Switzerland.

Abstract: As aircraft noise affects large areas around airports, noise exposure calculations need to be highly accurate. In this study, we compare noise exposure measurements with calculations of several thousand single flights at Zurich and Geneva airports, Switzerland, of three aircraft noise calculation programs: sonAIR, a next-generation aircraft noise calculation program, and the two current best-practice programs FLULA2 and AEDT. For one part of the flights, we had access to flight data recorder (FDR) data, which contain flight configuration information that sonAIR can account for. For the other part, only radar data without flight configuration information were available. Overall, all three programs show good results, with mean differences between calculations and measurements smaller than ±0.5 dB in the close range of the airports. sonAIR performs clearly better than the two best-practice programs if FDR data are available. However, in situations without FDR data (reduced set of input data), sonAIR cannot exploit its full potential and performs similarly well as FLULA2 and AEDT. In conclusion, all three programs are well suited to determine averaged noise metrics resulting from complex scenarios consisting of many flights (e.g., yearly air operations), while sonAIR is additionally capable to highly accurately reproduce single flights in greater detail.

Keywords: aircraft noise; simulation; sonAIR; FLULA2; AEDT; validation with measurements

Citation: Meister, J.; Schalcher, S.; Wunderli, J.-M.; Jäger, D.; Zellmann, C.; Schäffer, B. Comparison of the Aircraft Noise Calculation Programs sonAIR, FLULA2 and AEDT with Noise Measurements of Single Flights. *Aerospace* **2021**, *8*, 388. https://doi.org/10.3390/aerospace8120388

Academic Editors: Lothar Bertsch and Adrian Sescu

Received: 5 November 2021
Accepted: 8 December 2021
Published: 10 December 2021

1. Introduction

Aircraft noise usually affects large areas around airports. In populated regions, a large number of people may be exposed [1,2], entailing, among others, restrictions for land-use planning. Besides economic consequences, it can also cause various health effects [3,4]. In 2017, around 3.2 million people were estimated to be highly annoyed by aircraft noise and 1.4 million suffered from sleep disturbance around the 47 major European airports. In addition, more than 1 million people were exposed to more than 50 aircraft noise events per day exceeding 70 dB in 2017—this is 60% more than in 2005 [5].

Since aircraft noise has a large-scale impact, noise exposure cannot be assessed by measurements alone, but has to be established using model calculations. Due to their impact on land-use planning, aircraft noise calculations need to be highly accurate. Calculation uncertainties strongly depend on the modelling approach, model sophistication, traffic input data, available sound source data and airport peculiarities such as specific aircraft fleet or different flight procedures [6–8]. Systematic comparisons of simulations with independent measurements as a validation step are therefore essential.

In the past, model validations were conducted for different models such as ANCON in the United Kingdom [9], and FLULA2 [7,10] and sonAIR [11] in Switzerland. Another study compared calculations of the maximum A-weighted sound pressure level ($L_{A,max}$) with three different scientific simulation programs (PANAM, ANOPP, CARMEN) by means of benchmark tests with a reference and a low-noise vehicle [12]. Literature, however, does not provide any information on larger validation campaigns for AEDT. Previous research has focused on specific aspects, such as the accuracy of the AEDT noise input database, the noise power distance (NPD) curves and their potential for improvement [13]. Studies which compare different noise models together and against measurements are not available.

The aim of this study was therefore to conduct a comparison and systematic validation of three different aircraft noise calculation programs, namely sonAIR, FLULA2 and AEDT, using a large measurement data set. The three programs are based on inherently different modelling approaches. sonAIR [14,15], with its detailed spectral noise emission and propagation models, was recently developed to accurately reproduce single flights as well as (new) noise abatement flight procedures [16]. FLULA2 [17] is an empirical time-step model and an official tool to calculate yearly aircraft noise exposure in Switzerland. AEDT [18] is a Doc 9911 [19] compliant and Doc 29 [20] equivalent segmentation-based model. Both, FLULA2 and AEDT, are best practice programs, primarily developed to calculate complex scenarios such as yearly air traffic operations, which include various aircraft types and large numbers of flights. For the validation, we simulated 8785 single flights around the airports of Zurich and Geneva, Switzerland, and compared the results with corresponding measurements.

2. Materials and Methods

2.1. Concept Overview

We simulated approaches and departures around the airports Zurich (ZRH) and Geneva (GVA) with all three aircraft noise calculation programs and compared them with measured data at the corresponding microphone locations (cf. Section 2.3).

The acoustic quantity used for the comparisons is the A-weighted sound exposure level L_{AE} (total amount of sound energy of an event (single aircraft flyby), related to the duration of 1 s, also referred to as "SEL"). Specifically, for the measurements, the $L_{AE,tg}$ is used, with tg being the time period of an event with the instantaneous sound pressure level above a measurement location specific threshold (ZRH) or dynamic threshold (GVA). This quantity is used to have a sufficiently large signal-to-noise ratio to exclude measurement location specific or varying background noise. Further, the $L_{AE,t10}$ (total amount of sound energy of an event over the duration of t10) is used for the simulations with FLULA2 and sonAIR, where t10 is the 10 dB-down-time, i.e., the time period during which the sound pressure level is less than 10 dB below the $L_{A,max}$. The $L_{AE,10}$ is very similar to the $L_{AE,tg}$ and was thus used for the best comparability with the $L_{AE,tg}$ of the measurements. Finally, as AEDT does not yield the $L_{AE,t10}$ as output, we compare the L_{AE} obtained with AEDT. This is feasible since the sound exposure levels in the AEDT database (NPD curves) are based on 10 dB-down-time values (p. 28 in [21]). In addition, the difference between $L_{AE,t10}$ and L_{AE} is usually small (in the range of some 0.1 dB). Thus, all three quantities, the measured $L_{AE,tg}$, the calculated $L_{AE,t10}$ (FLULA2, sonAIR) and L_{AE} (AEDT) are highly comparable. We only considered large civil aircraft types, since they represent the majority of movements on public airports and dominate the aircraft noise exposure in these areas. The simulated aircraft types in the present study cover the most frequently operating commercial aircraft in Switzerland.

For 2126 of the overall 8785 simulated flights, Swiss International Air Lines provided us with flight data recorder (FDR) data. Besides time and position information, FDR data include Mach number, air density and N1 (rotational speed of the low-pressure compressor), which are mandatory input data for sonAIR. sonAIR is further able to include the airplane configuration for some aircraft types [15], in which case, the most detailed available sonAIR emission models can be applied. For flights where no FDR data were available, we used radar data, which include only time and position information. As N1 is a mandatory input parameter for sonAIR, N1 was in this case estimated based on the methodology of Schwab et al. [22]. Further, Mach number and air density, which are also mandatory input parameters, were deduced from local meteorological profiles in this particular case. However, the airplane configuration cannot be estimated from radar data. Therefore, these calculations were performed with a set of reduced sonAIR emission models [15] that do not require information on aircraft configuration.

We simulated three flight datasets separately. Flights with measurement data recorded close to the airport, once with and once without FDR data, and flights with measurement data recorded at greater distance to Zurich airport, for which FDR data were available. Table 1 lists the aircraft types (ICAO designation) used for simulation and distinguishes whether full or reduced models, i.e., with or without FDR data as input, were used for calculations with sonAIR. Note that several aircraft types are listed in both categories, as flights with and without FDR data were measured and analyzed for the corresponding aircraft types. In contrast to sonAIR, FLULA2 and AEDT do not make use of the additional information of FDR data and use the same modelling approaches for both, flights with and without FDR data available.

Table 1. Simulated aircraft types with and without FDR data available. For several aircraft types, both cases occurred.

Simulation Models	Aircraft Types (ICAO Designation)
Aircraft Types with FDR Data Available (ICAO Designation):	A319, A320, A321, A333, A343, B77W, BCS1, BCS3
Aircraft Types without FDR Data Available (ICAO Designation):	A319, A320, A321, A333, A343, A388, B733, B734, B735, B736, B737, B738, B739, B762, B763, B764, CRJ7, CRJ9, CRJX, E170, E190, E195, F100, FA7X

Note that the calculations with sonAIR were taken from [11], but had to be re-analyzed for the current purpose. The calculations with FLULA2 and AEDT were done specifically for this study.

2.2. Description of the Aircraft Noise Programs

The following subsections give an overview of the most important characteristics of the three noise calculation programs. Table 2 provides an overview of the model properties and their settings within the calculations of the current study.

Table 2. Properties of the three programs for the calculations in the scope of this paper.

Model Properties	sonAIR	FLULA2	AEDT
Model approach	Time step model	Time step model	Segmented model
Noise calculation method	Semi-empirical source model and separate sound propagation model	Combined sound emission and propagation in an empirical sound emission pattern	Combined sound emission and propagation using noise power distance (NPD) curves
Spectral calculation	Yes	No (A-weighted level)	No (A-weighted level)
Source model	3D (aircraft type specific), separate engine and airframe model	2D (longitudinal directivity)	3D (2D + engine installation effects)
Bank Angle considered	Yes	No (as 2D source model)	Yes
Thrust setting considered	Yes (with N1)	No, but separate emission models for departure with max and flex power and for approaches	Yes (calculated thrust)
Aircraft weight for thrust estimation	Implicitly in N1 estimation if no FDR data available, not used if FDR data available	D: different sound emission patterns (max and flex power) depending on weight	D: Depending on flight distance A: Max. landing weight
Departure Cut-Back	Yes (with N1)	Yes (sound level reduction)	Yes (with thrust)
Aircraft configuration	Yes (if FDR data available)	No	No
Approach reversed thrust	No	No	Yes

Table 2. *Cont.*

Model Properties	sonAIR	FLULA2	AEDT
Aircraft trajectory	Real 3D trajectory	Real 3D trajectory	2D (ground track) with standard profiles or 3D with fixed point profiles
Terrain	Yes	Yes	Yes
Buildings	Yes	No	No
Ground properties	Yes	No, soft ground	No, soft ground
Weather	Vertical profiles of pressure, temperature, humidity and wind speed	ISA atmosphere	ISA atmosphere or high fidelity weather data or average airport weather
Atmospheric attenuation	ISO 9613-1	ISO 9613-1	SAE-ARP-5534 or SAE-ARP-866A or SAE-AIR-1845

2.2.1. sonAIR

Details on sonAIR are given in [14,15]. In short, sonAIR is a spectral time step noise calculation program with three-dimensional sound emission, developed to accurately reproduce the sound exposure of single flights and/or noise abatement flight procedures [14]. The emission model describes airframe and engine noise separately through multiple regression models, derived from large measurement campaigns [15]. Separately calculated sound propagation accounts for geometric divergence, atmospheric absorption, shielding effects, ground reflection of uneven terrain and varying surface properties, foliage attenuation and meteorological effects [14]. If desired, also buildings can be taken into account to calculate corresponding barrier effects as well as reflections [23].

For emission calculations, sonAIR uses N1 as main parameter of the engine noise calculation. If FDR data are available, sonAIR is able to further account for airplane configuration, namely flaps, gears and speed brakes settings for certain aircraft types (cf. Table 1). The corresponding emission models are referred to as "3D" models [15]. If no FDR data are provided, sonAIR uses reduced emission models, which do not account for airplane configuration [15]. These models are referred to as "3Dred" ("red" for reduced) models. The sonAIR aircraft noise calculation program has been successfully validated in a rigorous validation exercise with roughly 20000 single flights around ZRH and GVA [11]. The simulations within the present study were conducted with the scientific MATLAB tool version 7 of sonAIR, using the sonAIR emission model version 1.1-0013 based on measurement campaigns in 2013, 2014, 2016 and 2017 at ZRH.

2.2.2. FLULA2

Details on FLULA2 are given in [24,25]. In short, FLULA2 is, like sonAIR, a time step noise calculation program. It uses two-dimensional sound emission patterns, rotationally symmetric with respect to the roll axis. The sound emission and propagation are combined in a fully empirical model description [25]. FLULA2 was developed for calculations of yearly air traffic scenarios and is one of three official models in Switzerland for aircraft noise calculations [26].

FLULA2 yields the A-weighted sound exposure level of single flights, accounting for real three-dimensional flight trajectories, as well as speed and terrain data. An additional term accounts for lateral attenuation, but no engine installation effects are considered. Emission models are available for approaches and departures with either maximum or reduced (flex) power [24]. For the current calculations, FLULA2 version 004 was used with the emission database RC_2019, based on measurement campaigns in 1996, 2013, 2014, 2016 and 2017 at ZRH. Note that the FLULA2 sound source data for the aircraft types examined in the present study were fully derived from sound source data of sonAIR, based on the latter four campaigns (internal Empa report No. 5214.019893).

2.2.3. AEDT

Details on AEDT are given in [18]. In short, AEDT is a software system to calculate fuel consumption, gaseous emissions and noise related to aircraft operations. The flight trajectories are divided in flight segments [27]. AEDT uses noise power distance (NPD) curves as noise calculation database, which contain A-weighted noise values depending on power setting and slant distances to the receiver for different aircraft and operational modes. The noise emission and propagation models are fully empirical and combined within the NPD curves. Additional terms account for lateral attenuation and engine installation effects. AEDT (previously Integrated Noise Model, INM) was developed for calculations of yearly air operations. Amongst other things, it is used by the U.S. government for regulatory studies, research and domestic as well as international aviation environmental policy analyses [15] and is fully compliant to ICAO Doc 9911 [19]. We used the software version AEDT 3c within this paper, with the corresponding underlying NPD database and airport specific default procedural profiles.

The aim of this study was to investigate how well AEDT performs with default settings, and therefore (unlike for FLULA2 and sonAIR), we did not make any adjustments to meet local conditions. This means that NPD data, (procedural) flight profiles and weather data are not adapted to conditions around ZRH and GVA. Instead, standard procedural profiles of ZRH and GVA, standard weather conditions (ISA atmosphere) and unadjusted NPD data were used. While adjusted NPD data and flight profiles are likely to improve results, adjustment of the weather data should only have minor effects on the noise calculations (for sonAIR, differences in calculations with standard atmospheric and local atmospheric conditions were found to be less than 0.1 dB [11]).

Simulation process and settings in AEDT for this study: The AEDT simulation process is not designed for calculating several thousand individual flights with one command, if the calculation results of every flight have to be evaluated separately. Therefore, a process workaround was done for the calculations in the present study. AEDT allows importing simulation specific input data for a single flight via AEDT standard input files (ASIF). The corresponding data are stored in an SQL server and can be manipulated via SQL commands. This allows defining all necessary calculation settings. The automation of input via ASIF and SQL commands, execution and output (L_{AE}) of the calculation of each single flight is realized by calling command line files.

2.3. Acoustic Measurements

Three existing acoustic measurement datasets are used for this study. One data set was obtained during a measurement campaign carried out in autumn of 2017, in the far range of up to 53 km from touchdown at ZRH, by the authors' institution, Empa [11]. The other two data sets were obtained from stationary noise monitoring terminals (NMT) in the close range to ZRH and GVA [11]. The measurement campaigns are described in [11]. For readers' convenience, we give a summary of the campaigns below.

Figures 1 and 2 show the NMT network in the close range to ZRH and GVA. The NMT network of ZRH consists of 14 terminals which are distributed around the airport to cover the major departure and approach routes. The NMT network of GVA consists of 4 terminals. The black dots in Figures 1 and 2 represent the locations of the microphones where the aircraft noise was measured. The black circles around each measurement location represent spatial gates, which the flight trajectories have to penetrate to be considered in this study. In the Figures, all simulated departure and approach trajectories are visualized. Figure 3 shows the measurement layout in the far range to ZRH. The black lines in this this Figure represent spatial gates, which the flight trajectories have to penetrate to be considered in this study. In this case, approaches on runways 28 (A28) and 34 (A34) were measured.

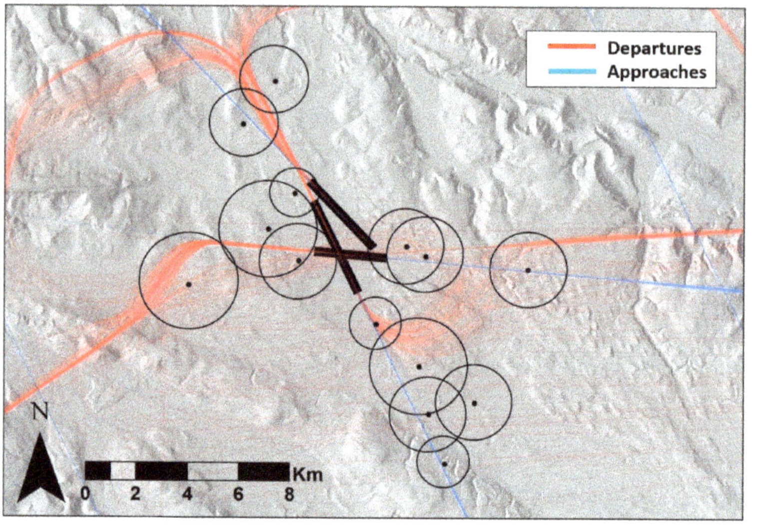

Figure 1. Noise monitoring terminals (NMT) in close range to ZRH with all flight trajectories used for this study, colored by procedure. The black circles around each terminal represent spatial gates, which the flight trajectories have to penetrate to be considered (basemap: swissALTI3D LV95, swisstopo; source: Federal Office of Topography).

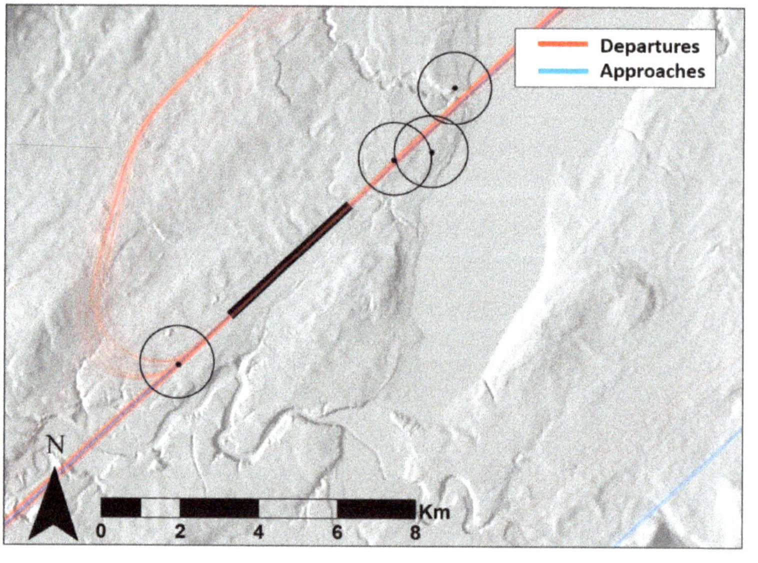

Figure 2. Noise monitoring terminals (NMT) in close range to GVA with all flight trajectories used for this study, colored by procedure. The black circles around each terminal represent spatial gates, which the flight trajectories have to penetrate to be considered (basemap: swissALTI3D LV95, swisstopo; source: Federal Office of Topography).

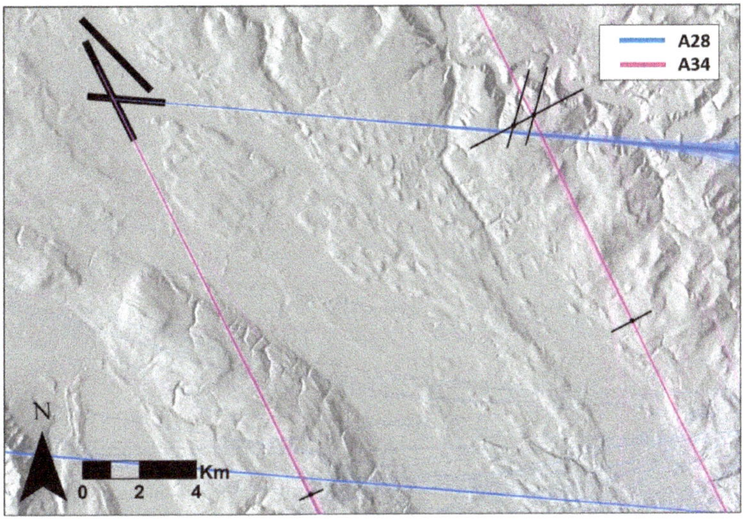

Figure 3. Measuring stations of the measurement campaign in the far range to ZRH with all flight trajectories used for this study. Approaches on runway 28 (A28) are depicted in blue and on runway 34 (A34) in purple. The black lines represent spatial gates, which the flight trajectories have to penetrate to be considered (basemap: swissALTI3D LV95, swisstopo; source: Federal Office of Topography).

As measurements conducted during bad weather conditions, such as strong wind situations with wind speeds exceeding 5 m/s and/or rain, generally have higher background noise, they were systematically excluded from the analysis.

2.4. Simulation Input Data

2.4.1. Flight Data

We simulated each individual flight separately using either FDR data or radar data (see Section 2.1). FDR data were provided by Swiss International Air Lines. If no FDR data were available, we used radar data provided by Flughafen Zürich AG (owner and operator of Zurich Airport). The trajectory data were processed as necessary (e.g., N1 estimation for sonAIR in case of radar data, see above) to serve as inputs for the single flight simulation in sonAIR, FLULA2 and AEDT. Table 3 provides an overview of the simulated flights that are compared with the corresponding measurements.

Table 3. Number of simulated flights and aircraft types.

Close/Far Range to Airports	FDR Available	# Flights	Airport (s)	# Different Aircraft Types
close	yes	1732	ZRH/GVA	8
close	no	6659	ZRH/GVA	24
far	yes	394	ZRH	5

2.4.2. Additional Input Data

Weather data for the sonAIR simulations were available as vertical profiles of air pressure, temperature, humidity and wind speed (COSMO-1 forecast data by the Federal Office of Meteorology and Climatology, MeteoSwiss). For FLULA2 and AEDT, standard atmospheric conditions for air pressure (1013.25 hPa) and temperature (15 °C) according to the International Standard Atmosphere (ISA) were used, with 70% relative humidity. Note that while AEDT may also consider specific (e.g., local) meteorological conditions, we used standard atmospheric conditions for AEDT, as we wanted to test how the program performs with default settings (cf. Section 2.2.3).

sonAIR and FLULA2 both by default perform simulations with terrain data (Digital Terrain Model DTM by the Federal Office of Topography, swisstopo). FLULA2 uses a resolution of 25 m × 25 m in the whole calculation area. sonAIR uses a resolution of 25 m × 25 m in the far range of the airport, but a finer resolution of 2 m × 2 m closer to the airports, to integrate buildings, on which the NMTs were installed, manually from the swissBUILDINGS3D vector based dataset by swisstopo. The terrain model was used in both FLULA2 and sonAIR, to account for receiver height (= microphone height), ground reflections from roofs of the NMT locations (sonAIR only), as well as possible shielding effects (FLULA2: by terrain only, sonAIR: also by buildings close to the airport), which, however, are negligible for the examined microphone locations in this study, given the present terrain situation. In AEDT, we did not model shielding by terrain, but accounted for the true microphone height.

Ground cover data were obtained from the Swiss Map Vector 25 by swisstopo in the far range and from cadastral surveying of the cantons in the close range to the airports, both with a resolution of 5 m × 5 m.

2.4.3. Sound Level Correction in AEDT

The aircraft and engine designations in the input data of the simulated flights had to be renamed to assign a suitable emission dataset in AEDT. If there is no corresponding emission model in the database, AEDT substitutes with a proxy type [20]. However, the sound emission levels (NPD curves) of the substitution model may differ from the actual aircraft, which needs to be corrected afterwards. The method of the sound level correction is given in Section 6.4 of Doc 29, Volume 1 [28]. The database for the sound level corrections is the EASA Jet aeroplanes noise database [29]. The list of aircraft types simulated with AEDT, their emission models and corresponding level corrections can be found in Table S1 in the Supplementary Materials. Note that for FLULA2 and sonAIR, no substitutions and corresponding level corrections were necessary as all studied aircraft types were available in their data bases.

2.5. Measurement and Calculation Uncertainties

Measurement uncertainties differ between accompanied and automated measurements [7,30]. In accompanied measurements, such as those in the far range to ZRH, local influencing factors such as ambient noise or reflections on buildings can largely be excluded. Here, the remaining uncertainty is due to the measurement devices and amounts to around 0.5 dB (68% standard uncertainty). Automated measurements, such as the NMT stations in the close range to ZRH and GVA, are characterized by less controlled measurement conditions. The locations might not be ideal from an acoustical point of view, and event trigger levels are used for recordings. In addition to device-related uncertainties, buildings and ambient noise may affect the accuracy of the measurement. Depending on the locations, the measurement uncertainty amounts to 0.5–0.9 dB [7,30].

Besides, calculations are also afflicted by input data and (sound emission and propagation) modelling uncertainties, which vary along the flight path [6–8]. The calculation uncertainties for FLULA2 were estimated to amount to between 0.5 dB (day) and 1.0 dB (night) for legally relevant yearly noise exposures $L_{Aeq} \geq 53$ dB (day) and $L_{Aeq} \geq 43$ dB (night), respectively [6,7]. These values are likely to be similar for sonAIR and AEDT.

When comparing the simulation results with the measurement and interpreting the differences, one should keep these uncertainties in mind [7,31].

3. Results

This section shows the comparisons between the simulations with the different programs and the measurements. The comparisons are separately done for each program and distinguished between events with FDR data (referred to as "FDR" in the following) and without FDR data available ("nonFDR"), as well as in close and far range to the airports. Note that the subsequent plots display multiple aircraft types and measurement locations with different local conditions. A list with mean values and standard deviation of the differences between calculations and measurements of the three models for each aircraft type can be found in Tables S2–S4 in the Supplementary Materials.

3.1. Close Range to Airports, FDR

The aircraft types with FDR data included ("FDR types") in the study in close range to the airports are given in Table 1. Figure 4 shows the differences between corresponding calculations and measurements with scatterplots, separately for all three programs.

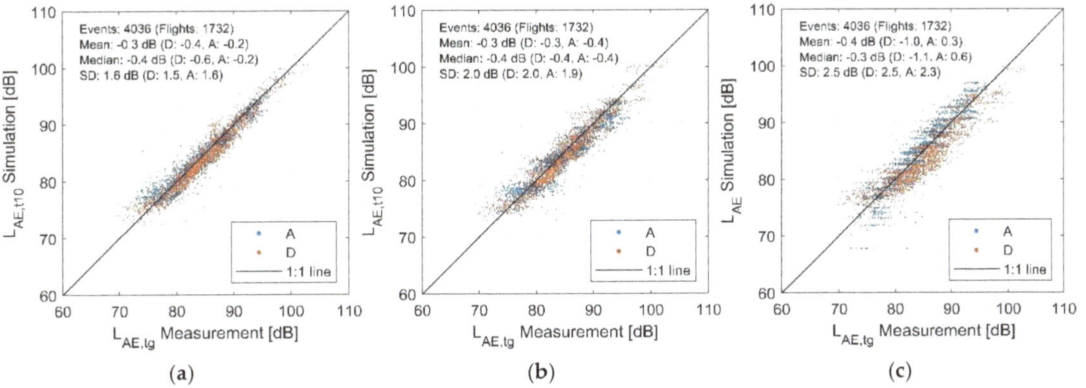

Figure 4. Scatter plots of simulation vs. measurements for aircraft with FDR data in the close range to the airports (ZRH and GVA combined), grouped by procedure (D: Departure, A: Approach, SD: Standard deviation). (**a**) sonAIR, (**b**) FLULA2, (**c**) AEDT.

Overall, there is a very good agreement between calculations and measurement in the close range with sonAIR. The single event level $L_{AE,t10}$ is slightly underestimated (−0.3 dB), with departures particularly contributing to this underestimation. The standard deviation for the $L_{AE,t10}$ is 1.6 dB.

The $L_{AE,t10}$ calculated with FLULA2 also agree well with measurements. Overall, calculations slightly underestimate measurements by −0.4 dB, but this difference is substantially smaller than the measurement and calculation uncertainties (Section 2.5). In the scatter plot, there are slight horizontal stratifications, which are due to the rigid directivity patterns of FLULA2. For sonAIR with dynamic sound source data, these stratifications are not visible.

The overall simulation results obtained with AEDT also agree well with measurements. The scattering is somewhat larger than for sonAIR and FLULA2. The L_{AE} for departures are generally underestimated and those for approaches generally overestimated. As in FLULA2, there are horizontal stratifications visible in the scatter plot. The reason in this case is that AEDT uses type-specific standard procedural profiles for the calculations in this study, in which the thrust and altitude profile is the same for a large number of flights. Especially for quieter approaches, some aircraft type-specific deviations can be identified, which appear as horizontally layered clusters.

sonAIR is able to represent all aircraft types with high accuracy, while FLULA2 and especially AEDT show more type-specific deviations. For more details on specific aircraft types, we refer to Table S2 of the Supplementary Materials.

3.2. Close Range to Airports, nonFDR

The aircraft types without FDR data included ("nonFDR types") in the study in close range to the airports are given in Table 1. Figure 5 shows the differences between corresponding calculations and measurements with scatterplots, separately for all three programs.

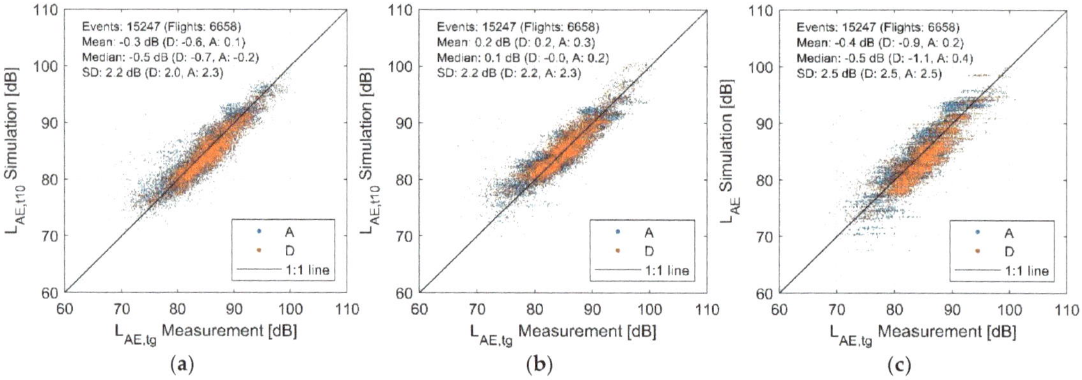

Figure 5. Scatter plots of simulation vs. measurements for aircraft without FDR data in the close range to the airports (ZRH and GVA combined), grouped by procedure (D: Departure, A: Approach, SD: Standard deviation). (**a**) sonAIR, (**b**) FLULA2, (**c**) AEDT.

On average, sonAIR reproduces the measurements as good as with the FDR types, with a mean difference of −0.3 dB, but with a larger standard deviation of 2.2 dB. Departures are again slightly underestimated, while the approaches show on average a very good match.

Overall, there is also a good agreement between calculations and measurements for FLULA2. The standard deviation is very similar for calculations with FDR data. Nevertheless, a larger number of outliers is visible, especially in cases of lower $L_{AE,t10}$. This is further discussed in Section 3.4.

The calculations of AEDT with nonFDR types also show on average a good agreement with the measurements. As with the FDR types, approaches are generally overestimated and departures are underestimated. The standard deviation of 2.5 dB is the same as for calculations with FDR data. This was expected, as AEDT and FLULA2, in contrast to sonAIR, do not include additional flight parameters from the FDR data. Again, horizontal stratifications are visible for both FLULA2 and AEDT.

As for the FDR types, the differences in aircraft type-specific differences are smallest for sonAIR. FLULA2 and especially AEDT show larger differences. The corresponding data can be found in Table S3 of the Supplementary Materials.

3.3. Far Range to Airports, FDR

The aircraft types with FDR data included ("FDR types") in the study in far range to the airports are A319, A320, A321, A333, A343. Figure 6 shows the differences between corresponding calculations and measurements with scatterplots, separately for all three programs.

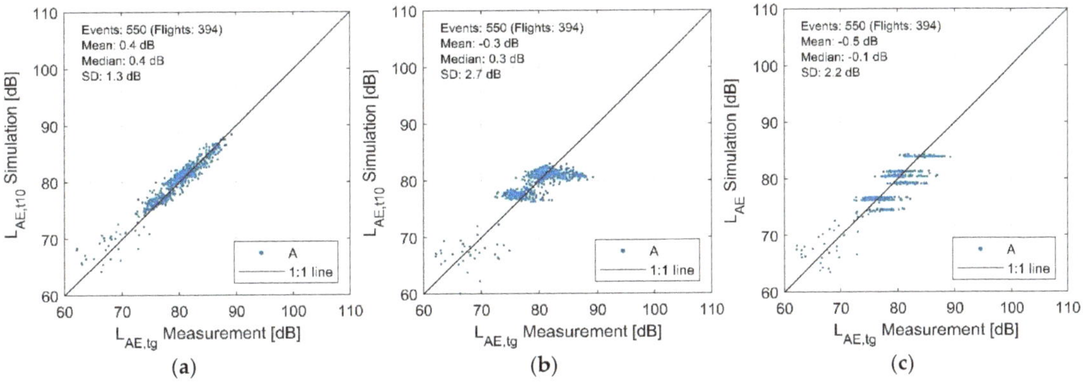

Figure 6. Scatter plots of simulation vs. measurements for approaches of aircraft with FDR data in the far range of ZRH. A: Approach, SD: Standard deviation. Note that no departures were recorded (Section 2.3). (**a**) sonAIR, (**b**) FLULA2, (**c**) AEDT.

sonAIR reproduced the measured noise exposure values well over all microphone locations and aircraft types. On average, there is a slight overestimation, with a mean difference of +0.4 dB and a standard deviation of 1.3 dB. In the area of lower sound pressure levels (corresponding to larger propagation distances/flight distances to touchdown), a slightly increased scattering can be observed. These differences are likely a consequence of turbulence induced scattering of the measured levels, which tends to be more pronounced at greater propagation distances. Another uncertainty arises from the fact that sonAIR has to estimate the sound emission outside of the verified model parameter range in these cases (extrapolation of model parameters).

The calculations with FLULA2 show, on average, a good agreement with the measurements. However, the differences are clearly larger than in the close range, especially for approaches on runway 34 with an underestimation of 3.2 dB (see Table 4). For FLULA2, the scattering is substantially larger in situations with $L_{AE,t10}$ below 70 dB. In addition, the horizontal stratifications in the scatter plot are more pronounced than in the close range. This is expected, as FLULA2 has fixed directivity patterns, which represent the configuration of the final approach state. Therefore, FLULA2 can represent flight conditions far away from the airport only to a limited extent, and larger differences to measurements, as observed here, may occur.

Table 4. Differences between simulation and measurements for all scenarios, in brackets separated into departures (D) and approaches (A).

Scenario	Model	Median (dB)	SD (dB)
ZRH & GVA, FDR, close range	sonAIR	−0.4 (D: −0.6, A: −0.2)	1.6 (D: 1.5, A: 1.6)
	FLULA2	−0.4 (D: −0.4, A: −0.4)	2.0 (D: 2.0, A: 1.9)
	AEDT	−0.3 (D: −1.1, A: +0.6)	2.5 (D: 2.5, A: 2.3)
ZRH & GVA, nonFDR, close range	sonAIR	−0.5 (D: −0.7, A: −0.2)	2.2 (D: 2.0, A: 2.3)
	FLULA2	0.1 (D: 0.0, A: +0.2)	2.2 (D: 2.2, A: 2.3)
	AEDT	−0.5 (D: −1.1, A: +0.4)	2.5 (D: 2.5, A: 2.5)
ZRH, FDR, far range	sonAIR	0.4 (A28: +0.5, A34: +0.2)	1.3
	FLULA2	0.3 (A28: +0.8, A34: −3.2)	2.7
	AEDT	−0.1	2.2

Overall, the calculations with AEDT show good agreement with the measurements. The standard deviation of 2.2 dB is similar to the close range. Similar to FLULA2, the horizontal stratification in the scatter plot is more pronounced, whereby two horizontal clusters are clearly to the right of the 1:1 line (at simulated L_{AE} of ~75 and 80 dB). These are the calculations of one specific aircraft type, which AEDT seems to generally underestimate (about 2 dB lower L_{AE} than FLULA2 and sonAIR). This might be explained by the fact that this type is poorly mapped by the NPD curves of its proxy type in AEDT.

The corresponding data can be found in Table S4 of the Supplementary Materials.

3.4. Overall Comparison of the Three Aircraft Noise Programs

Figure 7 shows the overall performance comparison (differences in calculations vs. measurements) of the three models with Box-Whiskers-plots, separately for approaches and departures. Table 4 numerically summarizes the results presented in Figure 7. In the close range to the airports, sonAIR and FLULA2 reproduce the measured values similarly well, with differences mostly smaller than 0.5 dB, while the calculations with AEDT show an overestimation for approaches (up to 0.6 dB) and a more pronounced underestimation for departures (up to −1.1 dB). In addition, for the FDR types, sonAIR has the smallest standard deviations (up to 1.6 dB) and AEDT has the largest (2.5 dB); FLULA2 lies in between (2.5 dB). For the nonFDR types, sonAIR calculations show similar standard deviations as FLULA2 and AEDT (Table 4).

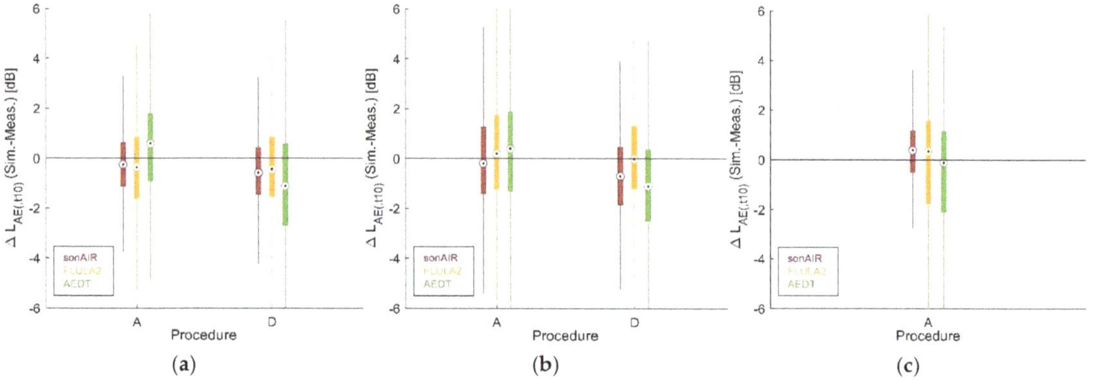

Figure 7. Box-whisker-plots of differences $\Delta L_{AE,t10}$ for sonAIR and FLULA2 and ΔL_{AE} for AEDT (simulation minus measurements) for all scenarios, grouped by procedure (D: Departure, A: Approach). (**a**) ZRH & GVA, FDR, close range; (**b**) ZRH & GVA, nonFDR, close range; (**c**) ZRH, FDR, far range.

In the far range, all three models show, on average, good agreement with the measurement data. The smallest standard deviations are obtained with the sonAIR calculations.

Figures 8 and 9 compare the calculated $L_{AE,t10}$ (L_{AE} for AEDT) of the three models for the datasets with and without FDR data in the close range to ZRH and GVA. This gives an indication of the overall agreement between the three model calculations (the results of the far range are not included in this comparison, as the number of flights are limited, and as only approaches of FDR types are available). The calculated event levels with sonAIR and FLULA2 agree well for both, FDR and nonFDR types, with a moderate standard deviation of 1.6 dB. This was expected, as FLULA2's sound source data for the aircraft types of the present study were derived from sound source data of sonAIR (Section 2.2.2). Thus, differences between sonAIR and FLULA2 are exclusively due to different underlying modelling approaches, but not different databases. The comparisons of AEDT with FLULA2 and sonAIR show somewhat larger deviations from the 1:1 line. The standard deviation varies between 2.2 and 2.5 dB. This may be due to aircraft type or measurement location specific differences in the calculated event levels. Larger differences between AEDT and sonAIR/FLULA2 were expected, as AEDT differs not only with respect to underlying modelling approaches, but also databases (NPD curves and standard procedural profiles). Overall, there are no systematic differences between the three programs, indicating that their calculation results are equivalent on average.

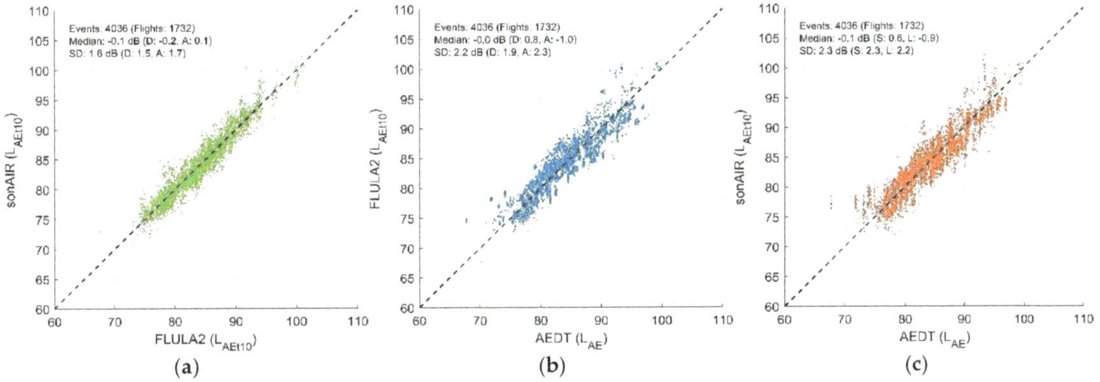

(a) (b) (c)

Figure 8. Scatter plots of the calculated event levels L_{AEt10} (L_{AE} for AEDT) between the three models for aircraft with FDR data in the close range (ZRH and GVA combined), grouped by procedure (D: Departure, A: Approach, SD: Standard deviation). (**a**) sonAIR vs. FLULA2; (**b**) FLULA2 vs. AEDT; (**c**) sonAIR vs. AEDT.

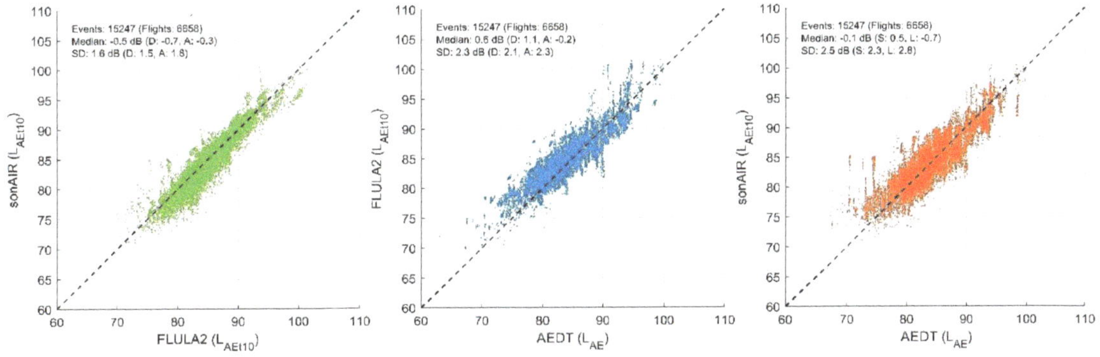

(a) (b) (c)

Figure 9. Scatter plots of the calculated event levels $L_{AE,\ t10}$ (L_{AE} for AEDT) between the three models for aircraft without FDR data in the close range (ZRH and GVA combined), grouped by procedure (D: Departure, A: Approach, SD: Standard deviation). (**a**) sonAIR vs. FLULA2; (**b**) FLULA2 vs. AEDT; (**c**) sonAIR vs. AEDT.

Larger outliers as observed in Figure 4 and especially Figure 5 mostly vanish. These deviations are therefore likely due to turbulence-induced sound propagation effects and/or ambient noise contaminations that neither of the examined calculation programs accounts for.

4. Discussion

In this study, a comparison and systematic validation of three inherently different aircraft noise calculation programs, namely sonAIR, FLULA2 and AEDT, was performed with a large measurement data set.

The calculations with sonAIR show a very good agreement with the measurements, both in the close and in the far range to the airports, although the $L_{AE,t10}$ of departures is generally slightly underestimated. If FDR data are available as input, sonAIR is particularly accurate, with smaller standard deviations of the differences to the measurements than those of FLULA2 and AEDT. This is particularly evident in the far range, where sonAIR calculates noise exposure with a standard deviation of 1.3 dB, compared to 2.7 dB of FLULA2 and 2.2 dB of AEDT. By considering airplane configuration and the separation of engine and airframe noise, approaches in particular can thus be more accurately reproduced by sonAIR. Only a few nonFDR types show somewhat larger mean differences: here, sonAIR performs similarly accurately as FLULA2. Also the far range calculations would be less accurate if no FDR data were available, with roughly 1 dB larger standard deviations [11].

In the close range, the $L_{AE,t10}$ calculated by FLULA2 also agree well with the measurements. However, FLULA2 is not able to reproduce the $L_{AE,t10}$ of all aircraft types equally well as sonAIR, and a larger number of aircraft type-specific differences larger than 1 dB result. In the far range, FLULA2 shows a strong underestimation of −3.2 dB for approaches on runway 34 (see Table 4). During the entire approach phase, FLULA2 uses static source model representing final approach configuration. This leads to an underestimation of the airframe noise far from the airport when airplane speeds are higher than close to touchdown. The FLULA2 emission patterns therefore only partially represent flight conditions far from airports, where, e.g., flaps and gears settings are completely different.

AEDT also overall yields a good agreement with measurements, in the near and far range to the airports, although in close range, approaches tend to be overestimated while departures are underestimated. The differences to measurements in the close range are generally greater than for sonAIR and FLULA2. In interpreting these results, however, one should keep in mind that AEDT's sound source database does not cover all aircraft engine types that we measured. Thus, they had to be substituted with comparable proxy types in some cases. In addition, contrary to sonAIR and FLULA2, neither sound source data nor flight profiles were specifically tuned to the local conditions of ZRH and GVA.

This study shows that sonAIR is clearly more accurate than the best practice models FLULA2 and AEDT if FDR data with flight configuration information are available for calculations. In this case, sonAIR can exploit its full potential, by using the best possible 3D source model with detailed consideration of the flight configurations. For calculations without FDR data, all three models show similar results. However, for all three programs, the accuracy of calculations varies between different aircraft types; here, the fleet mix determines how well the results are on average. For yearly aircraft noise calculations with a fleet consisting of many aircraft types, this point is not critical. While both FLULA2 and sonAIR are based on measurement data around ZRH and therefore adapted to local conditions such as the exact aircraft fleet, flight profiles, etc., in this region, this is not the case for AEDT. By adjusting the source data, AEDT in particular could be better tuned to local conditions which would improve modelling results. However, input data adjustment and corresponding calculations were beyond scope of this paper.

In conclusion, there are no systematic differences between the three models, indicating that their calculation results are equivalent on average. All three programs are well suited to reproduce measurements sufficiently accurate on average and thus also to determine averaged noise metrics resulting from complex scenarios consisting of many flights (e.g., yearly air operations), while sonAIR is additionally capable to highly accurately reproduce single flights in greater detail. sonAIR also shows the lowest aircraft type specific deviations, which makes it more robust with respect to single-flight simulations.

5. Conclusions

In this study, we calculated the A-weighted sound exposure level of several thousand single flights with three different aircraft noise calculation programs at receiver locations around Zurich and Geneva airports. We compared the results to measurements and between the programs. Overall, all three models yield results that achieve good agreement between each other and with the measurements. Using FDR data, sonAIR shows a substantially lower deviation from measurements than FLULA2 and AEDT, for both, departures and approaches, especially in far range to the airports. Without using FDR data, however, the deviations are similar to those of FLULA2 in the close range. However, substantially more aircraft type and measurement location-specific deviations larger than 1 dB occur with calculations of FLULA2 than with sonAIR, especially in the far range. This reveals the advantage of the sonAIR modeling approach with dynamic and separate sound source models for engine and airframe noise. AEDT also reproduced the measurements well, in both, the close and far range to the airports, with slightly larger mean differences and standard deviations than sonAIR and FLULA2. However, standardized procedural profiles and NPD curves were used without adaptions to the local conditions, while FLULA2 and sonAIR are adjusted to specific Swiss conditions. Consequently, AEDT can be attested a high modelling performance as well.

In conclusion, the present study shows that, on average, all three inherently different modelling approaches of the aircraft noise calculation programs are able to reproduce measurements in the close and far range sufficiently accurate, with mean differences below 1 dB. All three programs are therefore well suited to determine averaged noise metrics resulting from complex (e.g., yearly) airport scenarios. The strength of sonAIR is shown especially if FDR data are available and/or in the far range of airports, being able to reproduce individual flights very precisely (low scatter).

Supplementary Materials: The following are available online at https://www.mdpi.com/article/10.3390/aerospace8120388/s1, Table S1: List of aircraft types simulated with AEDT, their emission models and corresponding level corrections, Tables S2–S4: List of emission models with FDR data in the close range (ZRH and GVA combined), with mean values and standard deviations between calculations and measurements, separately for departures and approaches.

Author Contributions: Conceptualization, B.S.; data processing, J.M., S.S., D.J. and C.Z.; formal analysis, J.M., S.S. and D.J.; funding acquisition, B.S.; investigation, J.M., S.S., D.J. and C.Z.; project administration, B.S.; writing—original draft, J.M. and S.S.; writing—review & editing, J.-M.W., D.J., C.Z. and B.S. All authors have read and agreed to the published version of the manuscript.

Funding: This study (research project CompAIR) was funded by the Swiss Federal Office for the Environment FOEN (assignment No. 5211.01812.100.01). The manuscript preparation was funded by the authors' institution, Empa.

Institutional Review Board Statement: Not applicable.

Informed Consent Statement: Not applicable.

Data Availability Statement: Not applicable.

Acknowledgments: We would like to thank the Swiss International Airlines and Zurich and Geneva Airport for permission to use the data for this study.

Conflicts of Interest: The authors declare no conflict of interest.

References

1. Schäffer, B.; Thomann, G.; Huber, P.; Brink, M.; Plüss, S.; Hofmann, R. Zurich Aircraft Noise Index: An index for the assessment and analysis of the effects of aircraft noise on the population. *Acta Acust. United Acust.* **2012**, *98*, 505–519. [CrossRef]
2. Peris, E. *Environmental Noise in Europe—2020*; European Environment Agency (EEA): Luxembourg, 2020.
3. Jarup, L.; Babisch, W.; Houthuijs, D.; Pershagen, G.; Katsouyanni, K.; Cadum, E.; Dudley, M.-L.; Savigny, P.; Seiffert, I.; Swart, W.; et al. Hypertension and exposure to noise near airports: The HYENA study. *Environ. Health Perspect.* **2008**, *116*, 329–333. [CrossRef] [PubMed]
4. WHO. *Environmental Noise Guidelines for the European Region*; WHO: Copenhagen, Denmark, 2018.
5. EASA; EEA; EUROCONTROL. *European Aviation Environmental Report 2019*; EASA: Cologne, Germany, 2019.
6. Schäffer, B.; Plüss, S.; Thomann, G. Estimating the model-specific uncertainty of aircraft noise calculations. *Appl. Acoust.* **2014**, *84*, 58–72. [CrossRef]
7. Thomann, G. Uncertainties of Measured and Calculated Aircraft Noise and Consequences in Relation to Noise Limits (Mess-und Berechnungsunsicherheit von Fluglärmbelastungen und ihre Konsequenzen). Ph.D. Dissertation, Diss. ETH Nr. 17433, ETH Zurich, Zurich, Switzerland, 2007.
8. Bertsch, L.; Schäffer, B.; Guérin, S. Uncertainty Analysis for Parametric Aircraft System Noise Prediction. *J. Aircr.* **2019**, *56*, 529–544. [CrossRef]
9. Rhodes, D.; Ollerhead, J. Aircraft noise model validation. In Proceedings of the INTER-NOISE and NOISE-CON Congress and Conference Proceedings, The Hague, The Netherlands, 27–30 August 2001.
10. Buetikofer, R.; Thomann, G. Validation of FLULA, a time-step model for aircraft noise calculations. In *INTER-NOISE and NOISE-CON Congress and Conference Proceedings*; Institute of Noise Control Engineering: West Lafayette, IN, USA, 2001.
11. Jäger, D.; Zellmann, C.; Schlatter, F.; Wunderli, J.M. Validation of the sonAIR aircraft noise simulation model. *Noise Mapp.* **2020**, *8*, 95–107. [CrossRef]

12. Bertsch, L.; Sanders, L.; Thomas, R.H.; Le Griffon, I.; June, J.C.; Clark, I.A.; Lorteau, M. Comparative Assessment of Aircraft System Noise Simulation Tools. *J. Aircr.* **2021**, *58*, 4. [CrossRef]
13. Zaporozhets, O.; Levchenko, L. Accuracy of Noise-Power-Distance Definition on Results of Single Aircraft Noise Event Calculation. *Aerospace* **2021**, *8*, 121. [CrossRef]
14. Wunderli, J.M.; Zellmann, C.; Köpfli, M.; Habermacher, M.; Schwab, O.; Schlatter, F.; Schäffer, B. sonAIR—A GIS-Integrated Spectral Aircraft Noise Simulation Tool for Single Flight Prediciton and Noise Mapping. *Acta Acust. United Acust.* **2018**, *104*, 440–451. [CrossRef]
15. Zellmann, C.; Schäffer, B.; Wunderli, J.M.; Isermann, U.; Paschereit, C.O. Aircraft Noise Emission Model Accounting for Aircraft Flight Parameters. *J. Aircr.* **2017**, *55*, 682–695. [CrossRef]
16. Jäger, D.; Zellmann, C.; Wunderli, J.M.; Scholz, M.; Abdelmoula, F.; Gerber, M. Validation of an airline pilot assistant system for low-noise approach procedures. *Transp. Res. Part D Transp. Environ.* **2021**, *99*, 103020. [CrossRef]
17. Pietrzko, S.; Hofmann, R.F. Prediction of A-weighted aircraft noise based on measured directivity patterns. *Appl. Acoust.* **1988**, *23*, 29–44. [CrossRef]
18. U.S. Department of Transportation Volpe National Transportation Systems Center. *Aviation Environmental Design Tool (AEDT) Version 3c. Technical Manual*; U.S. Department of Transportation: Washington, DC, USA, 2020.
19. ICAO. *Recommended Method for Computing Noise Contours Around Airports. Doc 9911*, 1st ed.; International Civil Aviation Organization (ICAO): Montréal, QC, Canada, 2008.
20. ECAC. *ECAC.CEAC Doc 29: Report on Standard Method of Computing Noise Contours around Civil Airports, Volume 1: Applications Guide*, 4th ed.; European Civil Aviation Conference (ECAC): Neuilly-sur-Seine, France, 2016.
21. ECAC. *ECAC.CEAC Doc 29: Report on Standard Method of Computing Noise Contours around Civil Airports, Volume 2: Technical Guide*, 4th ed.; European Civil Aviation Conference (ECAC): Neuilly-sur-Seine, France, 2016.
22. Schwab, O.; Zellmann, C. Estimation of Flight-Phase-Specific Jet Aircraft Parameters for Noise Simulations. *J. Aircr.* **2020**, *57*, 1111–1120. [CrossRef]
23. Schlatter, F.; Wunderli, J.M.; Köpfli, M. Relevance of buildings in aircraft noise predictions. In Proceedings of the INTER-NOISE 2018—Impact of Noise Control Engineering, Chicago, IL, USA, 26–29 August 2018.
24. Empa. *FLULA2, a Method for the Calculation and Illustration of Aircraft Noise Exposure. Technical Program Documentation (FLULA2, Ein Verfahren zur Berechnung und Darstellung der Fluglärmbelastung. Technische Programm-Dokumentation), Version 4*; Empa, Swiss Federal Laboratories for Materials Science and Technology, Laboratory for Acoustics/Noise Control: Dübendorf, Switzerland, 2010.
25. Krebs, W.; Bütikofer, R.; Plüss, S.; Thomann, G. Sound source data for aircraft noise simulation. *Acta Acust. United Acust.* **2004**, *90*, 91–100.
26. Federal Office for the Environment (FOEN). Determination of Aircraft Noise. Available online: https://www.bafu.admin.ch/bafu/en/home/topics/noise/info-specialists/determination-and-evalution-of-noise/determination-of-aircraft-noise.html (accessed on 22 September 2021).
27. Isermann, U.; Vogelsang, B. AzB and ECAC Doc.29—Two best-practice European aircraft noiseprediction models. *Noise Control Eng. J.* **2010**, *58*, 455–461. [CrossRef]
28. ECAC. *DOC.29: Report on Standard Method of Computing Noise Contours around Civil Airports, Volume 1: Applications Guide*, 3rd ed.; European Civil Aviation Conference (ECAC): Neuilly-sur-Seine, France, 2005.
29. European Union Aviation Safety Agency (EASA). EASA Certification Noise Levels. Available online: https://www.easa.europa.eu/domains/environment/easa-certification-noise-levels (accessed on 27 October 2021).
30. Bütikofer, R.; Thomann, G. Uncertainty and level adjustments of aircraft noise measurements. In Proceedings of the 2009 International Congress and Exposition on Noise Control Engineering (Inter-Noise 2009), Ottawa, ON, Canada, 23–26 August 2009.
31. Thomann, G. Measurement and calculation uncertainty of aircraft noise exposure. In Proceedings of the 42nd International Congress and Exposition on Noise Control Engineering, Innsbruck, Austria, 15–18 September 2013.

aerospace MDPI

Article

Prediction of Aircraft Noise Impact with Application to Hong Kong International Airport

Chunhui Wu and Stephane Redonnet *

School of Mechanical & Aerospace Engineering, The Hong Kong University of Science & Technology, Hong Kong, China; cwuax@connect.ust.hk
* Correspondence: redonnet@ust.hk

Abstract: As part of a collective research effort towards greener aviation, the present study focuses on the noise impact of aircraft operations around major airports. To this end, an aircraft noise prediction platform is developed, which relies on state-of-the-art functionalities as well as more specific, innovative features. Originally built upon the Aircraft Noise and Performance (ANP) database and its Noise–Power–Distance (NPD) table, the method is further refined to alleviate most of their inherent limitations (e.g., standardized and simplified aircraft noise scenarios). The resulting aircraft noise prediction platform is validated against benchmark cases of increasing complexity, being then applied to real-life situations involving actual aircraft operations around Hong Kong International Airport (HKIA). Specific comparative analyses are conducted, which allow highlighting the variability of the noise impact by aircraft, depending on their type (A330, B777) and/or operational conditions (power settings, meteorological conditions, routes, banks, etc.). The study delivers insightful outcomes, whether phenomenological (aircraft noise impact) or methodological (aircraft noise prediction). As a by-product, it illustrates how noise prediction methods/platforms such as the present one may help in guiding the further expansion of airport operations and/or infrastructures (as is currently the case with HKIA).

Keywords: green aviation; airport operations; aircraft noise impact; noise prediction method

Citation: Wu, C.; Redonnet, S. Prediction of Aircraft Noise Impact with Application to Hong Kong International Airport. *Aerospace* **2021**, *8*, 264. https://doi.org/10.3390/aerospace8090264

Academic Editors: Adrian Sescu and Lothar Bertsch

Received: 6 August 2021
Accepted: 10 September 2021
Published: 15 September 2021

1. Introduction

The development of low-noise technologies and the establishment of more stringent regulations have led to a continuous decrease in the noise impact by air traffic over recent decades [1]. However, the environmental noise pollution caused by civil aviation is still a major societal concern, because of the many adverse effects it entails (annoyances, sleep disturbances, health issues, educational achievements, etc.) [2–4]. This makes aircraft noise a show-stopper for airports expansion [5], which may question the sustainable development of the aviation market [6–8] whose annual growth is about 5% [9]. Therefore, mitigating the noise impact of civil aviation is of utmost importance for all major countries, especially those which possess large aviation hubs located nearby highly dense cities, such as China [10,11]. Indeed, not only China is the most populous country in the world, but it also ranks second in the world in terms of passenger air traffic and it has now replaced the United States as the world's largest aviation market [12]. In particular, Hong Kong, which is the world's sixth most densely populated city, is also the busiest airport in terms of cargo volume, worldwide. To cope with the growing demand for air transportation, the Hong Kong International Airport (HKIA) is now expanding its capacity, through the deployment of a third runway [13]. This obviously raises legitimate concerns about how the noise impact by aircraft operations will affect the many residential areas surrounding HKIA.

Aircraft noise has been a matter of intense research over the past 50 years. Essentially, two main approaches exist to mitigate aircraft noise, namely at the aircraft level by reducing its sources [14] and at the operation level by reducing its perceived impact on the ground [15–17]. The former strategy seeks at developing quieter engines and low-noise aircraft designs [18], which however constitutes a daunting challenge given the multiple components involved (e.g., high-speed jets, combustion, turbine, fan, landing gear, high-lift devices, cavities, etc.) [19,20]. The latter strategy rather seeks at optimizing the aircraft operations nearby airports (e.g., through noise abatement flight procedures) [17,21,22]. This however requires properly assessing the noise impact induced by aircraft during their departure and approach flight phases, which can be achieved through field measurements and/or predictive models. Field measurements are relatively straightforward and reliable [23], but they can only be performed a posteriori, over certain predetermined locations (e.g., using in-situ microphones). Notably, they solely offer measuring the noise impact from actual aircraft operations, but they do not allow to extrapolate it to alternative scenarios (e.g., for exploring the impact of new flightpaths, runways, fleets, noise abatement procedures, etc.). Moreover, field measurements can be polluted by ambient sounds, which makes it more difficult to discriminate the actual aircraft noise [24]. All these issues can be overcome by using predictive models, whose principle is to virtually recreate the noise impact by aircraft operations. The major limitation of these models is their cost in terms of computational resources, which scale with the accuracy level that is sought after. Over recent decades, many aircraft noise prediction approaches were proposed, which can be subdivided into three categories [25,26], namely, empirical [27–30], semi-empirical [31–33] and analytical [34–36]. Falling in the first category, the so-called Integrated Noise Model (INM) [37] developed by the US Federal Aviation Administration (FAA) is used worldwide for assessing the noise impact around airports. This tool is commonly used for either (i) monitoring aircraft operations, (ii) studying their impact on the neighboring populations [38–40], (iii) guiding the further development of airports [22,41,42], or even (iv) establishing low-noise flight procedures [43,44]—all that whereas incorporating other constraints (e.g., fuel consumption and chemical emissions) [45,46]. In particular, the INM model is now the cornerstone of large prediction platforms aiming at optimizing airport operations, worldwide, such as the Aviation Environmental Design Tool (AEDT) in the U.S.A. The latter constitutes one of the major bricks of the FAA-led Next Generation Air Transport System [47], whose goal is to revamp America's aviation infrastructures and operations into a renewed, integrated, clean and efficient air system [47]. Similar large-scale initiatives exist worldwide, for instance in Europe with the Environmental Noise Directive [48], whose aircraft noise aspects are tackled using methods that also rely on the INM model. Notably, the baseline methodology underlying INM is best described in ref [49], which was issued by a concertation group of the European Civil Aviation Conference (ECAC) in the framework of the END initiative [48]. Since the latter initiative extends to other industries (civilian and military aviation, railway, road traffic, etc.), current efforts focus on unifying and standardizing the prediction tools for environmental noise mitigation, for instance by developing the so-called Common Noise Assessment Methods in Europe [50]. Regarding aviation noise, the CNOSSOS-EU approach also relies on the INM model whereas integrating additional databases, thereby offering to tackle not only civilian aircraft, but also military ones, as well as helicopters, or specific airport operations (e.g., engine run-up noise). This may be pivotal in implementing strategic noise maps [51,52], thereby helping policymakers in their efforts to improve the land-use planning around major airports, as advocated by ICAO for mitigating environmental noise by air traffic. All the above illustrates how aircraft noise prediction tools can be advantageously integrated into larger platforms, thereby allowing to tackle environmental issues that would be too challenging to handle otherwise.

The present study is part of a larger research effort, whose objective is to develop a flightpath optimization platform enabling to minimize both the fuel consumption and noise emission due to aircraft operations nearby airports [45]. As a starting point, we here develop a noise prediction approach that comes as a mix between the INM baseline methodology [49] and novel, specific features. Upon incremental validations, we apply the resulting approach/tool to actual aircraft operations, thereby highlighting its relevance for real-life situations. This paper is organized as follows. In Section 2, the noise prediction approach is outlined, with a special emphasis put on the specific improvements that were brought to it. The method and subsequent tool are then validated in Section 3, this being achieved using several benchmark cases of increasing representativeness. In Section 4, the approach is applied to realistic scenarios coming from actual aircraft operations nearby Hong Kong International Airport. Finally, some conclusions and perspectives are drawn.

2. Computational Methodology

This section outlines the present noise prediction approach, which incorporates specific supplements compared to the baseline methodology described in [49].

2.1. Overview

Similar to the INM-related baseline methodology it originates from, the present approach primarily relies on the so-called Aircraft Noise and Performance (ANP) database and its associated Noise–Power–Distance (NPD) subset. This ANP/NPD database can be seen as a semi-empirical tool that allows estimating the noise impact induced by a given aircraft, depending on what its power settings and distance to the ground are [53]. Here it is worth reminding that such a noise impact is commonly measured in terms of Sound Exposure Level (SEL) or Effective Perceived Noise Level (EPNL), both of which represent the overall sound energy of a single noise event once integrated over a given duration (one and ten seconds, respectively) [37,54] and tailored to the sensitivity of the human ear (e.g., the SEL is A-weighted and the EPNL is tone corrected). Both measures thus translate the loudness or noisiness of the noise event, whereas their cumulative effect over successive events can be averaged (e.g., Day–Night average sound Level, DNL, Equivalent Sound Level, Leq). Given its empirical nature, the ANP/NPD database/tool is a powerful resource for predicting the noise environment around airports. On the other hand, since it is built upon standardized scenarios of aircraft operations (aircraft types, flightpaths, power settings, atmospheric conditions, etc.), its application is limited to rather canonical situations of air transport exploitation. Moreover, since the NPD-based noise prediction kernel relies on a rather rudimentary scenario of noise emission by a simplistic source (isolated jet) within a homogeneous free-field medium, the approach somehow lacks accuracy. One way to mitigate these limitations is to improve both the source characteristics and the propagation features that the approach relies on, which can be partly achieved by incorporating additional correction factors. This is what was carried out here and is summarized below.

On one hand, the characteristics of the noise source can be improved through a more accurate description of both its intensity and directivity. First, the source intensity is directly driven by the aircraft power settings, which can be refined by using the actual ones instead of their standardized counterparts. To this end, Section 2.3.1 proposes a method for assessing the actual power settings of an aircraft, based on its flightpath characteristics. Second, the source directivity is related to the way the engines are installed within the airframe (e.g., wing-mounted or fuselage-mounted), with an installation effect that can be partly incorporated through the proper estimation of the aircraft bank angle (see Section 2.3.2). On the other hand, the propagation features can be refined through several means such as a more accurate description of the propagative path itself (e.g., incorporating *all* the flightpath sections that contribute to the noise impact, see Section 2.3.3), as well as of the noise phenomena to occur during the propagation phase (e.g., absorption due to the humidity, refraction due to the atmosphere inhomogeneities, reflection due to the ground, etc.). Some of these refinements are part of the INM-based method which is described in Section 2.2, whereas others are specific to the current work and are detailed in Section 2.3. Yet, not all of these effects can be incorporated in the approach, which would be too complex and costly otherwise. For instance, one can here remind that the sound refraction originates from the atmospheric heterogeneities, which either affect the sound speed (and thus the acoustic impedance) through the pressure and temperature variations, or add convection effects due to winds. Whereas the former effect can be incorporated in a rather straightforward manner, the latter raises more challenges in terms of practical implementations and CPU cost efficiency.

Figure 1 sketches the overall methodology, both summarizing the baseline approach [49] and highlighting the specific addendums brought to it in the present work.

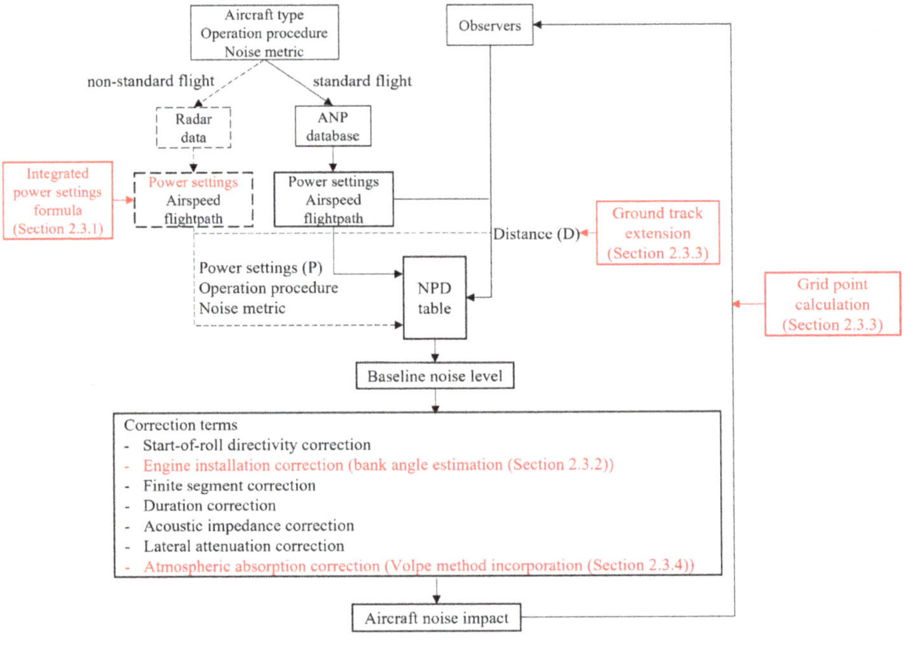

Figure 1. Aircraft noise prediction methodology, as built-up from the baseline approach [49] (in black) whereas incorporating specific addendums (in red).

2.2. Baseline Approach

This section gives a brief overview of the baseline approach, whose detail can be found in [49]. Recommended by the International Civil Aviation Organization (ICAO), this approach is widely adopted by airports and legislation authorities for assessing the noise impact of aircraft operations (at least for commercial aircraft) [55].

As detailed in ref [53], this methodology relies on the fragmentation of the flightpath into specific segments, each one being allotted the aircraft corresponding characteristics (position, power settings, airspeed, etc.). Those characteristics are then translated into suitable inputs (power settings, aircraft-to-ground distance) for the NPD-based noise prediction tool to assess the aircraft noise impact on the ground. More precisely, the 3D flightpath is decomposed into the so-called ground track (i.e., the vertical projection of the flightpath on the ground) and flight profile (i.e., the altitude changes of the aircraft along the ground track), both of which are made of a finite number of continuous straight-line segments. The detail of these flight segments (endpoints) and associated aircraft characteristics (operational data) can be inferred from the flightpath records (either radar or Flight Data Recorder, FDR), or estimated using specific formulas [49]. Using this, the aircraft power settings and aircraft-to-ground distance can be obtained, thereby allowing the aircraft noise impact to be assessed through proper (linear and logarithmic) interpolations of the NPD database.

As was said, however, the NPD database is built upon an ideal scenario following which the ground observer would be located right below the aircraft, with the latter flying at a constant speed and power settings along an infinite horizontal flightpath and under specific atmospheric conditions. For the method can be applied to more realistic scenarios, it is thus needed to enhance it with several correction terms, namely, start-of-roll directivity, engine installation, finite segment, duration, acoustic impedance, and lateral attenuation adjustments. The start-of-roll directivity correction reflects the highly directive radiation pattern of the engine noise during the take-off roll. The engine installation correction accounts for the modification of the aircraft noise source directivity due to the integration of the engines within the airframe (e.g., scattering by the wings, refraction effects in the near-field, etc.). The finite segment correction translates the fact that, different from the NPD assumption, the flightpath is made of finite segments. The duration correction accounts for the difference between the actual aircraft speed and that from the NPD standard (160 knots), such disparity in speed translating into shorter or longer sound exposures. The acoustic impedance adjustment accounts for the difference between the local atmospheric conditions (taken at the airport level [49]) and that from the NPD standard. The lateral attenuation translates all the noise interferences induced by the reflection (and, to a lesser extent, by the refraction) effects inherited from the ground presence. Here it is worth noticing that these refraction effects are incurred by either the temperature and/or wind gradients induced by the ground, depending on its characteristics (e.g., roughness, heat transfer). All these correction terms enable the NPD-based noise prediction tool to assess more accurately the noise impact of aircraft in different scenarios. As an illustration, Appendix B exemplifies how these various correction terms weigh on the prediction of the noise impact by a typical aircraft at take-off.

2.3. Specific Improvements Brought to the Methodology

The ANP database is built on a standardized scenario of aircraft operations (flight profile, power settings, airspeed, etc.). However, in reality, the actual operations often deviate from these standard cases, due to various factors (terrain, wind speed and direction, noise abatement flight procedure, etc.). For the overall methodology to be closer to reality, the NPD-based noise prediction must therefore be enhanced with various refinements—which was carried out here and is summarized hereafter.

2.3.1. Refinement of the Aircraft Noise Emission (Intensity), through the Incorporation of the Actual Power Settings

Ideally, the aircraft operational data should be inferred directly from actual records (either radar or Flight Data Recorder, FDR). Whereas FDR data offer the most comprehensive and accurate information, their proprietary nature makes them difficult to access. Conversely, radar data are open source and therefore readily available but they offer less detailed information. For instance, the flightpath characteristics of any aircraft can be easily accessed through their radar data, which can be accessed via public domain websites (e.g., FlightAware, Flightradar24, OpenSky Network, etc.). These data, however, do not provide any insights about the aircraft power settings, which makes them less straightforward to use in the present context. To overcome this issue, the present approach incorporates a functionality that offers assessing the aircraft power settings directly from the flightpath characteristics (flight profile and airspeed). To this end, we derive a simplified model for the thrust to be delivered by the engines when the aircraft is in dynamic equilibrium, whether it is considered at roll or aloft. By developing the corresponding equations of motion (cf. Appendix A), one can express the aircraft thrust via the following integrated formula (which is valid whether the flight segment is, i.e., either ground roll or aloft):

$$T = m\left[a + g\left(\frac{\mu_F}{cos\gamma + \mu_F sin\gamma}\right)\right] + L\left[\frac{C_D}{C_L} + cos\phi\cdot\left(\frac{sin\gamma - \mu_F cos\gamma}{cos\gamma + \mu_F sin\gamma}\right)\right] \quad (1)$$

In the above, T indicates the thrust (in lbf), L is the lift force, C_D/C_L is the drag-to-lift ratio, m is the aircraft mass (in lbm), a is the aircraft inertial acceleration (in average), g is the gravitational acceleration. Besides, γ and ϕ, respectively, stand for the flightpath and bank angles (which are zero when the aircraft is on its ground roll). On the other hand, μ_F is the friction coefficient [56], which accounts for the additional drag effect due to the runway (and is zero when the aircraft is aloft). From there, two scenarios can be distinguished:

When the aircraft is aloft ($\mu_F = 0$), Equation (1) can be re-expressed as (cf. Appendix A)

$$T = m\left[a + g\left(\frac{C_D}{C_L}\cdot\frac{cos\gamma}{\cos\phi} + sin\gamma\right)\right] \quad (2)$$

Oppositely, when the aircraft is on the ground roll ($\gamma = \phi = 0$), and by introducing a lift model [57], one can show that (cf. Appendix A)

$$T = m\left\{a + g\left[\frac{C_D}{C_L}\cdot\left(\frac{V_{TAS}}{V_{LOF}}\right)^2 + \mu_F\left(1 - \left(\frac{V_{TAS}}{V_{LOF}}\right)^2\right)\right]\right\} \quad (3)$$

where V_{TAS} and V_{LOF} stand for the aircraft true airspeed (TAS) and its value at lift-off, respectively.

All parameters appearing in the above Equations (2) and (3) are readily available from the flight data and/or ANP database, for each aircraft type and flight setting. One can thus get the actual thrust directly from the aircraft and flight parameters, in all flight phases (whether the aircraft is aloft or on its ground roll). For being readily exploitable, however, the actual thrust must be scaled back to the standardized value upon which the NPD database is built upon, namely the corrected net thrust at sea level. To do so, the actual thrust is then adjusted through a gross-to-net thrust factor, which translates the fact that the thrust delivered at a given altitude deviates from its sea-level counterpart. This factor can be expressed through δ, the ratio of the ambient air pressure around the aircraft to the standard air pressure at mean sea level [23]. Once scaled by the number of engines, N, one finally gets the corrected net thrust per engine, to be used as input for the NPD database.

$$F_n = \frac{T}{N\delta} \quad (4)$$

2.3.2. Refinement of the Aircraft Noise Emission (Directivity), through the Incorporation of the Engine Installation Effects (Bank Angle)

The way the engines are mounted within the airframe affects the directivity of the aircraft noise source. These so-called acoustic installation effects can be partially incorporated through the aircraft bank angle [49], which relates to the depression angle (φ) that is directly linked to the engine installation (see Figure 2 below, as well as Figure A1 in Appendix A). Indeed, when installed on an aircraft, engines see their noise directivity importantly altered because of the multiple scattering effects induced by the airframe elements (e.g., wings) [23]. This directivity, which is measured through the depression angle, mostly depends on the way the engines are positioned within the airframe (e.g., wing-mounted, fuselage-mounted). From a ground observer perspective, the noise intensity is thus to be driven by the combination of the depression and bank angles.

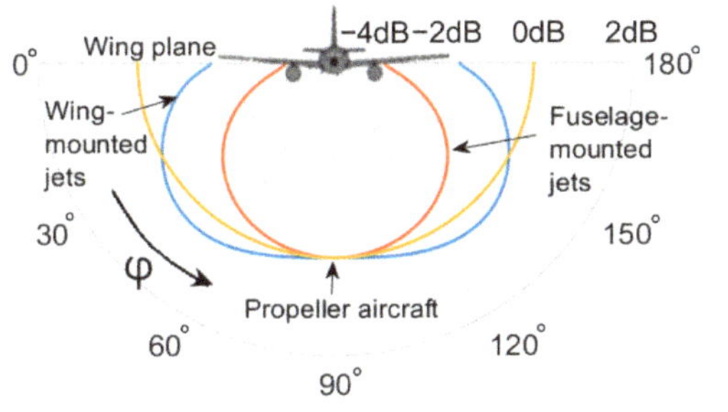

Figure 2. Engine installation effects on aircraft noise directivity. Noise level correction (dB) as a function of depression angle (deg).

In the absence of any specific formulas provided in Ref [49], the aircraft bank angle is here estimated through the flightpath turn radius, using a geometrical approach [37]. To do so, the ground track is first approximated (using cubic spline interpolation) as a succession of evenly spaced points, each one being then inferred its corresponding turn radius (which is determined from the circumradius of the triangle formed with its two neighboring points). From there, using the aircraft airspeed along with the gravitational acceleration, the bank angle is derived [49]. Once filtered out from all possible irregularities coming from the flight characteristics inaccuracies (especially the aircraft position and airspeed), the bank angle values are interpolated back to the original ground track. Section 4.3.1 and Appendix C illustrate the sensitivity of the noise prediction towards these bank effects.

2.3.3. Refinement of the Aircraft Noise Propagation (Distance), through the Dynamic Specification of the Ground Track and Observers

Dynamic Extension on the Ground Track

By definition, the departure flight sequence begins when the aircraft initiates its ground roll and terminates once it has climbed up to an altitude of 10,000 ft. Oppositely, the approach flight sequence starts when the aircraft has descended down to an altitude of 6000 ft and finishes at the end of the landing ground-roll [37]. Because of this somehow arbitrary definition of the departure and approach procedures, any flightpath segment that would be beyond their altitude ranges is usually discarded in the aircraft noise assessment process. This may lead to underestimating the noise levels on the ground, whose discarded flight segments may still contribute to, depending on the situation (e.g., aircraft settings/flightpath ⇔ ground observer relationship). In order to correct such a potential bias in the noise estimation, some authors proposed to automatically enhance the default flight sequences with an extra flight segment, of arbitrary length [49]. Given its arbitrary nature, however, this fix may be partially conclusive (e.g., extra flight segment of insufficient length). Therefore, we here propose to rather incorporate *all* the flightpath segments that actually contribute to the noise impact on the ground, on a case-by-case basis. To do so, the flight sequence is extended in a dynamic fashion, being automatically added all the flight segments that are beyond its standard altitude range—this being carried out until their contribution to the aircraft noise impact is negligible, i.e., the noise levels on the ground are converged. More detail about this process is provided in Section 4.3.2.

Dynamic Specification of Ground Observers

Classically, the aircraft noise impact is characterized using a set of ground observer locations, over which the noise contribution by the various flight segments is summed up over time—all this leading to integrated noise contour maps. The overall efficiency (i.e., accuracy and rapidity) of the noise prediction is directly driven by the number and distribution of these prediction points. Therefore, past researches focused on how to optimize the latter, for instance using irregular grids [58,59]. Different from the baseline methodology, the present method incorporates an advanced dynamic grid refinement functionality [37], which allows increasing locally the density of observers wherever and whenever needed. This dynamic, local grid refinement process is conducted automatically during the noise prediction, which can be seen as a multi-stage calculation. More precisely, the noise prediction is first conducted on a coarse mesh, being then bi-linearly interpolated on a twice finer grid. The interpolated result is then compared to that of the direct calculation, which is repeated for each new grid point created. Whenever the two results fall within a certain threshold, the calculation is considered converged, and the local grid refinement stops. Should this not be the case, the process is repeated, until a local convergence is reached. Besides the tolerance (i.e., interpolation accuracy threshold), the degree of refinement is controlled through two other criteria, namely, the maximum refinement level allowed (with respect to the initial, coarse grid), and some pre-defined extrema of noise levels. Thanks to such a dynamic grid refinement functionality, the calculation process is optimal, offering the best trade-off in terms of accuracy and cost.

2.3.4. Refinement of the Aircraft Noise Propagation (Attenuation), through the Incorporation of Meteorological Effects

Aside from its emission characteristics (intensity, directivity), the aircraft noise is also driven by the properties of the atmospheric medium through which it propagates to the ground. It is not rare, however, that the actual atmospheric conditions deviate from the standard values upon which the NPD database is built. In this case, the noise prediction must be corrected from the specific biases induced by these offsets in atmospheric properties [55]. This can be achieved through two correction terms, namely, the acoustic impedance adjustment and the atmospheric absorption adjustment. The former correction accounts for the offset in the sound speed (which is driven by the atmospheric pressure and temperature values) whereas the latter accounts for the offset in the noise absorption effect (which is primarily driven by the temperature and relative humidity).

Deviation of the Sound Speed Characteristics (Acoustic Impedance Adjustment)

Acoustic impedance adjustment accounts for the deviation of the medium sound speed with respect to its NPD standard value. This correction term [49] is classically built as a function of the atmospheric pressure and temperature (e.g. using Equation (4–6) and Equation (4–7) from Ref. [49]), both of which have a monotonic effect on the propagation. As exemplified in Figure 3, a rise in pressure (resp. temperature) shall result in a higher (resp. lower) correction required. The latter corrections, however, are usually moderate, with a variability that remains in the order of a decibel (dB). For instance, a pressure increase of 50 kPa entails a difference of +2.3 dB (regardless of the temperature), whereas a temperature increase of 40 °C induces an effect of −0.3 dB, regardless of the pressure.

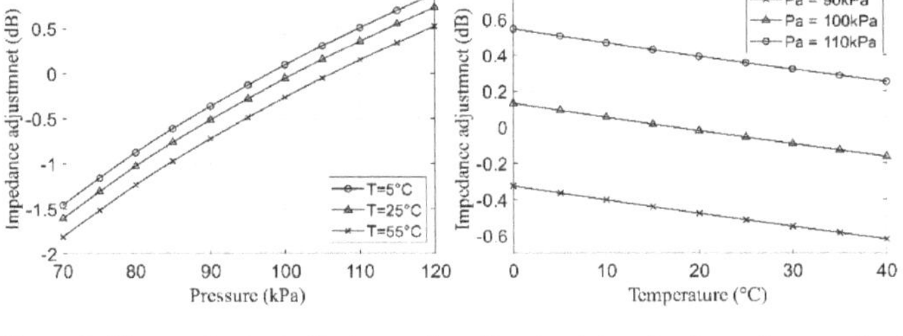

Figure 3. Meteorological effects on aircraft noise intensity. Impedance adjustment as a function of the atmospheric pressure and temperature.

Deviation of the Sound Absorption Characteristics (Atmospheric Absorption Adjustment)

The atmospheric absorption relates to the energy dissipation that sound waves experience when interacting with air molecules. It is mostly driven by the temperature and relative humidity, upon which the atmospheric absorption adjustment is thus usually built [60]. As an illustration, Figure 4 depicts the difference in the noise levels that are radiated on the ground by a given sound source located at an altitude of 25,000 ft, depending on if the propagation medium corresponds to a standard or a non-standard atmosphere. As can be seen, these differences are nonlinear and non-monotonic, which makes it uneasy to predict how a given deviation in temperature and/or relative humidity shall impact the noise prediction. Notably, these effects may cancel out each other—as highlighted in Figure 4 by the red line, which depicts those atmospheric conditions for which the difference is nil. In the present approach, the atmospheric absorption adjustment not only incorporates the deviations in temperature and/or relative humidity (as classically carried out [60]), but it also includes those in atmospheric pressure. Relying on the so-called Volpe method [61], this correction is thus expected to yield still more accurate results, especially for what concerns long-distance propagation. For more detail about this correction term, the interested reader is referred to Ref. [49] (Equations (D-1) to (D-4)) and Ref. [61] (Equations (1) to (6), Equations (16) and (17)). These atmospheric absorption effects are further discussed in Section 4.2.

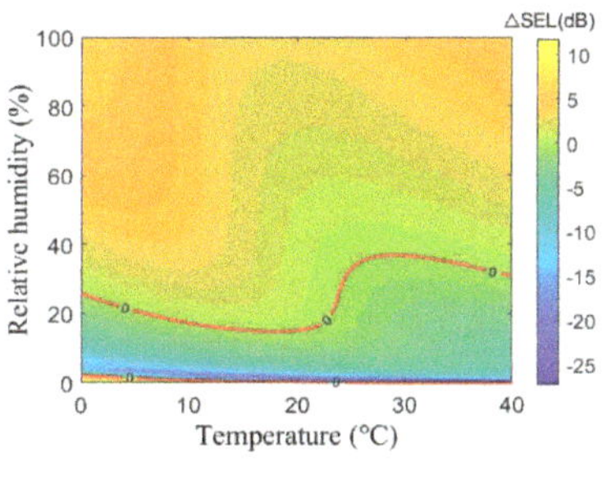

Figure 4. Meteorological effects on aircraft noise intensity. Absorption adjustment (difference in SEL with respect to that of a standard atmosphere).

3. Validation of the Methodology Using Standardized Cases

In this section, the present noise prediction approach is validated using benchmark cases, which correspond to either virtual standard scenarios or to real-life but well-known situations (noise certification tests).

3.1. Standardized Scenarios

Here, we consider a total of 12 standardized scenarios, which were previously documented in [62]. These cases describe the noise impact induced by an aircraft, depending on its type and/or flightpath. More precisely, we here consider three different types of aircraft, namely, a jet-powered aircraft with either fuselage-mounted (JETF) or wing-mounted (JETW) engines, as well as a propeller-powered aircraft (PROP). Each aircraft is taken under either a departure (D) or an approach (A) flight, which both follow either a straight (S) or a curved (C) route. Please, see Ref. [62] for a detailed description of these benchmark cases, some aspects of which are nevertheless further documented in Appendix B (e.g., aircraft routes, observers locations).

Figures 5 and 6, respectively, depict the noise contour maps associated with all three aircraft at departure and approach (only the curved routes are considered, for the sake of conciseness). Of note, the noise metric used here is Sound Exposure Level (SEL). For each case, the differences between the present prediction and the reference one [62] are quantified using error maps (obtained by subtracting the reference values from the calculated results). As one can see, the agreement is good, with absolute error levels that are typically less than 0.01 dB. Notably, compared to some of the past works [23], the present calculations rely on a coarser distribution of ground observers, whereas delivering more accurate results (compare for instance the error map of JETFDC case with its counterpart from Ref. [23] whose visual rendering was here duplicated on purpose.).

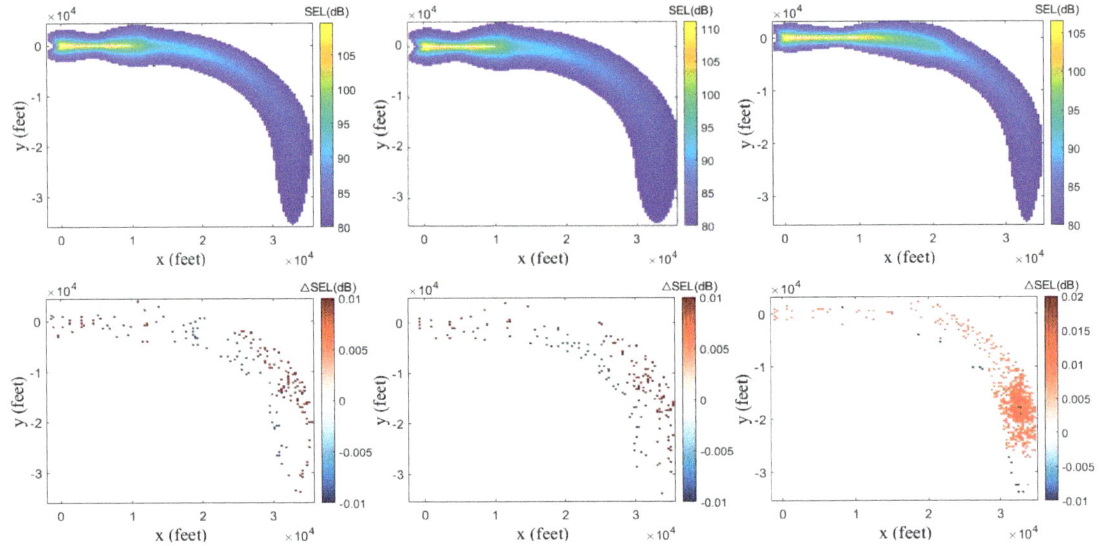

Figure 5. Noise contour map (**top**) and corresponding error (**bottom**) for JETFDC (**left**), JETWDC (**center**), PROPDC (**right**) cases.

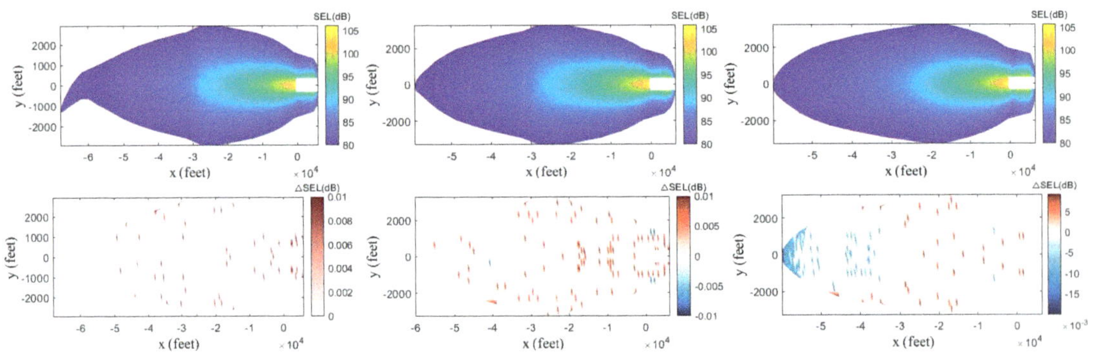

Figure 6. Noise contour map (**top**) and corresponding error (**bottom**) for JETFAC (**left**), JETWAC (**center**), PROPAC (**right**) cases.

This good agreement is verified for all 12 benchmark cases, as shown in Table 1, which quantifies the mismatches between the present predictions and the reference values. For each case, this mismatch is characterized via both the maximum absolute error (δ_{max}) and the average (root-mean-square) error δ_{RMS}, as recorded throughout the ground observers. Notably, for all cases, the average error, δ_{RMS}, is much less than the threshold of 0.01 dB, which is the indicator of a reliable prediction [62]. Moreover, this accuracy level appears to be higher than the one reached by previous similar studies which focused on the same benchmark [23].

Table 1. Errors between the calculated noise levels and the reference results.

Reference Case	δ_{max} (dB)	δ_{RMS} (dB)
JETFAC	0.0125	0.00135
JETFAS	0.0125	0.00149
JETFDC	0.0125	0.00237
JETFDS	0.0125	0.00160
JETWAC	0.0125	0.00183
JETWAS	0.0125	0.00179
JETWDC	0.0125	0.00215
JETWDS	0.0125	0.00157
PROPAC	0.0100	0.00218
PROPAS	0.0016	0.00162
PROPDC	0.0200	0.00423
PROPDS	0.0124	0.00159

All in all, these benchmark cases provide a validation of the present noise prediction platform and underlying methodology. Appendix B provides a further illustration (along with a more incremental validation) of the overall prediction process, exemplifying each calculation step in the particular JETWDS case.

3.2. Noise Certification Cases

To further assess the accuracy of the methodology as well as to lean towards real-life situations, the present section focuses on actual flight scenarios whose characteristics are however close enough from the ANP standard ones. To this end, we consider several noise certification flight tests, for which field measurements were made available by the European Union Aviation Safety Agency (EASA) [63]. These flights tests were performed for two different aircraft (B737-800 and A320-211), delivering for each the Effective Perceived Noise Level (EPNL) over a set of certification points. Indeed, aircraft noise certification tests follow a very specific protocol [64], which consists in flying the airplane along its standard approach and take-off flightpaths whereas measuring the resulting noise at three ground locations, namely the approach, lateral full-power and flyover reference points [64] (see Figure 7). Notably, these ground measurements are performed such that the ground reflections are minimized at best (e.g., displaying the microphones over a flat terrain of known impedance, with no surrounding buildings, etc.). The two considered certification flights were here virtually duplicated using the noise prediction platform, thereby delivering the EPNL results provided in Table 2.

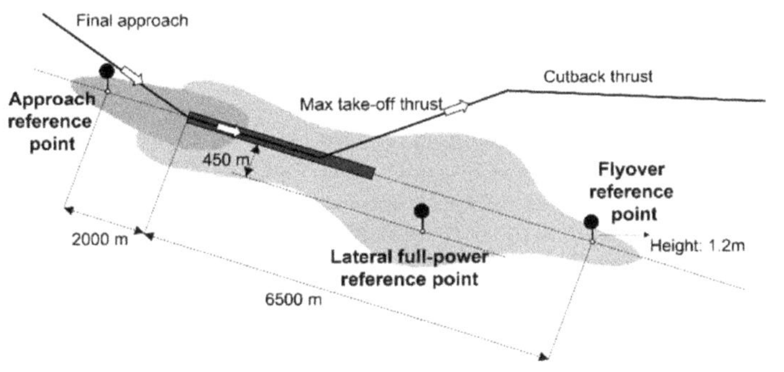

Figure 7. Reference measurement points in noise certification tests.

Table 2. Noise levels for B737-800 and A320-211 aircraft at three certification points, as measured and calculated (in EPNdB) with associated errors (in percentage).

	Approach		Lateral		Flyover	
	A320-211	**B737-800**	**A320-211**	**B737-800**	**A320-211**	**B737-800**
Measurement	96.10	96.30	93.70	93.90	87.40	86.40
Present study	96.32	94.80	94.41	94.71	94.40	94.00
Error	0.229	1.558	0.758	0.863	8.009	8.796

For what concerns the approach procedure, the noise levels predicted at the approach reference point compare favorably with their flight test counterpart, entailing an error of 0.229% and 1.558% for the A320-211 and B737-800 aircraft, respectively. Besides modeling aspects, these slight discrepancies between the predicted and measured results may be attributed to the multiple uncertainties coming from the field test (e.g., slight deviations in the aircraft trajectory, variations in the atmospheric conditions, variable impedance effects by the ground, etc.). For what concerns the departure procedure, the EPNL perceived on the lateral full-power reference point is first considered. Here too, the prediction matches well the flight tests, with an error of less than 1% between the predicted and measured noise levels. Notably, this good agreement also holds when the two aircraft are considered under different atmospheric conditions (as proven by successfully retrieving the prediction results from Ref. [65], which are not reproduced here for the sake of conciseness). The assessment is then conducted for the flyover reference point (see Figure 7), leading to an agreement that is less favorable. Indeed, whatever the aircraft is, the predicted noise levels are about 8% higher than their measured counterpart. Such discrepancies are likely to be due to the mismatches between the flight procedure used in the prediction (ANP database standard scenario) and that adopted in the certification (flight test scenario). Indeed, compared to the former, the latter comes with an earlier reduction in the power settings (thrust is decreased before the flyover reference point [64]), thereby leading to lower noise levels, overall. Rather than questioning the validity of the noise prediction approach (which was demonstrated through all previous cases), these discrepancies rather advocate for basing it on actual flightpaths rather than on standard ANP-based ones—as usually carried out. This point is discussed in the following section.

4. Further Illustration of the Methodology Using Realistic Scenarios

The present section further illustrates the capacity of the aircraft noise prediction methodology to tackle real-life situations by considering actual flight scenarios coming from one of the major international airports, worldwide.

4.1. Context

Hong Kong International Airport (HKIA) is among the busiest airports in the world, ranking 1st and 13th in terms of cargo and passenger traffic, respectively. On the other hand, Hong Kong is the 6th densest city worldwide, and many of its highly populated residential areas are prone to be impacted by air traffic pollution, whether chemical or acoustical. Since HKIA's current two-runway system has almost reached its maximum operational capacity, a third runway is now being developed so as to cope with the expected growth of air traffic (of about 5% per year). This makes it critical to accurately assess the noise impact by aircraft take-off and landing operations at HKIA on local communities. Such an assessment must account properly for the specificities that characterize the local Hong Kong aviation scene, whether these concern aircraft operations (power settings, flight trajectories) or meteorological conditions. Indeed, because of both the complicated airspace (crowded air traffic, topography, densely populated areas, available routes) and the adoption of specific procedures (e.g., continuous descent approach, CDA [66]), the flight trajectories around HKIA are pretty diverse, and definitely non-standard. In addition, due to its unique geographical location, Hong Kong undergoes important seasonal changes over the year (namely, the summer is hot and very humid, whereas the winter is rather cool and dry.).

To further illustrate the ability of the present approach to tackle real-life scenarios, the present section assesses the noise impact entailed by actual aircraft flying in and out from HKIA, a special focus being put on those Hong Kong's densely populated areas that are likely to be more exposed to it. The left side of Figure 8 plots the spatial distribution of Hong Kong's population in 2020 (as provided by the Hong Kong Census and Statistics Department) whereas its center side depicts some typical flightpaths of representative aircraft flying to/from Taipei in 2020 (as provided by the major airline in Hong Kong, Cathay Pacific). The areas depicted in black (namely, D1-D7) are densely populated districts with many aircraft flying over, which the focus is put on here. In that regard, a representative set of aircraft and flightpaths are selected, their noise footprint on the ground being then simulated using the noise prediction platform.

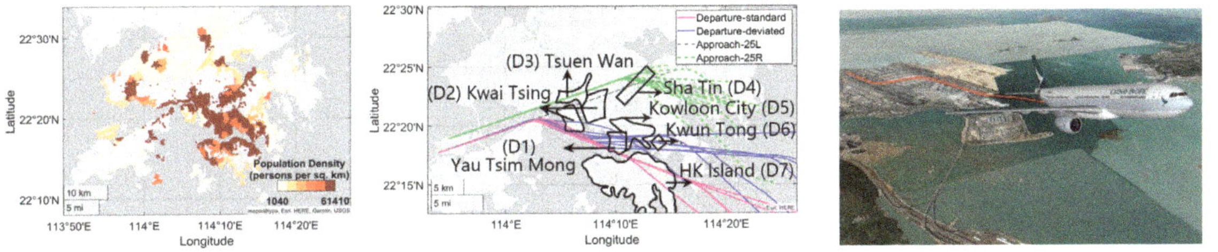

Figure 8. Noise impact by aircraft operation around Hong Kong. Left: Distribution of Hong Kong's population (2020 year). Center: flightpaths of aircraft flying from/to HKIA whereas passing over densely populated districts in Hong Kong (D1 to D7). Right: A330 aircraft departing from HKIA following a standard departure route.

A quick analysis of the air traffic around HKIA in 2021 reveals that the most frequently used aircraft types are A330-343, B787-9 and B777-3ER. We thus here consider both A330-343 and B777-3ER aircraft flying in/out HKIA from/to Taipei (the B787-9 being discarded since its characteristics are not yet incorporated in the ANP database and are rather close to those of A330-343). For each aircraft considered, the operational characteristics and associated flightpaths are excerpted from actual data provided by the airline (Cathay Pacific). Among all available possibilities, specific flights are selected, and their noise impact (SEL) on Hong Kong residential areas is predicted using the present approach. The latter incorporates most features, e.g., it accounts for the atmospheric properties (humidity, temperature, pressure), the installation effects (bank angle), the dynamic grid functionality. On some occasions, additional methodological ingredients are used, e.g., the ground track extension (cf. Section 4.3.2), the thrust approximation (cf. Section 4.3.3). Unless stated otherwise (e.g., Section 4.2.2), the atmospheric conditions correspond to the so-called ISA+10 atmosphere, which is classically used (including for noise certification tests) and is rather close to the yearly averaged meteorological data recorded in Hong Kong. Finally, for helping the reader identifying more easily the various scenarios considered hereafter, each flight is labeled using a specific nomenclature, namely, XYn, where X refers to the aircraft type ("A" for A330-343, "B" for B777-3ER) whereas Y relates to the flight type ("A" for approach, "D" for departure) and n is the subsequent scenario index (1, 2, etc.).

Regarding the aircraft noise exposure in Hong Kong, the solely available data are the official records from HKCAD. The latter come as *averaged* ground noise levels, which are recorded annually over a few specific locations and integrated over time (irrespective of aircraft movements). In the absence of suitable validation means coming from field tests, all the noise predictions presented hereafter are thus used for illustration purpose. They however deliver insightful information, whether the latter relates to phenomenological aspects (see Section 4.2) or to methodological considerations (see Section 4.3). More precisely, here, we assess the noise impact sensitivity towards various key factors, namely the flight characteristics (flightpath, power settings), the meteorological effects (atmospheric absorption, acoustic impedance), the aircraft type, but also the way to infer them better (flightpath extension, trajectory and/or thrust approximation), etc. Of note, these predictions incorporate the ground effects regardless of the specificities coming from the actual terrain (e.g., landscape, buildings, vegetation, or water), which would be too cumbersome to account for. These specificities are nevertheless expected to be of secondary importance compared to the effects that are primarily sought after, here.

4.2. Phenomenological Aspects

4.2.1. Noise Impact Variability upon the Flight Scenarios

Aircraft operations around HKIA rarely comply with standardized flightpaths, being rather characterized by scattered flight profiles. In that regard, the present section illustrates the variability of aircraft noise impact towards the diversity in flight routes.

Departure Scenarios

We consider those departure routes followed by A330-343 aircraft flying from Hong Kong to Taipei, focusing on those Hong Kong densely populated areas that are located underneath flight corridors (namely, D1, D5, D6, and D7 in Figure 8). Among all corresponding flights available from the database, two specific A330-343 departure flights are chosen upon the high disparity between their characteristics. Indeed, the first aircraft (flight AD1) flies over D1 and D7 districts following a standard route, while the second aircraft (AD2) rather flies over D5 district with a much lower rate of climb (ROC). The left and center sides of Figure 9, respectively, depict the ground track and flight profile of the corresponding routes, which are here plotted up to the nominal maximum altitude of 10,000 ft [37]. The right side of Figure 9 plots the observer grid, as generated using the dynamic meshing technique (cf. Section 2.3.3).

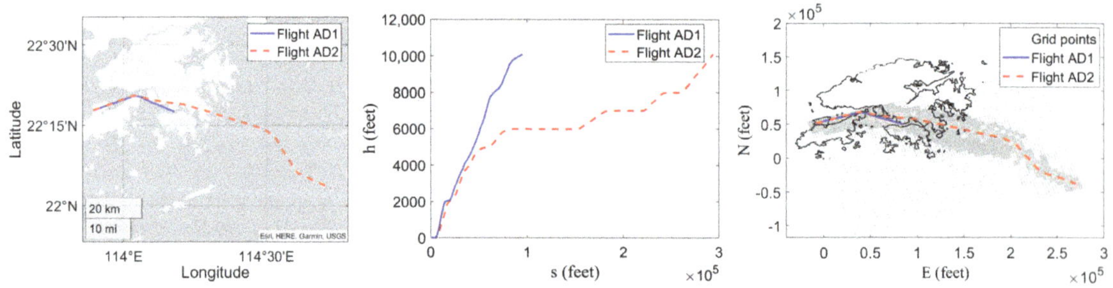

Figure 9. A330 aircraft departing from HKIA. Flights AD1 and AD2 respective routes (**left**: ground track, **center**: flight profile) and corresponding ground observers (dynamic grid, **right**).

The left and right sides of Figure 10 depict the SEL contours generated by flights AD1 and AD2, respectively. As can be seen, each flight exhibits a rather specific noise signature on the ground, thereby impacting Hong Kong in its own way. This disparity between both flights' impacts is quantified on the Figure 11, which depicts the differences between the two noise maps (obtained here by subtracting the former from the latter).

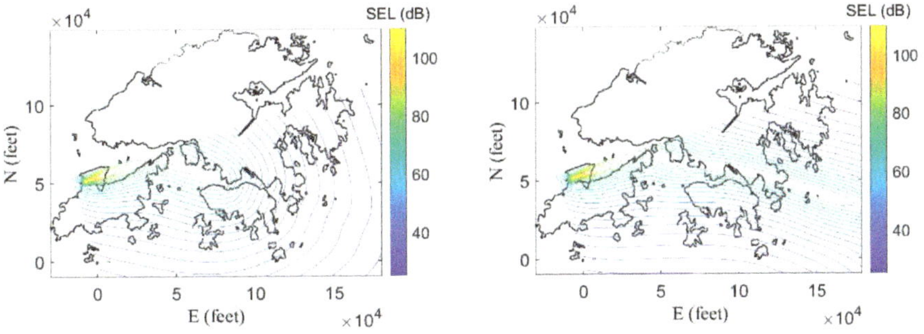

Figure 10. A330 aircraft departing from HKIA. Noise impact (SEL) generated by flight AD1 **(left)** and AD2 (**right**) over Hong Kong city.

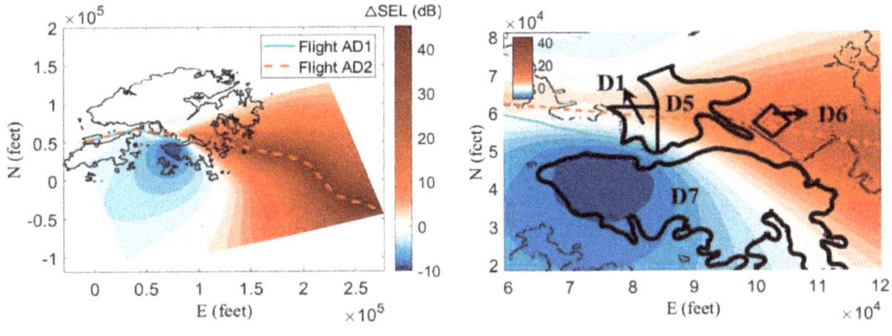

Figure 11. Difference in the noise impact (ΔSEL) by flights AD1 and AD2 (as obtained by subtracting the former from the latter), with a closer view provided on the right side.

Compared to flight AD1 (standard route, higher ROC), flight AD2 (deviated route, lower ROC) has a much higher impact on some of the populated areas, namely districts D1, D5 and D6 (see the right side of Figure 11). More precisely, the excess in SEL raises up to 5dB in districts D1 and D5, whereas it can reach up to 15dB in district D6, which is the most densely populated area in Hong Kong. On the other hand, the opposite holds for other city areas—such as district D7, where the impact of flight AD1 is about 10dB higher than that of flight AD2.

At this stage, it is worth noting that the higher altitude of flight AD1 does not necessarily translate into a lower noise impact on the ground areas it flies over, compared to what happens for flight AD2. For instance, the noise impact by flight AD1 on the D7 district is comparatively higher than that of flight AD2 on the D5 district, although the latter flight travels at an altitude that is half of the former. This can be explained by the difference in the power settings adopted by both aircraft along their respective routes. Indeed, the power settings associated with both flights are depicted in Figure 12, which delineates (in dash-dotted lines) the particular flight phase during which both aircraft pass nearby districts D1 and D5. During this phase, flight AD1 exhibits an engine power that is more than twice that of flight AD2 (whose ROC is suddenly and drastically reduced at that time). This excess of propulsive power translates into a more important noise emission, whose effect cannot be completely mitigated by the longer propagation distance entailed from flight AD1 higher altitude. Independently of the engines power (i.e., noise source amplitude), the flightpath angle (i.e., noise source directivity) may also explain such a comparatively higher noise impact by flight AD1. Indeed, during this flight phase, flight AD2 adopts a lower climb angle than that of flight AD1, as revealed by the flight profiles on the center side of Figure 9. Considering the highly directive patterns of jet noise (whose maximum radiation is aligned with the aircraft axis), this lower flightpath angle by flight AD2 is likely to induce a lower noise impact on the ground. This being said, these more favorable characteristics of flight AD2 (less power, reduced climb angle) do not prevent it to generate a quite substantial noise impact on these areas it flies over rather closely, such as districts D1 and D5.

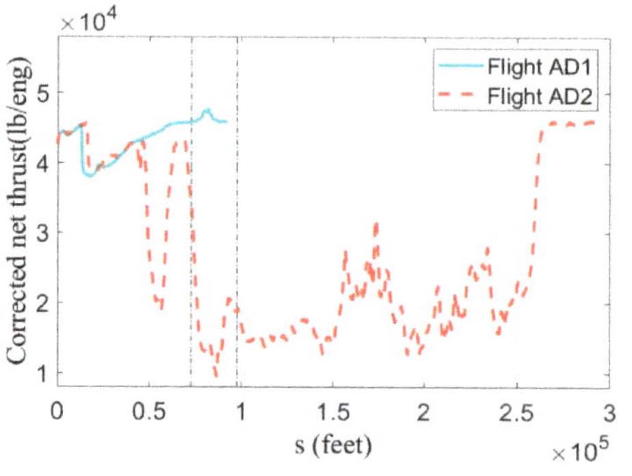

Figure 12. Power settings (engine thrust) evolution of flights AD1 and AD2.

A similar analysis was conducted for B777-3ER aircraft flying along both the standard and an alternative deviated route, leading to the same observations (see Appendix C).

All the above illustrate well how the variability in departure procedures around HKIA may have a quite large impact on specific, highly densely populated areas of Hong Kong city. From a methodological viewpoint, this underlines the necessity to account as accurately as possible for all flight characteristics when assessing the noise impact by air traffic operations.

Approach Scenarios

Whereas the previous section focused on A330 aircraft departing from HKIA (with their B777 counterpart being documented in Appendix C), this section rather focuses on B777-3ER aircraft at approach. More precisely, we consider the two approach routes to HKIA (namely 25R and 25L), which both pass through three densely populated districts (namely D2, D3, and D4 in Figure 8). Among all B777-3ER aircraft flying from Taipei to Hong Kong in 2020, two specific flights are selected for their high disparity in terms of characteristics (see Figure 13). Whereas the first one (flight BA1) approaches the 25L runway at a higher altitude, the second one (flight BA2) follows the route 25R at a lower altitude. As can be seen, these flight characteristics differences are more prominent at the beginning of the flight sequence, since both aircraft adopt similar trajectories as they get closer to HKIA, to comply with the noise abatement procedures enforced in Hong Kong, namely the Continuous Descent Approach.

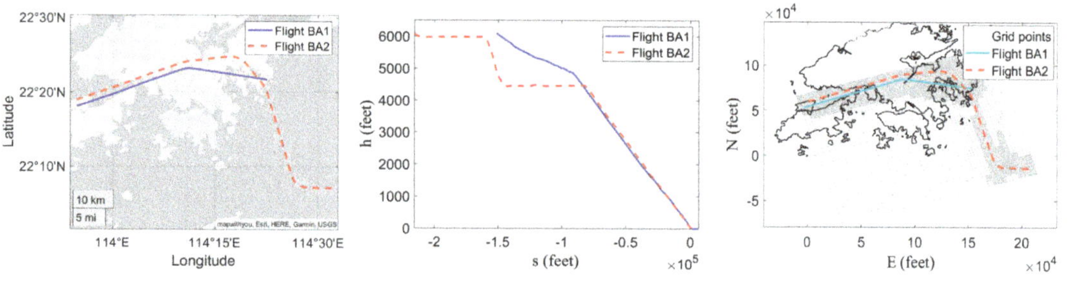

Figure 13. B777 aircraft approaching HKIA. Flights BA1 and BA2 respective routes (**left**: ground track, **center**: flight profile) and corresponding ground observers (dynamic grid, **right**).

The left and right sides of Figure 14 depict the SEL contours generated by these two flights BA1 and BA2, respectively. As for what could be observed with the departure scenarios (cf. Section 4.2.1), each flight impacts differently Hong Kong residential areas. This disparity between their respective noise impact is quantified on the Figure 15, which depicts the differences between the two noise maps (obtained here by subtracting the former from the latter).

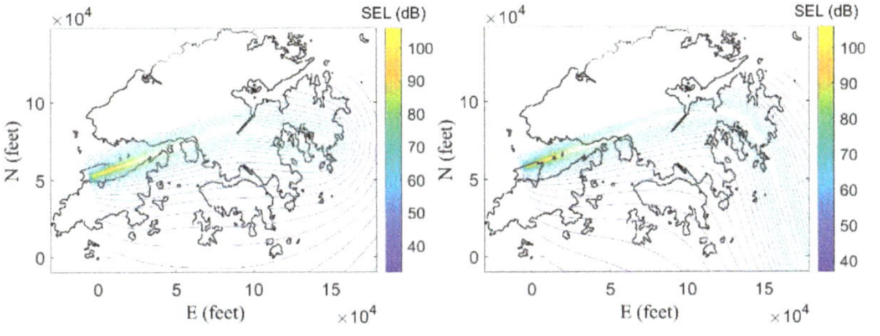

Figure 14. B777 aircraft approaching HKIA. Noise impact (SEL) generated by flight BA1 (**left**) and BA2 (**right**) over Hong Kong city.

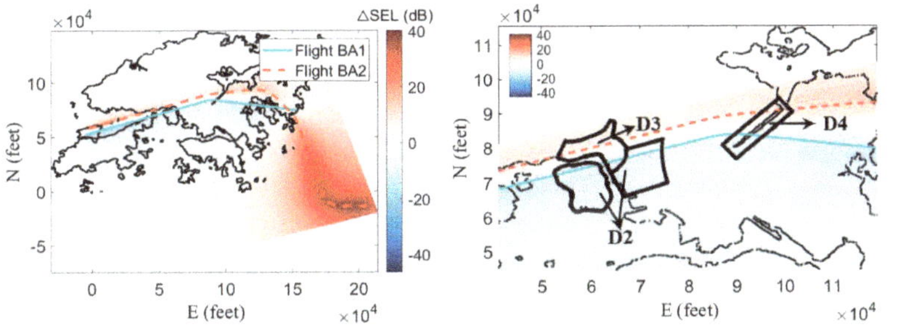

Figure 15. Difference in the noise impact (ΔSEL) by flights BA1 and BA2 (as obtained by subtracting the former from the latter), with a closer view provided on right side.

Being sandwiched between 25L and 25R routes, the populated D3 district sees its northwestern part impacted more by flight BA2 than by flight BA1 (with a difference in SEL of up to 7.6 dB), whereas the opposite occurs on D3 southern part (where flight BA1 impact exceeds that of flight BA2 by up to 7.3 dB in SEL)—see the right side of Figure 15.

Considering that most of D3 residents are concentrated in its central and southern areas, flight BA1 thus appears to be less environmentally friendly for this district. The same observation holds for the D2 district, whose population concentration is four times higher than that of its D3 counterpart. Being located right underneath 25L route, this district is impacted more by flight BA1, with an excess in SEL of up to 9dB (compared to that of flight BA2). On the other hand, the D4 district appears to be equally exposed to both flights, each inducing an excess of 5–7 dB compared to the other for those observers that are located underneath their respective flightpath. All in all, it appears that these densely populated districts are more exposed to flight BA1, overall. Therefore, the 25R approach route should be privileged whenever the conditions permit (operational, meteorological, etc.).

A similar analysis was conducted for an A330-343 aircraft flying along these two approach routes, leading to the same observations (see Appendix C).

As for the departure scenarios (cf. Section 4.2.1), the above results illustrate further how the approach procedures around HKIA may impact very differently those highly densely populated areas of Hong Kong city.

4.2.2. Noise Impact Variability upon the Meteorological Conditions

As said, because Hong Kong usually undergoes important seasonal changes, the local atmospheric conditions may present large deviations compared to the standard ones. For instance, according to the meteorological data records acquired over the past decade by the Hong Kong Observatory and Iowa State University, Hong Kong is characterized by large yearly variations in temperature and relative humidity (which can range from 3 °C to 37 °C and 6.91% to 100% throughout the year, respectively). Moreover, these fluctuations may be important on a much shorter timescale, with large deviations occurring throughout a single day.

The present section illustrates how such a meteorological variability may affect the aircraft noise impact on the ground, for instance, because of the deviations in the atmospheric absorption (which focus is put on, here). To do so, we consider the particular case of an A330-343 aircraft flying from Hong Kong to Taipei along a standard departure route, this flight being successively operated under four specific meteorological scenarios, which reproduce the extrema in temperature or relative humidity one can get in Hong Kong, yearly (see Table 3). For each one of the four scenarios (respectively labeled as AD3, AD4, AD5, AD6), Figure 16 depicts the corresponding SEL on the ground, whose differences are quantified and discussed in the next subsections.

Table 3. Yearly extrema in atmospheric conditions recorded in Hong Kong (sea level pressure) and meteorological scenarios considered for the four flights.

		Temperature (°C)	
		7	**35**
Relative humidity (%)	13	AD3	AD4
	100	AD5	AD6

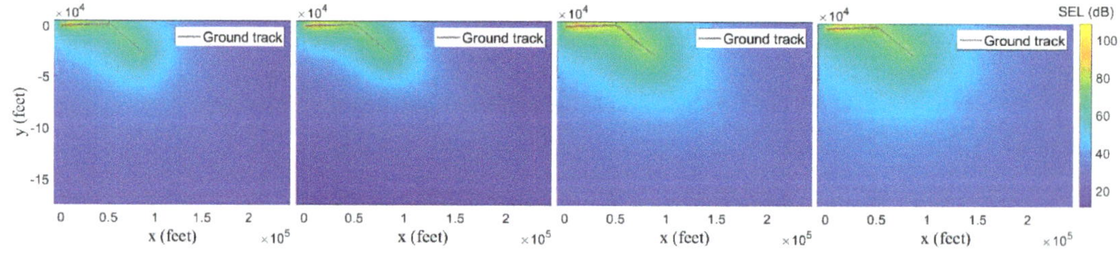

Figure 16. A330 aircraft departing from HKIA. Aircraft noise impact on the ground, as depicted in terms of SEL for cases AD3, AD4, AD5, AD6 (from left to right, respectively).

Effect of Temperature

By comparing either cases AD3 and AD4 or cases AD5 and AD6, we first highlight how the noise impact may be altered by variations in temperature (alone), for a given relative humidity. For each pair of cases, Figure 17 depicts the differences in ground noise levels, as obtained by subtracting the SEL map associated with case AD3 (resp. AD5) from that of case AD4 (resp. AD6). As can be seen, such a 28 °C rise in temperature noticeably impacts the ground noise levels, whatever the relative humidity is. This effect on the noise impact appears to be driven by both the observer location and the actual level of relative humidity. When the latter is low (13%), the noise impact is seen to increase (resp. decrease) for those observers that are located close to (resp. away from) the ground track (cf. left side of Figure 17). Oppositely, when the relative humidity is high (100%), a lower (resp. higher) noise impact is recorded for those observers that are positioned close to (resp. away from) the ground track (cf. right side of Figure 17).

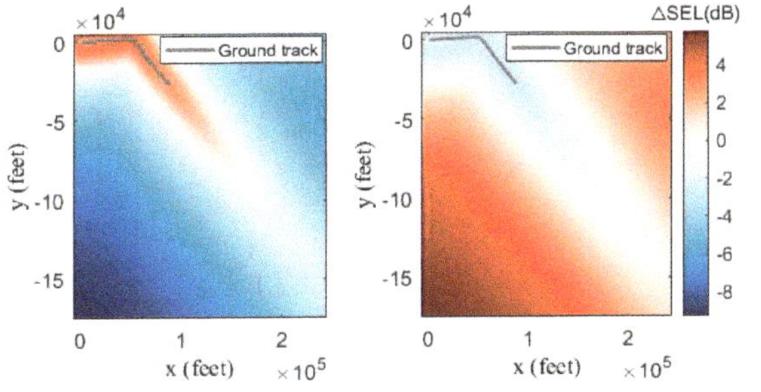

Figure 17. Effect of the atmospheric properties on the aircraft noise impact. Differences in SEL induced by a variation in temperature, as obtained for a given relative humidity (**left**: AD4—AD3, **right**: AD6—AD5).

Effect of Relative Humidity

Similarly, by comparing either cases AD3 and AD5 or cases AD4 and AD6, we highlight how the noise impact may be altered by the variations in relative humidity (alone), for a given temperature. Figure 18 depicts the differences in ground noise levels, as obtained by subtracting the SEL map associated with case AD3 (resp. AD4) from that of case AD5 (resp. AD6). Whatever the temperature is, such an 87% variation in relative humidity significantly alters the ground noise levels, which now increase everywhere. Again, this effect on the noise impact appears to depend on both the observer location and the actual temperature. When the latter is low (7 °C), the impact is seen to be more important for those observers that are located close to the ground track (cf. left side of Figure 18). Oppositely, under a high temperature (35 °C), the impact is higher for those observers that are positioned further away from the ground track (cf. right side of Figure 18).

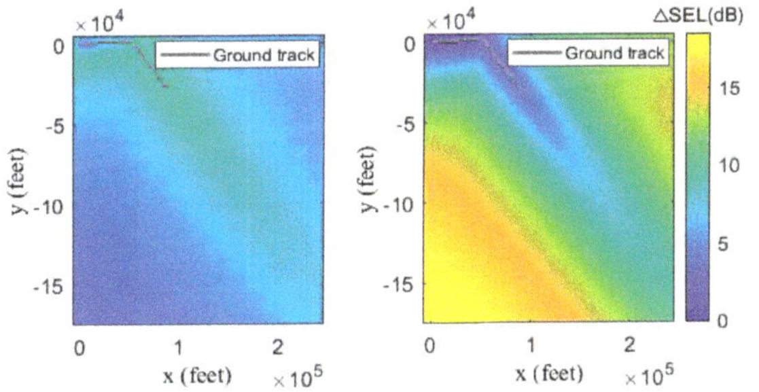

Figure 18. Effect of the atmospheric properties on the aircraft noise impact. Differences in SEL induced by a variation in relative humidity, as obtained for a given temperature (**left**: AD5—AD3, **right**: AD6—AD4).

Combined Effect of Both Temperature and Relative Humidity

Finally, by comparing cases AD3 and AD6, we illustrate the combined effect of temperature and relative humidity on the noise impact. Figure 19 depicts the differences in ground noise levels, as obtained by subtracting the SEL map associated with case AD3 from that of case AD6. As can be seen, when both temperature and relative humidity increase, the ground noise impact goes higher. For instance, a variation of +9.6 dB is recorded for that observer located at the left/bottom corner of the grid. Notably, this difference represents the arithmetic sum of the ones recorded separately for a variation in temperature alone (−9.1 dB, cf. left side of Figure 17/AD3 → AD4) and in relative humidity alone (+18.7 dB, cf. right side of Figure 18/AD4 → AD6). This illustrates well the dual dependency of the noise impact onto both the temperature and humidity, whose respective effects are cumulative (and may mitigate each other, depending on the situation, cf. Section 2.3.4).

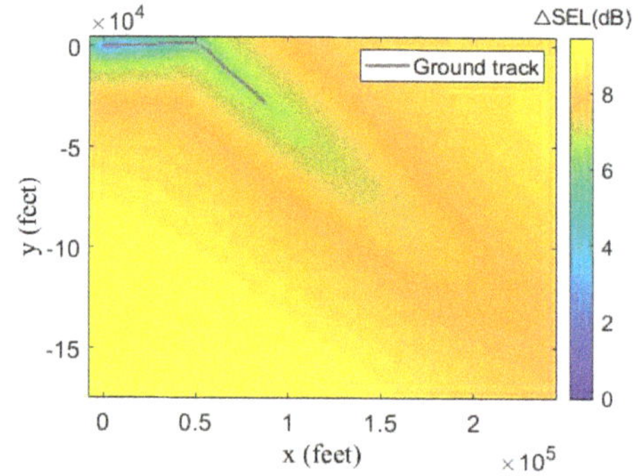

Figure 19. Effect of the atmospheric properties on the aircraft noise impact. Differences in SEL induced by a variation in both temperature and relative humidity (AD6—AD3).

At this stage, it is worth observing that the above results (both trends and numbers) are specific to the present HKIA-based scenario and should not be extrapolated readily to other situations. They however highlight well how important the atmospheric properties may alter the noise impact by air traffic operations, from a phenomenological viewpoint. From a methodological perspective, they underline the crude necessity of accounting properly for the local atmospheric conditions when assessing the noise impact of air traffic operations, especially in those regions that are known to undergo important seasonal variations, such as Hong Kong.

4.2.3. Noise Impact Variability upon the Aircraft Type

The previous sections highlighted how far a given aircraft may see its noise impact varying, depending on its operational conditions (flight route and power settings, meteorological conditions, etc.). Here, we rather explore how, for identical operational conditions, two distinct aircraft may entail different noise impacts. To do so, we consider two of the previous flights, whereas allotting them with a different aircraft type than previously carried out.

Focusing first on departure scenario (cf. Section 4.2.1), we repeat the noise prediction associated with flight AD2 (deviated route, lower ROC), the previous aircraft (A330-343) being now replaced with a (virtual) B777-3ER. Of note, it was checked that such a virtual scenario is realistic, e.g., the B777 engines can deliver the power required by the corresponding A330 flight settings. Figure 20 depicts the SEL maps induced by both cases (respectively labeled AD2 and BD1), along with their difference. As can be seen, B777-3ER aircraft impacts Hong Kong less than its A330-343 counterpart, with noise levels that are lower by about 3 dB over most residential areas. This trend is however inverted further away from the ground track, with an excess noise by the B777 of up to 5 dB recorded over some more remote locations (which are nevertheless sparsely populated).

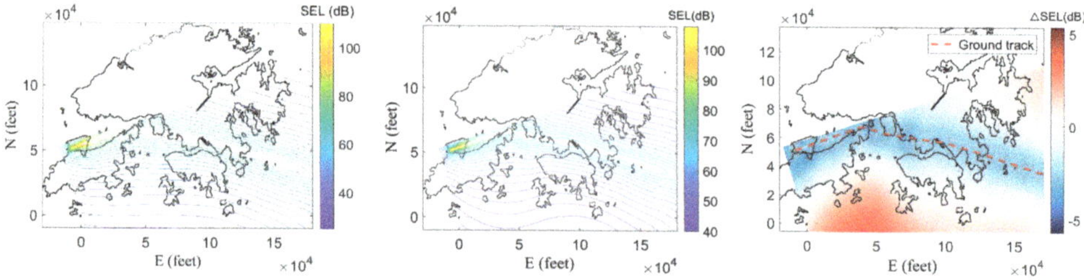

Figure 20. A330 and (virtual) B777 aircraft departing from HKIA. Noise levels (SEL) generated by flight AD2 (**left**) and BD1 (**center**) over Hong Kong city, along with their difference (obtained by subtracting the former from the latter, **right**).

At this stage, one could wonder about the exact reason for such an inverted trend, in terms of noise impact by both aircraft with respect to the distance to the ground track. Indeed, both AD2 and BD1 scenarios differ only by the aircraft type (i.e., noise source), all other parameters being strictly identical (flightpath, power settings, atmospheric absorption, etc.). However, as can be inferred from the NPD database (see Figure 21), both aircraft see their noise levels varying (i.e., decreasing) differently along the propagation distance. In particular, although they are less than the A330 ones at short distances, the B777 noise levels emerge more further away. This can be related to the atmospheric attenuation effects, which are to impact differently both noise signatures depending on their respective spectral content.

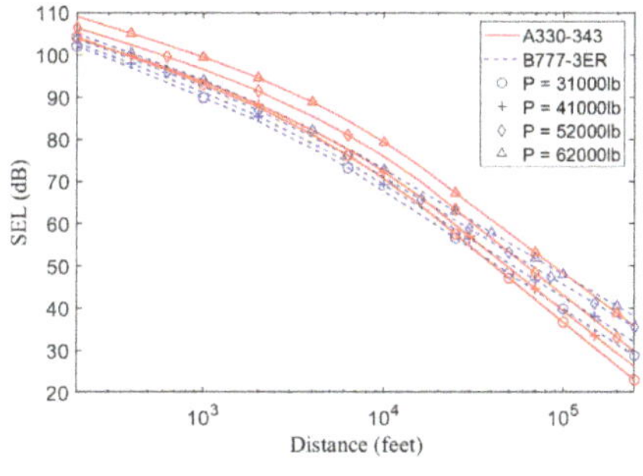

Figure 21. A330 and B777 aircraft noise impact (SEL) as recorded for various power settings (corrected thrust) and propagation distances. Excerpted from NPD database.

Focusing now on approach scenarios (cf. Section 4.2.1), we repeat the noise prediction associated with flight BA1 (25L runway, lower altitude), the previous aircraft (B777-3ER) being now replaced with a (virtual) A330-343 counterpart. Notably, here too, it was checked that such a virtual scenario is realistic, e.g., the A330 engines can deliver the power required by the corresponding B777 flight settings. Figure 22 depicts the SEL maps induced by both flights (respectively labeled BA1 and AA1), along with their difference. As can be seen, compared to its B777-3ER counterpart, A330-343 aircraft impacts Hong Kong less, with noise levels that are systematically lower—wherever the observer location is. In particular, the populated D6 district sees its noise impact reduced by as much as 5 dB when flown over by an A330 instead of a B777.

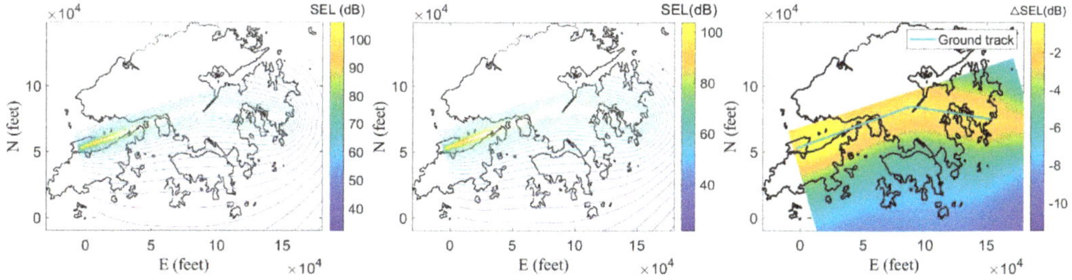

Figure 22. B777 and (virtual) A330 aircraft approaching HKIA. Noise levels (SEL) generated by flight BA1 (**left**) and AA1 (**center**), along with their difference (obtained by subtracting the former from the latter, **right**).

The above results illustrate how aircraft noise prediction methods such as the one presented here constitute not only a useful diagnostic tool but also a powerful predictive means. Indeed, beyond simply allowing to measure the noise impact by existing aircraft operations around major airports, such kind of tool can also help better planning them. This is of critical importance when new airport developments are on their way, as is currently the case with HKIA.

4.3. Methodological Aspects

Whereas the previous sections primarily focused on phenomenological aspects of aircraft noise impact, we here discuss the prediction process under a more methodological angle.

4.3.1. Accounting for the Bank Angle

In the previous sections, all noise predictions accounted for the engine installation effects (cf. Section 2.3.2), which were incorporated through the bank angle. One can however wonder about how big their influence on the noise impact is, that is, how critical it is to integrate them in the calculation process. To do so, we repeat the noise prediction of flight AD2 (which undergoes more banked turns than flight AD1), the bank angle effect being either incorporated or neglected in the calculation. Figure 23 compares the SEL maps obtained with (left) and without (center) the bank angle effect, along with their difference (right).

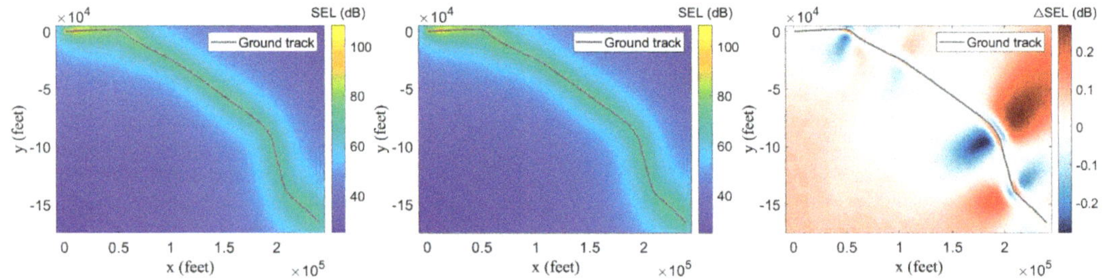

Figure 23. A330 aircraft departing from HKIA. Noise impact (SEL) generated by flight AD2 with (**left**) and without (**center**) the incorporation of the installation effect (bank angle), along with the difference between both results (obtained by subtracting the former from the latter, **right**).

As can be seen, when the aircraft is engaged into a banked turn, most of the observers that are located on the same (resp. opposite) side of the turn are exposed to lower (resp. higher) noise levels (see what happens on the right side of Figure 23, for instance). This trend, however, is inverted for the small fraction of observers that are located very close to the ground track. All this can be easily explained upon the respective way each observer perceives the *relative* depression angle, which is either increased or decreased because of the bank (see Figure 2, wing-mounted engines polar). From another hand, this effect of the bank angle appears to be rather modest, with noise differences of less than 0.3 dB overall. This number appears to be even lower for flight AD1, whose straighter flightpath naturally leads to a less important effect of the bank angle (cf. Appendix C).

A similar analysis was conducted for B777-3ER aircraft flying along different departure routes, leading to the same trends (see Appendix C).

On one hand, the above results confirm that the noise impact by an aircraft does not vary much during a banked turn, which explains why this aspect is often neglected in the prediction [62]. On the other hand, however, these results illustrate how one can further refine the accuracy of the noise prediction by incorporating still more additional effects, even though they are of secondary importance.

4.3.2. Extending the Flightpath Cut-off Limit

In the previous sections, all noise predictions relied on the classical approach of incorporating only those flightpath segments that would fall under an (arbitrary) reference altitude of 10,000 ft. As discussed previously (see Section 2.3), this is likely to induce a bias in the noise prediction because some of the contributing flight segments would then be ignored. In Section 4.2.1, for instance, flight AD1 (standard route, higher ROC) reaches the limit altitude much earlier than flight AD2 (deviated route, lower ROC), which may translate into a predicted noise impact that becomes less than the actual one. To illustrate this point, the noise prediction of flight AD1 is here refined, being now conducted using the automatic flightpath extension proposed in Section 2.3.3 (i.e., all contributing flight segments are incorporated in the prediction). As revealed by Figure 24, the noise impact by such an extended flight (labeled hereafter AD1-ext) goes on increasing beyond its former (10,000 ft limited) value until it reaches a plateau (around an altitude that is almost twice, i.e., 18,670 ft). This leads to a ground SEL that is higher, with differences of up to 30 dB at some locations (see right side of Figure 24).

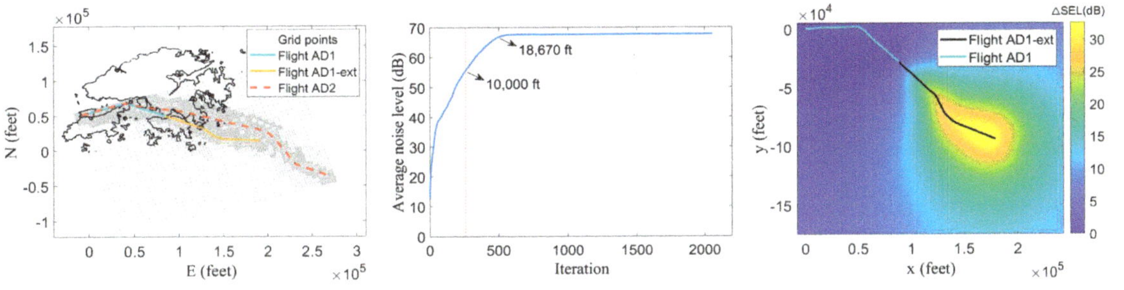

Figure 24. A330 aircraft departing from HKIA. Impact of the flightpath extension, as revealed by comparing flight AD1-ext to AD1. **Left**: respective routes and ground observers. **Center**: history of the ground noise impact by the AD1-ext flight, plotted as an averaged (over all observers) and cumulative (over all flight segments) measure. **Right**: Differences between the ground noise levels recorded for AD1-ext and AD1 flights.

The above result illustrates well how aircraft noise predictions can be biased by the arbitrary limitation of the route beyond a given reference altitude. In turn, it shows how to avoid such a bias by using a dynamic flightpath extension, such as the one proposed here.

4.3.3. Approximating the Flightpath and/or Power Settings

As said, aircraft operations around HKIA rarely comply with standardized flightpaths, being rather characterized by scattered flight profiles (cf. Figure 25). In the absence of any FDR (Flight Data Recorder) inputs, modeling these flights requires using radar information and inferring power settings, for instance using the model proposed in Section 2.3.1. Radar data, however, are of limited accuracy [67,68], and it is not rare that they noticeably differ from the actual (FDR) ones. These discrepancies may question the validity of either the thrust estimation (which requires knowing accurately both the acceleration and attitude of the aircraft, cf. Section 2.3.1) or of the noise prediction (which relies on the correct knowledge of the aircraft-to-ground distance, i.e., flight trajectory). Assessing this point is the matter of the present section.

We here consider two pairs of specific flights, each relating to an identical aircraft and mission (B777-3ER or A330-343 aircraft flying from Hong Kong to Taipei) whereas presenting rather different characteristics in terms of flight profile (cf. Figure 25) and/or engine power evolution (cf. top right of Figures 26–29). Here, we solely consider departure scenarios since those are primarily concerned with pure propulsive noise, for which a correct estimation of the thrust is vital. We first assess the validity of the thrust estimation alone, this being performed by applying Equations (2)–(4) to accurate trajectory inputs coming from FDR data. The latter are then replaced with their (less accurate) radar data counterparts, which are extracted from FlightRadar24 website (business version). The objective is to assess how far the radar data inaccuracies may impact the thrust estimation, and ultimately, the noise prediction. Of note, since the present radar data solely provide the aircraft ground speed, we infer its airspeed counterpart by adding to it a headwind component (whose value is arbitrarily taken as 8 kt [49], for a lack of better information on wind profiles). Besides, here, the atmosphere characteristics are based on the average values recorded in Hong Kong for 2020.

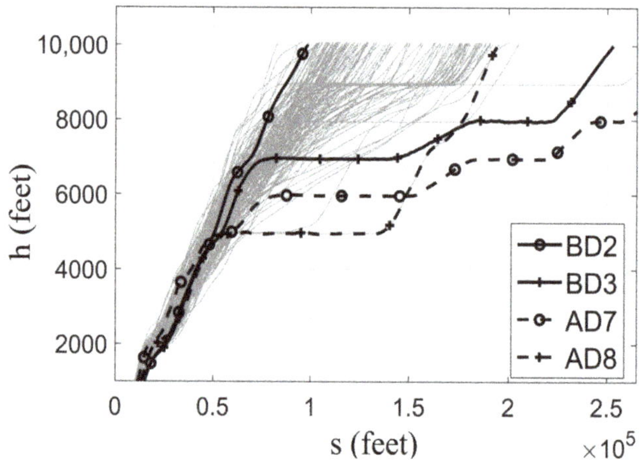

Figure 25. Flight profiles of B777 and A330 aircraft departing from HKIA (grey curves), with two pairs of flights selected (namely, BD2 and BD3, AD7 and AD8).

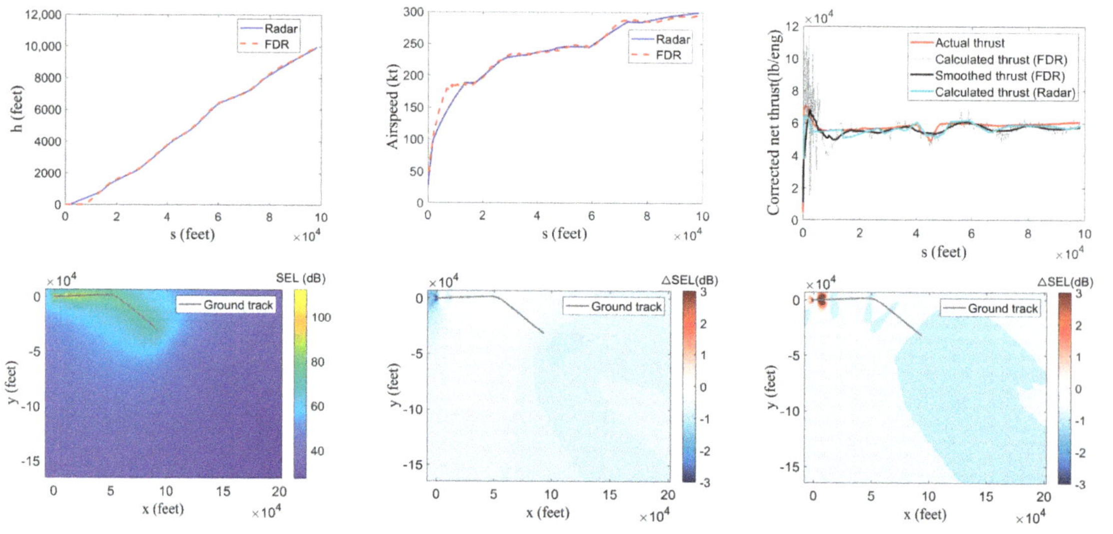

Figure 26. B777 aircraft departing from HKIA (flight BD2). **Top**: Comparison between the flight profile (**left**), airspeed (**center**), calculated and actual thrust (**right**) inferred from the FDR and radar data. **Bottom**: Noise contour map, as obtained using FDR data and actual thrust (**left**). Error entailed by an approximation of the thrust (**center**: smooth calculated thrust) plus the trajectory estimation (**right**: radar data, calculated thrust).

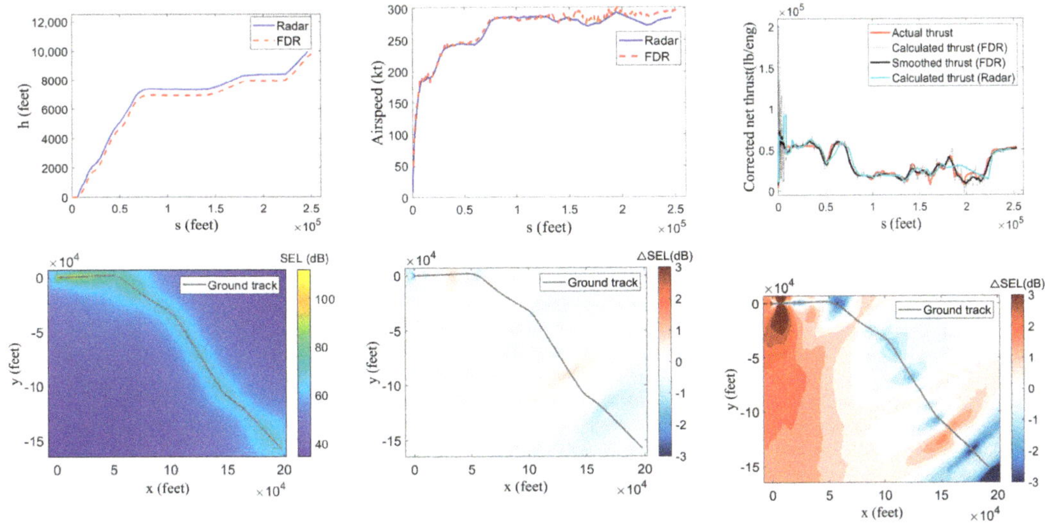

Figure 27. B777 aircraft departing from HKIA (flight BD3). **Top**: Comparison between the flight profile (**left**), airspeed (**center**), calculated and actual thrust (**right**) inferred from the FDR and radar data. **Bottom**: Noise contour map, as obtained using FDR data and actual thrust (**left**). Error entailed by an approximation of the thrust (**center**: smooth calculated thrust) plus the trajectory estimation (**right**: radar data, calculated thrust).

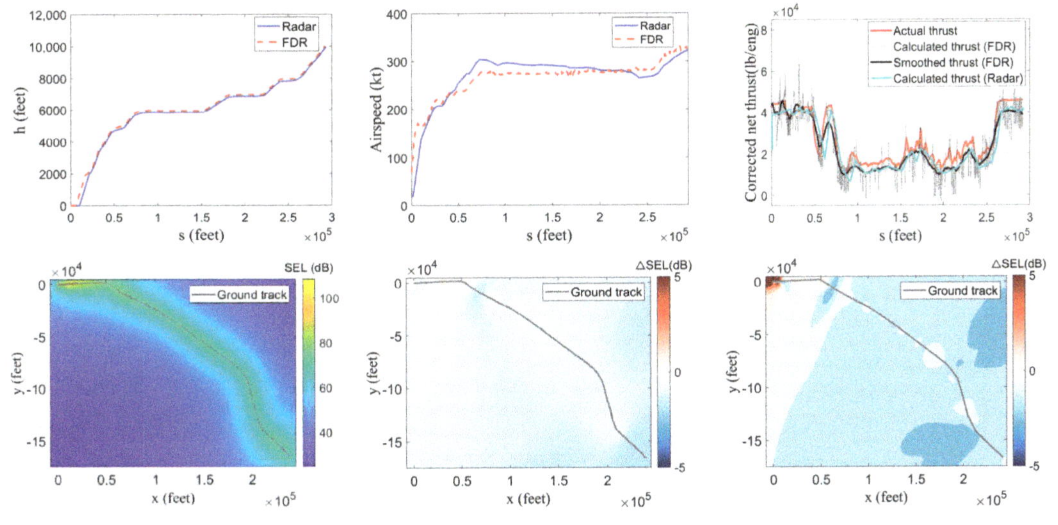

Figure 28. A330 aircraft departing from HKIA (flight AD7). **Top**: Comparison between the flight profile (**left**), airspeed (**center**), calculated and actual thrust (**right**) inferred from the FDR and radar data. **Bottom**: Noise contour map, as obtained using FDR data and actual thrust (**left**). Error entailed by an approximation of the thrust (**center**: smooth calculated thrust) plus the trajectory estimation (**right**: radar data, calculated thrust).

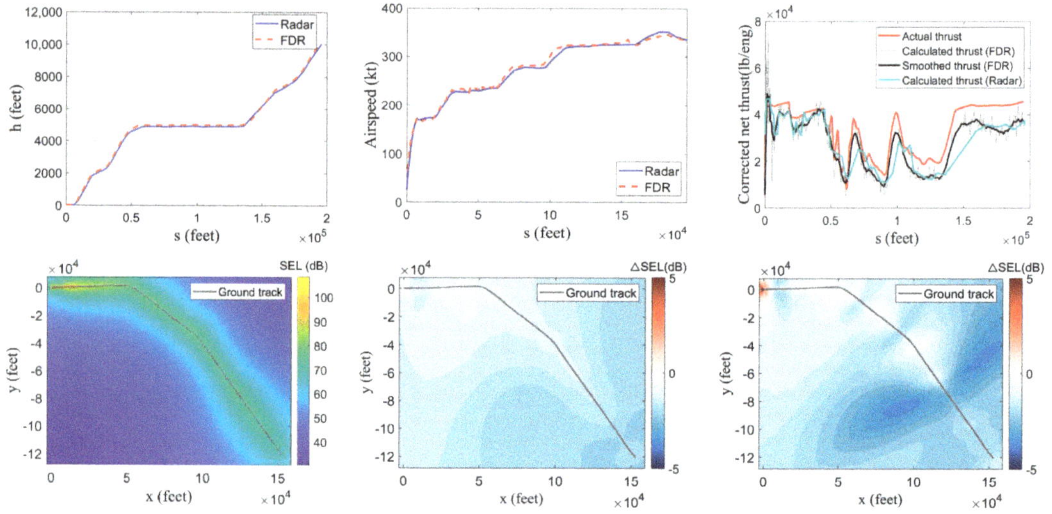

Figure 29. A330 aircraft departing from HKIA (flight AD8). **Top**: Comparison between the flight profile (**left**), airspeed (**center**), calculated and actual thrust (**right**) inferred from the FDR and radar data. **Bottom**: Noise contour map, as obtained using FDR data and actual thrust (**left**). Error entailed by an approximation of the thrust (**center**: smooth calculated thrust) plus the trajectory estimation (**right**: radar data, calculated thrust).

For both pairs of flights (hereafter labeled as BD2, BD3 and AD7, AD8), the thrust evolution is first estimated using the actual (FDR) data, its averaged value (labeled as "smoothed thrust") being then compared with its actual counterpart (labeled as "actual thrust"). As can be seen on the top right of Figures 26–29, the agreement is fairly good, whatever the scenario is. In particular, for both B777 flights (BD2, BD3), the estimation is very satisfactory, with an estimated thrust that matches the actual one almost perfectly (compare black and red curves on the top right of Figures 26 and 27, respectively). This good match slightly degrades for both A330 flights (AD7, AD8), whose actual and predicted thrust differ in some places (compare again the black and red curves on the top right of Figures 28 and 29, respectively). These discrepancies come from the much less monotonic character of the flight (and, thus, power) settings, whose important variations make the thrust estimation more challenging. Despite such a fussy timeline of the power settings (which turned out to be the more complicated ones, among all the considered cases), the thrust estimation is still very favorable, delivering trends that are quite close to the actual result. This fair character of the thrust estimation naturally translates into a fair prediction of the aircraft noise impact, as shown on the bottom of Figures 26–29: whereas the left image plots the actual SEL map (predicted using the actual thrust), its center counterpart depicts the error incurred by the thrust estimation (obtained by subtracting the former SEL map from the one predicted using the estimated thrust). Whatever the considered flight is, this error on the SEL prediction (ΔSEL) is mostly in the range of $\pm$ 1dB: whereas it can reach up to $\pm$2.5 dB at the beginning (start-of-roll) and/or the end of the flight sequence (standard limit altitude of 10,000 ft), the ΔSEL never exceeds $\pm$ 1dB in the rest of the flight sequence. Such discrepancies, which primarily reflect the accuracy limitations of FDR data (e.g., lack of discretization, absence of information on flap settings, etc.), are deemed to be acceptable for such low-order modeling of the thrust.

Independent of the thrust model itself, the uncertainties that weigh on the flight settings (trajectory, airspeed) provided by the radar data may however entail additional errors. To illustrate this point, we here re-assess the thrust estimation, its FDR-originated inputs being now replaced with their radar data counterpart. The latter may indeed importantly differ from the former, whether this concerns the flight profile (cf. top/left images of Figures 26–29) or the airspeed (cf. top/center images of Figures 26–29). Notably, only the actual ground track appeared to be accurately reproduced by the radar data, across all cases. For each flight, the cumulative effect of all these discrepancies onto the aircraft noise is again quantified in terms of ΔSEL maps (see bottom right image of Figures 26–29). The latter are obtained by subtracting the actual ground noise levels (predicted using the FDR inputs, among which is the actual thrust) from their approximated counterpart (predicted using the radar-originated inputs and subsequently estimated thrust). Overall, the fidelity of the noise prediction appears to be driven by the radar data accuracy. Considering for instance the BD2 flight, one can see that the radar-originated flight profile, airspeed and thrust do not devoid much from their actual counterparts (the early take-off excepted). The resulting bias on the noise prediction is fairly low (cf. bottom right of Figure 26), with deviations of less than 0.5 dB for most of the flight sequences (except at the take-off start, where radar uncertainties entail noise discrepancies of about 15 dB). For BD3 flight, the radar-estimated altitude is higher than its actual value throughout the entire flight sequence (so much that the 10,000 ft limit altitude is reached earlier), which results in slightly higher error levels on the noise prediction ($\Delta SEL \sim 1$ dB, overall). Notably, the radar-based estimated thrust still agrees well with both its actual and its FDR-estimated counterparts (compare the blue, red and black curves on the top right of Figure 27). Only at the start and at the end of the flight sequence, this agreement on both the thrust estimation and the noise impact appears to be less satisfactory. This may tend to indicate that, here, the radar data inaccuracies primarily weigh on the thrust prediction, more than on the propagation aspects (e.g., the aircraft-to-ground distance). Notably, because their respective discretization levels differ, FDR and radar data exhibit slightly different start-of-roll point, lift-off point and endpoint. Similar trends are observed for the AD7 flight, whose radar-originated airspeed appears to be less accurate than its flight profile counterpart (cf. left and center images on top of Figure 28, respectively). Combined with the fuzzy evolution of the power settings, these radar data inaccuracies challenges still more the thrust estimation (compare the blue, red and black curves on the top right of Figure 28). Besides, such a variability of the radar data airspeed is very likely to incur a different exposure duration to the noise associated with each flight segment, thereby further affecting the entire prediction, ultimately. Despite all these discrepancies, however, the noise impact is predicted rather accurately, with error levels that fall in the range of $[-2; 0]$ dB (except again near the start-of-roll point, where differences reach 40 dB). Finally, AD8 flight reveals how radar data may resemble their actual counterparts (cf. left and center images on top of Figure 29) whereas entailing still more bias on the thrust estimation (compare the blue, red and black curves on the top right of Figure 29). At this stage, however, it is believed that the lesser discretization of the radar data (compared to their FDR counterpart) might explain such discrepancies on the thrust, whose estimation might then be still more challenged by the fuzzy timeline of the power settings. Whatever the reason for such biases on the thrust estimation, the overall impact on the noise prediction is still acceptable, with ΔSEL that never goes under -3 dB (cf. bottom right of Figure 29).

All in all, the above results illustrate the ability of the noise prediction method (and its underlying thrust model) to solely rely on radar data, thereby allowing to handle flightpaths that differ from standard ones whenever FDR information is unavailable.

5. Conclusions and Perspectives

The present study focuses on the noise impact by aircraft operations around major airports. To this end, an aircraft noise prediction platform was developed, which relies on state-of-the-art functionalities as well as more specific, innovative features. More precisely, the method classically relies on the Aircraft Noise and Performance (ANP) database and its Noise–Power–Distance (NPD) table. The latter, however, are known to suffer from restrictive assumptions which limit their application to simplified situations. For instance, the ANP database is built upon standardized scenarios of aircraft operations (aircraft types, flightpaths, power settings, atmospheric conditions, etc.). Besides, its NPD component relies on the simplistic scenario of a punctual noise source radiating within a homogeneous free field. Last but not least, the ground observer is assumed to be located right underneath the aircraft, with the latter flying with a constant speed and power settings along an infinite horizontal flightpath. To alleviate all these limitations, the present method incorporates several functionalities, some of which are specific to the present study. These features aim at (i) refining the noise emission stage (e.g., source intensity and directivity) as well as (ii) improving the propagation phase. Regarding for instance the refinement of the aircraft noise source, an innovative method is proposed, which allows inferring the engine power solely from the aircraft flightpath characteristics (whose access is much easier than that of Flight Data Recorder). Besides this, a functionality is introduced which allows extending automatically and dynamically the flightpath, thereby avoiding an arbitrary truncation of the aircraft noise emission as well as ensuring an accurate representation of its subsequent propagation. The latter propagation is also refined through the proper incorporation of the noise attenuation effects induced by any realistic, non-standard atmosphere. Other specific features are incorporated into the method, which aim at improving its efficiency (accuracy and execution speed). For instance, a recursive, dynamic grid refinement technique offers to optimize the number and distribution of ground observers, thereby maximizing the results accuracy whereas minimizing the prediction time.

The noise prediction method and subsequent computational platform are successfully validated using several benchmark cases of increasing representativeness. They are then applied to several realistic scenarios coming from actual aircraft operations around Hong Kong International Airport (HKIA). Specific comparative analyses are conducted, which allow highlighting the variability of the noise impact by aircraft, depending on their type (A330, B777) and/or operational conditions (power settings, meteorological conditions, routes, banks, etc.). From a phenomenological viewpoint, results allow discriminating the more prominent drivers, namely, the aircraft flightpath characteristics, the engine power settings, etc. It is also shown how important the meteorological conditions can be, because of the specific atmospheric attenuation effects they entail. Oppositely, the engine installation effects (assessed through the aircraft bank effects) appear to be of secondary importance. From a methodological perspective, the results showcase the capacity of the present approach to handling real-life situations, including when the latter greatly differ from the standardized scenarios underlying the ANP database. They also illustrate how noise prediction methods/platforms such as the present one may help in guiding the further expansion of airport operations and/or infrastructures (as is currently the case with HKIA).

With this end in view, it is planned to further improve the present approach, whether it is by refining still more its noise emission stage (e.g., assessing better the aircraft power settings using the Base of Aircraft Data database, BADA [69–71]), or by strengthening further its propagation phase (e.g., incorporating variable atmospheric conditions such as wind effects [72–74]). Last but not least, it is envisioned to extend the tool's capacities from an aircraft to a fleet level [75] so as to allow better optimizing air traffic operations.

Supplementary Materials: The following are available online at https://www.mdpi.com/article/10.3390/aerospace8090264/s1.

Author Contributions: Conceptualization, S.R. and C.W.; methodology, S.R. and C.W.; software, C.W. and S.R.; validation, C.W. and S.R.; formal analysis, S.R. and C.W.; investigation, C.W. and S.R.; resources, C.W. and S.R.; data curation, S.R. and C.W.; writing—original draft preparation, C.W.; writing—review and editing, S.R.; visualization, S.R. and C.W.; supervision, S.R.; project administration, S.R.; funding acquisition, S.R. Both authors have read and agreed to the published version of the manuscript.

Funding: This research received no external funding.

Institutional Review Board Statement: Not applicable.

Informed Consent Statement: Not applicable.

Data Availability Statement: The data presented in Section 3 are openly available online: https://www.ecac-ceac.org/images/documents/ECAC-Doc_29_4th_edition_Dec_2016_Volume_3_Part_1.pdf. Besides, publicly available datasets used in Sections 3 and 4 can be found here: https://www.easa.europa.eu/document-library/type-certificates and https://www.flightradar24.com/. Some of these data are made available to the reader, as Supplementary Material (namely the radar data of flights BD2, BD3, AD7 and AD8 in Section 4). The FDR data used in Section 4 cannot be made publicly available, due to their proprietary nature (Cathay Pacific airline).

Acknowledgments: The present research is a byproduct of a cooperative effort between Cathay Pacific airline (CX) and the Hong Kong University of Science and Technology (HKUST). The authors are grateful to Cathay Pacific for sharing with HKUST some of the flight data excerpted from CX aircraft operations around HKIA airport. Moreover, the authors are thankful to Rhea Liem and Dajung Kim (HKUST) for their support in accessing the data.

Conflicts of Interest: The authors declare no conflict of interest.

Appendix A. Integrated Thrust Equation for the Aircraft at Roll and Aloft

Appendix A.1. Integrated Thrust Equation

We here derive an integrated expression of the engine(s) thrust required for the aircraft to remain in *dynamic* equilibrium, either aloft (in-flight segments) or on the ground (ground-roll segments). To do so, we consider a virtual situation in which the aircraft is exerted all typical forces (including the ground reaction, which shall vanish automatically in the air). Assuming that the aircraft undergoes a straight accelerated motion, we derive the longitudinal equations of motion (EOM). Once projected along the so-called stability axes (i.e., parallel and perpendicular to the airflow direction, see Figure A1), these EOM are respectively given by:

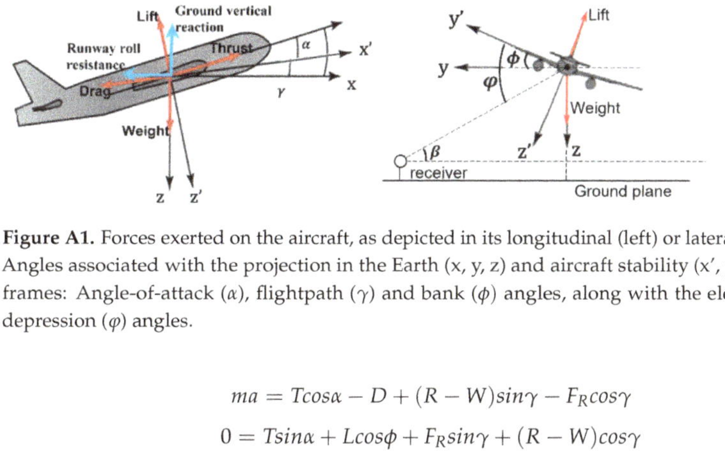

Figure A1. Forces exerted on the aircraft, as depicted in its longitudinal (left) or lateral (right) planes. Angles associated with the projection in the Earth (x, y, z) and aircraft stability (x′, y′, z′) reference frames: Angle-of-attack (α), flightpath (γ) and bank (ϕ) angles, along with the elevation (β) and depression (φ) angles.

$$ma = Tcos\alpha - D + (R - W)sin\gamma - F_R cos\gamma \tag{A1}$$

$$0 = Tsin\alpha + Lcos\phi + F_R sin\gamma + (R - W)cos\gamma \tag{A2}$$

In the above equations, m and W, respectively, stand for the mass and weight of the aircraft, whose (average) acceleration is a. On the other hand, α, γ and ϕ stand for the aircraft's angle of attack, flightpath and bank angles, respectively. Of note, the bank angle refers to the angle between the Earth reference frame's horizontal axis (y) and that of the aircraft's stability reference frame (y'), whose angle with the aircraft-to-observer noise propagation path corresponds to the depression angle (φ). Aside from that, T, L and D are the propulsive (thrust) and aerodynamic (lift and drag) forces. R and F_R are the two components of the runway reaction, namely the vertical reaction and the roll resistance, which are related through the friction coefficient characterizing the runway, μ_F

$$F_R = \mu_F \cdot R \tag{A3}$$

As a reminder, a typical value for this friction factor is 0.02 (resp. 0.4) for a dry concrete runway without (resp. with) brakes applied.

By combining Equations (A3) and (A2), one can express the ground vertical reaction as

$$R = \frac{W cos\gamma - L cos\phi - T sin\alpha}{cos\gamma + \mu_F sin\gamma} \tag{A4}$$

When introducing the lift and drag coefficient (C_L, C_D), one can then express the drag in terms of the lift alone

$$D = L \cdot \frac{C_D}{C_L} \tag{A5}$$

One can legitimately assume that the aircraft's angle of attack is small (that is $cos\alpha \sim 1$ and $sin\alpha \sim 0$), which further simplifies the above relationships. After a few developments, the thrust can be expressed as

$$T = m\left[a + g\left(\frac{\mu_F}{cos\gamma + \mu_F sin\gamma}\right)\right] + L\left[\frac{C_D}{C_L} + cos\phi \cdot \left(\frac{sin\gamma - \mu_F cos\gamma}{cos\gamma + \mu_F sin\gamma}\right)\right] \tag{A6}$$

Appendix A.2. Particular Case of in-Flight Segments

When the aircraft is aloft, the ground reaction is nil ($\mu_F = 0$), and Equation (A6) becomes

$$T = m[a] + L\left[\frac{C_D}{C_L} + cos\phi \cdot \left(\frac{sin\gamma}{cos\gamma}\right)\right] \tag{A7}$$

Considering that the lift equals the weight corrected from the simultaneous projection of flightpath and bank angles

$$L = W\frac{cos\gamma}{\cos\phi} \tag{A8}$$

Equation (A6) then becomes

$$T = m\left[a + g\left(\frac{C_D}{C_L} \cdot \frac{cos\gamma}{\cos\phi} + sin\gamma\right)\right] \tag{A9}$$

Of note, all parameters appearing in the above equation are readily available from the flight data and/or ANP database.

Appendix A.3. Particular Case of Ground-Roll Segments

On the other hand, when the aircraft is on the ground roll ($\gamma = \phi = 0$), Equation (A6) becomes

$$T = m(a + g\mu_F) + L\left(\frac{C_D}{C_L} - \mu_F\right) \tag{A10}$$

Expressing the lift force in these conditions is less straightforward, and requires developing an alternative lift model [57]. To do so, we start from the definition of lift

$$L = \left(\frac{1}{2}\rho S C_L\right) V_{TAS}^2 \tag{A11}$$

where ρ is the air density, V_{TAS} indicates the true airspeed of the aircraft, S is the wing area and C_L is the lift coefficient. Assuming that the wing can be considered as a thin airfoil, the lift coefficient can be classically related to the angle of attack, α (which is supposed to be small enough).

$$C_L = C_{L,0} + C_{L,\alpha}\alpha \tag{A12}$$

where $C_{L,0}$ stands for the zero-lift coefficient and $C_{L,\alpha}$ is the lift coefficient slope. Assuming that the aerodynamic settings (α, S) and the atmospheric conditions (ρ) do not vary during the ground roll (i.e., $C = \frac{1}{2}\rho S C_L = const$), Equation (A11) turns into a parabolic equation of the aircraft speed (alone)

$$L = C \cdot V_{TAS}^2 \quad \text{(A13)}$$

At the lift-off point, the ground reaction vanishes and the lift strictly equals the aircraft weight, W

$$W = L = C \cdot V_{LOF}^2 \quad \text{(A14)}$$

where V_{LOF} is the lift-off speed of the aircraft. By neglecting any variation of the aircraft weight during the ground roll, one can thus re-express the lift force as

$$L = W \cdot \left(\frac{V_{TAS}}{V_{LOF}}\right)^2 \quad \text{(A15)}$$

All in all, Equation (A6) then becomes

$$T = m\left\{a + g\left[\frac{C_D}{C_L} \cdot \left(\frac{V_{TAS}}{V_{LOF}}\right)^2 + \mu_F\left(1 - \left(\frac{V_{TAS}}{V_{LOF}}\right)^2\right)\right]\right\} \quad \text{(A16)}$$

Here too, all parameters appearing in the above equation are readily available from the flight data and/or ANP database.

Appendix B. Further Illustration of the Noise Prediction Process and of Its Validation against Reference Cases

This section further illustrates the validation cases provided in Section 3.1, one of which is selected and documented more thoroughly.

Figure A2 depicts the routes (departure and approach, either straight or curved) as well as the set of observers (namely R1–R17) used for all cases [62]. For more details about the other parameters (flightpath, meteorological conditions, power settings, airspeed, etc.), the reader is referred to Ref. [62]. Among the various cases listed, we here specifically focus on the JETWDS configuration (i.e., a wing-mounted jet engines aircraft flying along a straight departure route), whose relevant observers as those labeled R1–R5. Figure A3 depicts the time trace of various key quantities, as recorded for each observer along the flight evolution, which is broken down in terms of flight segments, either ground-roll (1–9) or airborne (10–29). These key quantities constitute the main ingredients of the noise prediction process, namely the baseline noise level perceived by each observer (Figure A3a) and its further adjustment through the cumulative corrections introduced in Section 2.2 (Figure A3b–g). The final result is provided in Figure A3h, which also quantifies the error made with the reference solution [62]. This breakdown of the noise prediction process is summarized hereafter, for illustration and validation purposes.

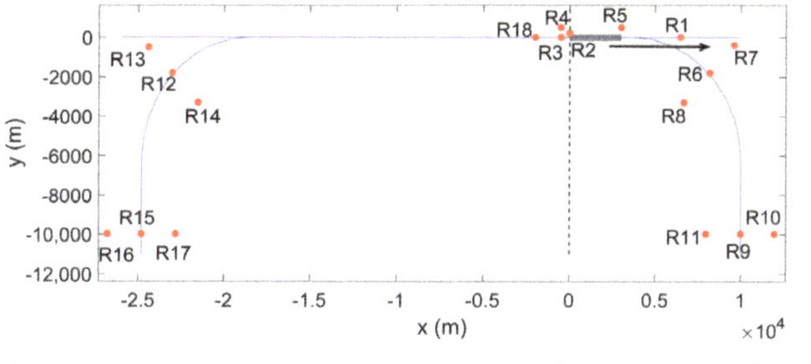

Figure A2. Departure and approach routes for reference cases, with the associated observers.

- *Baseline noise level* (cf. Figure A3a): Originated from the NPD database, this quantity represents the noise that would be perceived by each selected observer if the aircraft was flying according to a standard scenario (cf. Section 2.2). It was checked that the associate prediction results agree well with the reference values [62], the maximum errors being less than $10^{-5}\%$. Since the JETWDS case corresponds to a non-standard flight, these baseline noise levels must be refined through successive correction terms (see below).
- *Start-of-roll directivity adjustment* (cf. Figure A3b): This correction accounts for the highly directive pattern of the jet noise at take-off, which impacts more especially those observers that are located upstream the ground roll segments (cf. Section 2.2). Logically, the effect is nil for these observers that are located downstream (i.e., R1 and R5). Here too, the predicted results match well the reference values, with a maximum error of less than $10^{-5}\%$.
- *Engine installation correction* (cf. Figure A3c): This correction accounts for the lateral directivity patterns induced by the engine installation effects, which depend on the relative depression angle seen by each observer (cf. Section 2.3.2). Logically, the effect is nil for those observers that are located right underneath the airborne segment (e.g., R1). Again, the predicted values compare favorably against their reference counterparts, with an error of less than $10^{-9}\%$.
- *Finite segment correction* (cf. Figure A3d): This correction accounts for both (i) the finite nature of each concerned flight segment and (ii) its relative position with respect to the observers. The effect is more important for those observers that are located away from the flight segment, whose perceived noise is then corrected from the loss incurred by the respective propagation distance. Here too, the calculation results are in good agreement with the reference ones, leading to a maximum error of less than 1%.
- *Duration correction* (cf. Figure A3e): This correction translates the variation in the noise exposure duration, which logically varies with the fly-over time and, thus, the aircraft speed. In the present case where the aircraft continuously accelerates (from 0kt at the start-of-roll point) and eventually exceeds the standard value of 160kt (for which the correction is nil), this effect decreases monotonically. Here too, the prediction result is very close to the reference one, with errors in the range of $10^{-12}\%$–$10^{-2}\%$.
- *Acoustic Impedance adjustment* (cf. Figure A3f): This correction translates the impact of atmospheric effects (cf. Section 2.3.4), which are solely driven by the local meteorological conditions, thereby affecting all observers equally. Here, no error is recorded.
- *Lateral attenuation* (cf. Figure A3g): This correction accounts for the lateral attenuation induced by the ground presence (cf. Section 2.2). Logically, this effect does not impact those observers that are located right beneath the flightpath (here, R1 and R3 for airborne segments 10-29). The other observers are impacted differently, depending on their respective (lateral) distance to the aircraft, as well as on the nature of the flight segment considered (airborne or ground roll [49], e.g., R2-R4 for segments 9 and 10, respectively). Again, the predicted values match the reference ones, with maximum errors of less than $10^{-1}\%$.
- *Single event noise level* (cf. Figure A3h): This quantity represents the cumulative noise level perceived by each observer, once the contributions coming from all segments are summed up into a single noise event. As can be seen, the prediction is very close to the reference, with errors in the order of zero (except for the R1 observer, for which an error of about 0.011% is recorded).

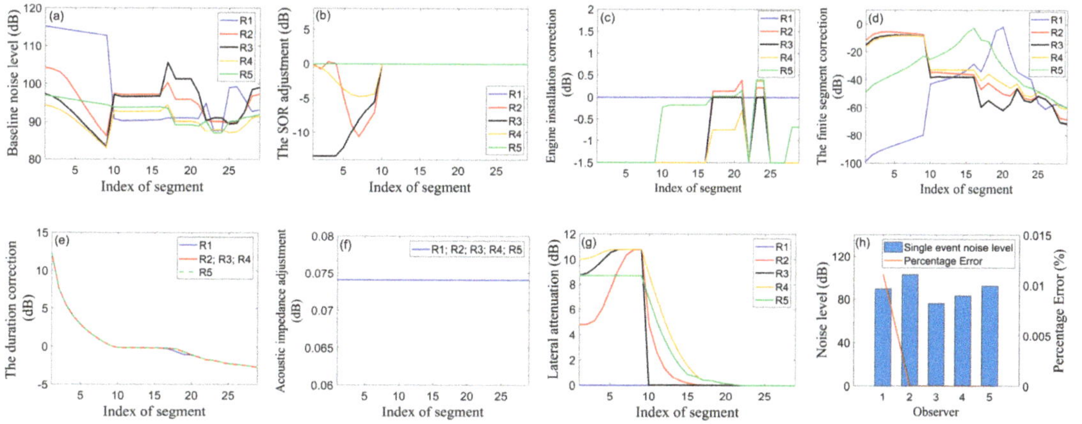

Figure A3. Breakdown of the noise prediction process, as illustrated for the JETWDS validation case (all in dB). Baseline noise level (**a**), start-of-roll directivity adjustment (**b**), engine installation correction (**c**), finite segment correction (**d**), duration correction (**e**), acoustic impedance adjustment (**f**), lateral attenuation (**g**), single event noise level (**h**).

Appendix C. Further Illustration of the Noise Impact Dependency towards Real-Life Operations

This section first focuses on the noise impact incurred by a B777-3ER (resp. an A330-343) aircraft departing from (resp. approaching) HKIA, thereby mirroring the discussion of Section 4.2.1. In a second time, the effect of banked turn is assessed on these various flights.

Appendix C.1. Noise Impact Incurred by B777-3ER Aircraft Flying along Two Departure Routes

Regarding the departure scenario, we select two specific B777-3ER flights with very distinct characteristics, and we assess their respective noise impact on three densely populated districts (namely D1, D5, and D7) that they both fly over. Whereas the first flight (labeled BD4) flies over districts D1 and D7 along a standard route with a high rate of climb (ROC), the second one (BD5) passes over district D1 at a lower altitude (see Figure A4).

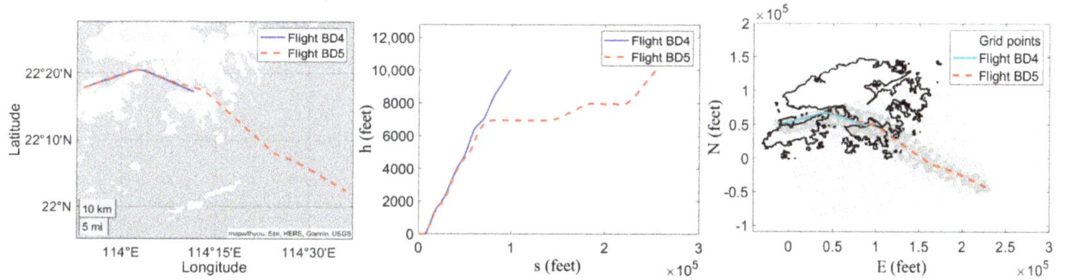

Figure A4. B777 aircraft departing from HKIA. Flights BD4 and BD5 respective routes (**left**: ground track, **center**: flight profile) and corresponding ground observers (dynamic grid, **right**).

Figure A5 depicts the SEL contours generated by flights BD4 (left) and BD5 (center) along with their difference, which is obtained by subtracting the former from the latter (right).

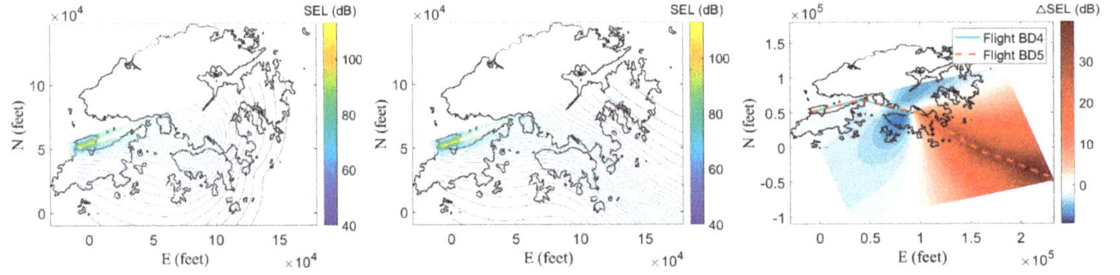

Figure A5. B777 aircraft departing from HKIA. Noise impact (SEL) generated by flight BD4 (**left**) and BD5 (**center**) over Hong Kong city, along with difference (ΔSEL) between both results (as obtained by subtracting the former from the latter, **right**).

Compared to flight BD4 (standard route, higher ROC), flight BD5 (deviated route, lower ROC) impacts less some of the populated areas, with noise levels that are lower by around 4 dB in D1 and D5 districts, and by about 7.5 dB in the western part of D7 district. The opposite trend is observed to occur in the eastern part of D7 district, where flight BD5′s impact exceeds that of BD4 by up to 13 dB in SEL (see Figure A6).

As was observed in Section 4.2.1 for Airbus-related flights, the lower altitude of a given flight (here BD5) does not automatically entail a higher noise impact on the ground. For instance, even though flight BD5 passes over D1 district at a lower altitude than its BD4 counterpart, it results in a smaller noise footprint over this area. Here too, this is likely to be explained by the difference in the power settings (i.e., engine thrust) characterizing both aircraft along their respective routes. For instance, Figure A7 compares the time history of the thrust delivered by the two aircraft, with a focus put on that specific flight phase where they both pass nearby D1 and D6 districts (highlighted with vertical dash-dotted lines). At that time, flight BD5 exhibits a propulsive power that is almost one-third of its BD4 counterpart. This is likely to incur a comparatively much lower noise emission for BD5 flight, whose lower noise attenuation (short propagation distance) is then favorably balanced. In addition, here again, the smaller ROC exhibited by flight BD5 is likely to diminish its noise footprint, compared to its BD4 counterpart whose higher flightpath angle somehow exacerbates its noise impact (jet noise directivity). All in all, these two features of flight BD5 (lower power setting, smaller ROC) suffice to offset the detrimental effect entailed by its lower altitude (reduced noise attenuation), when flying over D1 and D6 districts.

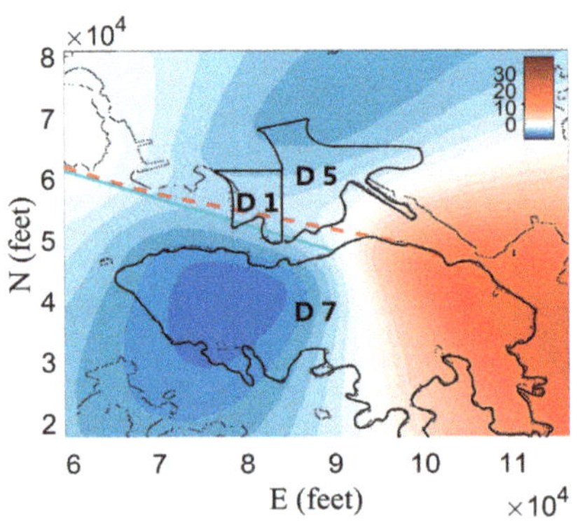

Figure A6. Difference in SEL (ΔSEL) between flights BD4 and BD5 (closer view).

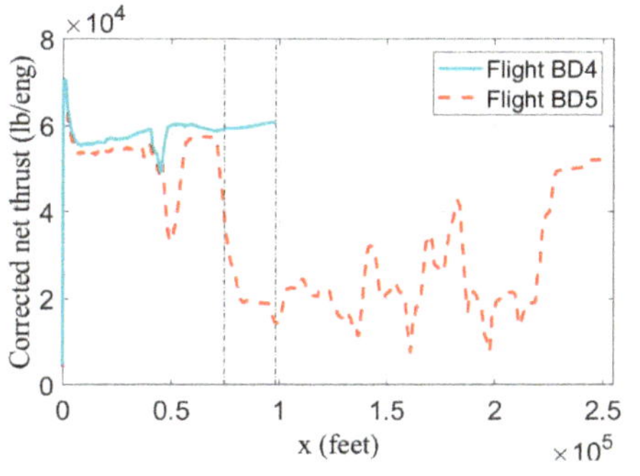

Figure A7. Power settings (engine thrust) evolution of flights BD4 and BD5.

As for what could be observed with the A330 aircraft (cf. Section 4.2.1), the above analysis further illustrates how HKIA departure procedures may impact very differently those highly densely populated areas of Hong Kong.

Appendix C.2. Noise Impact Incurred by A330-343 Aircraft Flying along Two Approach Routes

Mirroring Section 4.2.1, we here compare the noise impact induced by two A330-343 aircraft approaching HKIA, with a special focus on those densely populated areas (namely D2, D3, and D4) that are located beneath the flight corridors. The two flights are selected for their rather different flightpaths, so as to highlight how the latter may alter the noise footprint on the ground. More precisely, the first aircraft (flight AA2) approaches the 25L runway via a higher altitude route, while the second one (flight AA3) flies towards the 25R runway at a lower altitude. These differences hold to a certain point, after which both aircraft adopt a similar flightpath, following the Continuous Descent Approach protocol enforced in Hong Kong. Figure A8 depicts the ground track and flight profile of both flights.

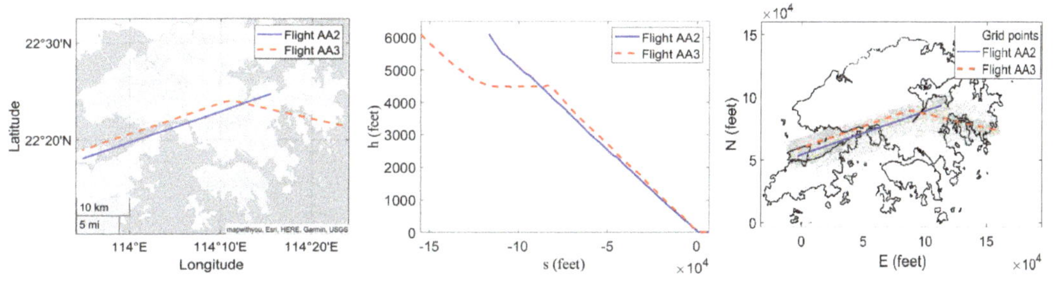

Figure A8. A330 aircraft approaching HKIA. Flights AA2 and AA3 respective routes (**left**: ground track, **center**: flight profile) and corresponding ground observers (dynamic grid, **right**).

Figure A9 depicts the SEL contours incurred by both flight AA2 (left) and flight AA3 (center), along with their difference (right). The densely populated D3 district which is sandwiched between 25L and 25R routes sees its northwestern part impacted more importantly by flight AA3 (with noise levels exceeding AA2 ones by about 10.6 dB, see Figure A10), while the opposite occurs in the southeastern part (with an excess noise by flight AA2 of up to 7.8 dB). Given the distribution of residents in the D3 district (whose central and southern areas are more populated, see Figure 8), flight AA2 (lower altitude) appears to be less environmentally friendly than its AA3 counterpart (higher altitude). This observation also holds for the four times denser D2 district, which is flown over directly by flight AA2 whose noise impact exceeds the AA3 one by about 9.1 dB. Finally, both flights impact equally the D4 district, each incurring a noise excess of 1–2 dB underneath its respective route. To sum up, flight AA2 appears to be less environmentally friendly for these densely populated districts, thereby revealing that the 25R approach route should be given priority whenever the conditions permit (operational, meteorological, etc.). Notably, all these observations are fully consistent with the ones recorded for Boeing aircraft (cf. Section 4.2.1).

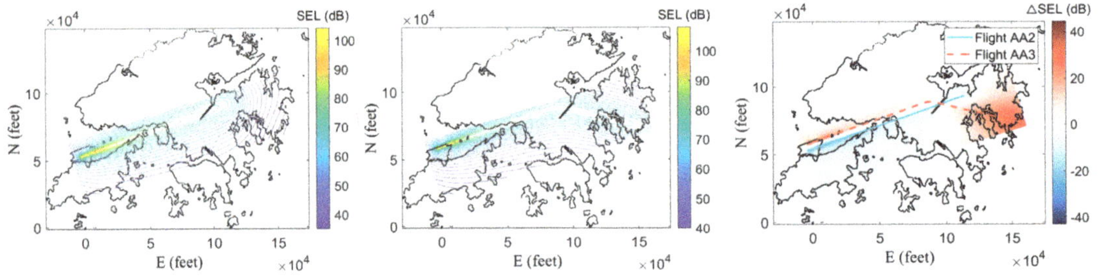

Figure A9. A330 aircraft approaching HKIA. Noise impact (SEL) generated by flight AA2 (**left**) and AA3 (**center**) over Hong Kong city, along with difference (ΔSEL) between both results (as obtained by subtracting the former from the latter, **right**).

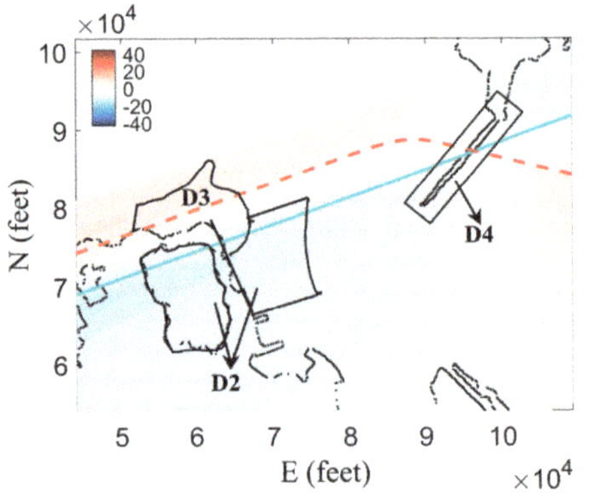

Figure A10. Difference in SEL (ΔSEL) between flights AA2 and AA3 (closer view).

Appendix C.3. Specific Effect of the Bank Angle on the Various Routes and Aircraft

This section further illustrates the sensitivity of the noise prediction towards the bank angle (i.e., engines installation) effect, thereby extending the investigation conducted in Section 4.3.1. Here, we perform the noise prediction for alternatives flights, with the bank angle being either incorporated or neglected in the SEL evaluation.

Focusing first on departure scenarios, we consider the flights AD1 (cf. Section 4.2.1), BD4, and BD5 (cf. Appendix C.1) departing from HKIA. Similar to what happened for flight AD2 (cf. Section 4.3.1), when the aircraft performs a banked turn, most of the observers located on the same (resp. opposite) side of the turn experience less (resp. more) noise impacts (see Figure A11). The opposite nevertheless holds for those observers that are located very close to the ground track. All the previous observations derive logically from the relative depression angle perceived by each observer (see Section 2.3.2). This being said, the bank effect on the SEL levels appears to be fairly subtle, with differences of less than 0.4 dB overall (see Figure A11).

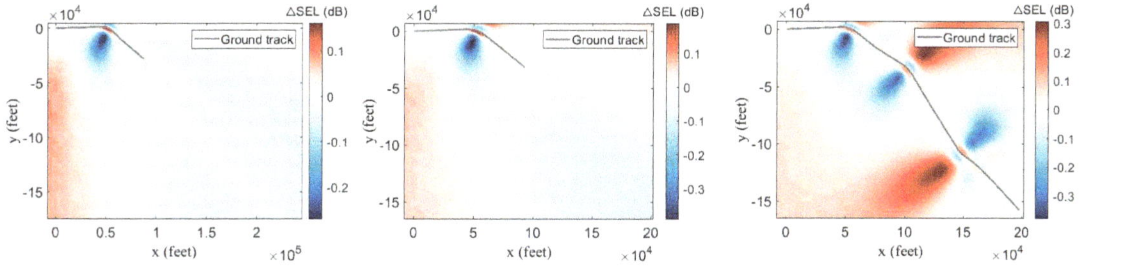

Figure A11. Various aircraft departing from HKIA. Difference in the noise impact (SEL) generated by flights AD1 (**left**), BD4 (**center**) and BD5 (**right**), due to the incorporation of bank angle effect.

Focusing then on approach scenarios, we repeat the noise prediction associated with flights AA3 (cf. Appendix C.2), BA1 and BA2 (cf. Section 4.2.1). Here too, the impact of the bank angle is rather modest, with no more than 0.5dB differences overall (see Figure A12). These differences are again positive or negative, depending on how each observer perceives the relative depression angle.

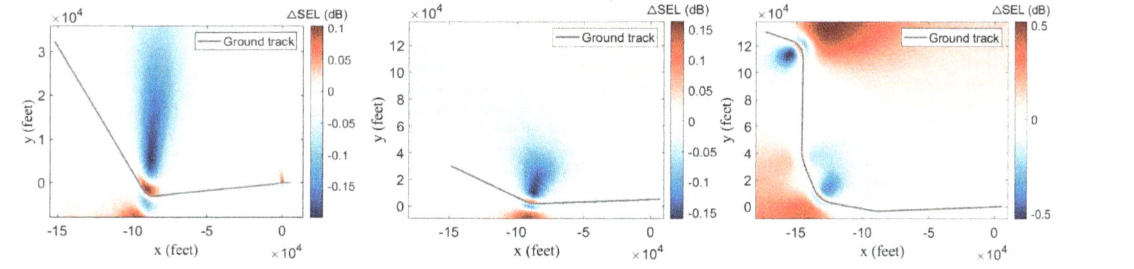

Figure A12. Various aircraft approaching HKIA. Difference in the noise impact (SEL) generated by flights AA3 (**left**), BA1 (**center**) and BA2 (**right**), due to the incorporation of bank angle effect.

All the above observations, which are fully consistent with those obtained in Section 4.3.1, further confirm that the noise impact is not that sensitive to the bank effect of turning flight segments. Even though of secondary importance, this effect however helps the noise prediction to better reflect real-life aircraft operations.

References

1. International Civil Aviation Organisation (ICAO). Environmental Report 2010. Available online: https://www.icao.int/environmental-protection/Documents/Publications/ENV_Report_2010.pdf (accessed on 7 September 2021).
2. Baudin, C.; Lefèvre, M.; Champelovier, P.; Lambert, J.; Laumon, B.; Evrard, A.-S. Aircraft Noise and Psychological Ill-Health: The Results of a Cross-Sectional Study in France. *Int. J. Environ. Res. Public Health* **2018**, *15*, 1642. [CrossRef] [PubMed]
3. Peters, J.L.; Zevitas, C.D.; Redline, S.; Hastings, A.; Sizov, N.; Hart, J.E.; Levy, J.I.; Roof, C.J.; Wellenius, G.A. Aviation Noise and Cardiovascular Health in the United States: A Review of the Evidence and Recommendations for Research Direction. *Curr. Epidemiol. Rep.* **2018**, *5*, 140–152. [CrossRef]
4. Nassur, A.-M.; Léger, D.; Lefèvre, M.; Elbaz, M.; Mietlicki, F.; Nguyen, P.; Ribeiro, C.; Sineau, M.; Laumon, B.; Evrard, A.-S. Effects of Aircraft Noise Exposure on Heart Rate during Sleep in the Population Living Near Airports. *Int. J. Environ. Res. Public Health* **2019**, *16*, 269. [CrossRef]
5. *European Aviation Environmental Report 2019*; European Environment Agency; European Aviation Safety Agency; Eurocontrol: Copenhagen, Denmark, 2019. [CrossRef]
6. Thomas, C.; Hume, K.; Hooper, P. Aircraft Noise, Airport Growth and Regional Development. In Proceedings of the 10th AIAA/CEAS Aeroacoustics Conference, Manchester, UK, 10–12 May 2004; p. 2806. [CrossRef]
7. Upham, P.; Thomas, C.; Gillingwater, D.; Raper, D. Environmental capacity and airport operations: Current issues and future prospects. *J. Air Transp. Manag.* **2003**, *9*, 145–151. [CrossRef]
8. Zaporozhets, O.; Tokarev, V.; Attenborough, K.; Miller, N.P. Aircraft Noise Assessment, Prediction and Control. *Noise Control. Eng. J.* **2012**, *60*, 222. [CrossRef]

9. Port Authority of New York and New Jersey. Airport Traffic Report. 2019. Available online: https://www.panynj.gov/airports/en/statistics-general-info.html (accessed on 7 September 2021).
10. Wei, N.; Zhu, F.; ChaoGao, Z. Research on Airport Noise Prediction Method Based on Noise Model INM. In Proceedings of the 2020 12th International Conference on Machine Learning and Computing, Shenzhen, China, 15–17 February 2020; pp. 489–494. [CrossRef]
11. Ziyao, X.; Xisheng, H. A Study on the Legislation Issues of Airport Noise Abatement in China. *J. Beijing Univ. Aeronaut. Astronaut. Soc. Sci. Ed.* **2011**, *24*, 38. [CrossRef]
12. Centre for Aviation. China Becomes the Largest Aviation Market in the World. 2020. Available online: https://centreforaviation.com/analysis/reports/china-becomes-the-largest-aviation-market-in-the-world-521779 (accessed on 7 September 2021).
13. Law, C.K.; Fung, M.; Law, J.; Tse, D.; Chan, K.Y. HKIA's Third Runway—The Key for Enhancing Hong Kong's Aviation Position. *Aviat. Policy Res. Cent. Novemb.* **2007**, *28*. Available online: https://citeseerx.ist.psu.edu/viewdoc/download?doi=10.1.1.493.3575&rep=rep1&type=pdf (accessed on 7 September 2021).
14. Leylekian, L.; Lebrun, M.; Lempereur, P. An Overview of Aircraft Noise Reduction Technologies. *AerospaceLab* **2014**, 1–15. [CrossRef]
15. Alam, S.; Nguyen, M.H.; Abbass, H.; Lokan, C.; Ellejmi, M.; Kirby, S. Multi-Aircraft Dynamic Continuous Descent Approach Methodology for Low-Noise and Emission Guidance. *J. Aircr.* **2011**, *48*, 1225–1237. [CrossRef]
16. Clarke, J.-P.B.; Ho, N.T.; Ren, L.; Brown, J.A.; Elmer, K.R.; Tong, K.-O.; Wat, J.K. Continuous Descent Approach: Design and Flight Test for Louisville International Airport. *J. Aircr.* **2004**, *41*, 1054–1066. [CrossRef]
17. Alam, S.; Nguyen, M.H.; Abbass, H.A.; Lokan, C.; Ellejmi, M.; Kirby, S. A dynamic continuous descent approach methodology for low noise and emission. In Proceedings of the 29th Digital Avionics Systems Conference, Salt Lake City, UT, USA, 3–7 October 2010; p. 1-E. [CrossRef]
18. Bertsch, L.; Dobrzynski, W.; Guérin, S. Tool Development for Low-Noise Aircraft Design. *J. Aircr.* **2010**, *47*, 694–699. [CrossRef]
19. Molin, N.; Piet, J.-F.; Chow, L.C.; Smith, M.; Dobrzynski, W.; Seror, C. Prediction of Low Noise Aircraft Landing Gears and Comparison with Test Results. In Proceedings of the 12th AIAA/CEAS Aeroacoustics Conference, Cambridge, MA, USA, 8–10 May 2006; p. 2623. [CrossRef]
20. Bertsch, L.; Heinze, W.; Lummer, M. Application of an Aircraft Design-To-Noise Simulation Process. In Proceedings of the 14th AIAA Aviation Technology, Integration, and Operations Conference, Atlanta, GA, USA, 16–20 June 2014; p. 2169. [CrossRef]
21. Nero, G.; Black, J.A. A critical examination of an airport noise mitigation scheme and an aircraft noise charge: The case of capacity expansion and externalities at Sydney (Kingsford Smith) airport. *Transp. Res. Part D Transp. Environ.* **2000**, *5*, 433–461. [CrossRef]
22. Arafa, M.H.; Osman, T.; Abdel-Latif, I.A. Noise assessment and mitigation schemes for Hurghada airport. *Appl. Acoust.* **2007**, *68*, 1373–1385. [CrossRef]
23. Koster, R. *Using NOMOS Measurements to Assess Improvements of ECAC Doc. 29 Aircraft Noise Calculations*; Delft University of Technology: Delft, The Netherlands, 2020; Available online: http://resolver.tudelft.nl/uuid:a8ba2f3c-7289-4aff-ad8a-5b358cd4ffbe (accessed on 7 September 2021).
24. Raimbault, M.; Dubois, D. Urban soundscapes: Experiences and knowledge. *Cities* **2005**, *22*, 339–350. [CrossRef]
25. Filippone, A. Aircraft noise prediction. *Prog. Aerosp. Sci.* **2014**, *68*, 27–63. [CrossRef]
26. Bertsch, L.; Schäffer, B.; Guérin, S. Uncertainty Analysis for Parametric Aircraft System Noise Prediction. *J. Aircr.* **2019**, *56*, 529–544. [CrossRef]
27. Pietrzko, S.; Rudolf, B. FLULA-Swiss Aircraft Noise Prediction Program. *Proc. Acoust.* **2002**, 13–15. Available online: https://www.dora.lib4ri.ch/empa/islandora/object/empa:11080 (accessed on 7 September 2021).
28. Ollerhead, J.B. The CAA Aircraft Noise Contour Model: ANCON Version 1. *Civ. Aviat. Auth. Dep. Transp.* **1992**. Available online: https://publicapps.caa.co.uk/docs/33/ERCD9120.PDF (accessed on 7 September 2021).
29. Ollerhead, J.; Sharp, B. MAGENTA-assessments of future aircraft noise policy options. *Air Space Eur.* **2001**, *3*, 247–249. [CrossRef]
30. Roof, C.; Hansen, A.; Fleming, G.; Thrasher, T.; Nguyen, A.; Hall, C.; Dinges, E.; Grandi, F.; Kim, B.; Usdrowski, S. Aviation Environmental Design Tool (AEDT) System Architecture. *Fed. Aviat. Adm. Off. Environ. Energy* **2007**. Available online: https://rosap.ntl.bts.gov/view/dot/12254 (accessed on 7 September 2021).
31. Wunderli, J.M.; Zellmann, C.; Köpfli, M.; Habermacher, M. sonAIR—A GIS-Integrated Spectral Aircraft Noise Simulation Tool for Single Flight Prediction and Noise Mapping. *Acta Acust. United Acust.* **2018**, *104*, 440–451. [CrossRef]
32. Bertsch, L.; Sébastien, G.; Looye, G.; Pott Pollenske, M. The Parametric Aircraft Noise Analysis Module-Status Overview and Recent Applications. In Proceedings of the 17th AIAA/CEAS Aeroacoustics Conference, Portland, OR, USA, 5–8 June 2011; p. 2855. [CrossRef]
33. Jäger, D.; Zellmann, C.; Schlatter, F.; Wunderli, J.M. Validation of the sonAIR aircraft noise simulation model. *Noise Mapp.* **2021**, *8*, 95–107. [CrossRef]
34. Malbéqui, P.; Rozenberg, Y.; Bulté, J. Aircraft Noise Prediction in the IESTA Program. In Proceedings of the 3rd European Conference for Aerospace Sciences, Versailles, France, 6–9 July 2009; Available online: https://www.onera.fr/sites/default/files/Departements-scientifiques/DCPS/AircraftnoisepredictionintheIESTAprogram.pdf (accessed on 7 September 2021).
35. Lopes, L.; Burley, C. Design of the next Generation Aircraft Noise Prediction Program: ANOPP2. In Proceedings of the 17th AIAA/CEAS Aeroacoustics Conference, Portland, OR, USA, 5–8 June 2011; p. 2854. [CrossRef]

36. Tuinstra, M. A Fast Atmospheric Sound Propagation Model for Aircraft Noise Prediction. *Int. J. Aeroacoustics* **2014**, *13*, 337–361. [CrossRef]
37. Boeker, E.R.; Dinges, E.; He, B.; Fleming, G.; Roof, C.J.; Gerbi, P.J.; Rapoza, A.S.; Hermann, J.; United States. Federal Aviation Administration; Office of Environment and Energy. Integrated Noise Model (INM) Version 7.0 Technical Manual. 2008. Available online: https://rosap.ntl.bts.gov/view/dot/12188 (accessed on 7 September 2021).
38. Mato, R.R.; Mufuruki, T. Noise pollution associated with the operation of the Dar es Salaam International Airport. *Transp. Res. Part D Transp. Environ.* **1999**, *4*, 81–89. [CrossRef]
39. Eriksson, C.; Bluhm, G.; Hilding, A.; Östenson, C.-G.; Pershagen, G. Aircraft noise and incidence of hypertension—Gender specific effects. *Environ. Res.* **2010**, *110*, 764–772. [CrossRef]
40. Babisch, W.; Pershagen, G.; Selander, J.; Houthuijs, D.; Breugelmans, O.; Cadum, E.; Vigna-Taglianti, F.; Katsouyanni, K.; Haralabidis, A.S.; Dimakopoulou, K.; et al. Noise annoyance—A modifier of the association between noise level and cardiovascular health? *Sci. Total. Environ.* **2013**, *452–453*, 50–57. [CrossRef]
41. Ignaccolo, M. Environmental capacity: Noise pollution at Catania-Fontanarossa international airport. *J. Air Transp. Manag.* **2000**, *6*, 191–199. [CrossRef]
42. El Fadel, M.; Chahine, M.; Baaj, M.; Mezher, T. Managing Noise Emission Impacts of Airport Traffic. *INTER-NOISE NOISE-CON Congr. Conf. Proc.* **2000**, *2000*, 1395–1399. Available online: http://www.conforg.fr/internoise2000/cdrom/data/articles/000863.pdf (accessed on 7 September 2021).
43. Hebly, S.J.; Visser, H.G. Advanced noise abatement departure procedures: Custom-optimised departure profiles. *Aeronaut. J.* **2015**, *119*, 647–661. [CrossRef]
44. Clarke, J.-P.; Bennett, D.; Elmer, K.; Firth, J.; Hilb, R.; Ho, N.; Johnson, S.; Lau, S.; Ren, L.; Senechal, D. Development, Design, and Flight Test Evaluation of a Continuous Descent Approach Procedure for Nighttime Operation at Louisville International Airport. Partnership for Air Transportation Noise and Emissions Reduction. 2006. Available online: https://rosap.ntl.bts.gov/view/dot/28416 (accessed on 7 September 2021).
45. Kim, D.; Lyu, Y.; Liem, R.P. Flight Profile Optimization for Noise Abatement and Fuel Efficiency during Departure and Arrival of an Aircraft. *AIAA Aviation 2019 Forum* **2019**, 3622. [CrossRef]
46. Wijnen, R.; Visser, H. Optimal departure trajectories with respect to sleep disturbance. *Aerosp. Sci. Technol.* **2003**, *7*, 81–91. [CrossRef]
47. U.S. Department of Transportation. NextGen Annual Report. Available online: https://www.faa.gov/nextgen/media/NextGenAnnualReport-FiscalYear2020.pdf (accessed on 7 September 2021).
48. Environmental Protection Agency, Ireland. Guidance Note for Strategic Noise Mapping For the Environmental Noise Regulations 2006. Available online: https://www.epa.ie/publications/monitoring--assessment/noise/EPA-Guidance-Note-for-Strategic-Noise-Mapping-(version-2).pdf (accessed on 7 September 2021).
49. Report on Standard Method of Computing Noise Contours around Civil Airports. *Eur. Civ. Aviat. Conf.* **2016**, *2*. Available online: https://www.ecac-ceac.org/images/documents/ECAC-Doc_29_4th_edition_Dec_2016_Volume_2.pdf (accessed on 7 September 2021).
50. Kephalopoulos, S.; Paviotti, M.; Anfosso-Lédée, F. Common Noise Assessment Methods in Europe (CNOSSOS-EU). *Publ. Off. Eur. Union* **2012**, 180. [CrossRef]
51. Civil Aviation Authority. Strategic Noise Maps for Heathrow Airport 2016. 2018. Available online: https://publicapps.caa.co.uk/docs/33/MappingHeathrow.pdf (accessed on 7 September 2021).
52. Vogiatzis, K.; Remy, N. Strategic Noise Mapping of Herakleion: The Aircraft Noise Impact as a factor of the Int. Airport relocation. *Noise Mapp.* **2014**, *1*. [CrossRef]
53. Procedure for the Calculation of Airplane Noise in the Vicinity of Airports. *Soc. Automot. Eng. Comm. A-21* **2012**. [CrossRef]
54. Federal Aviation Administration. Fundamentals of Noise and Sound. 2020. Available online: https://www.faa.gov/regulations_policies/policy_guidance/noise/basics/ (accessed on 7 September 2021).
55. International Civil Aviation Organization. Recommended Method for Computing Noise Contours Around Airports. 2008. Available online: https://global.ihs.com/doc_detail.cfm?document_name=ICAO9911&item_s_key=00520247 (accessed on 7 September 2021).
56. Vos, E.; Groenendijk, J.; Do, M.T.; Tyre and Road Surface Optimisation for Skid Resistance and Further Effects. D05 Report on Analysis and Findings of Previous Skid Resistance Harmonisation Research Projects. 2009. Available online: https://trimis.ec.europa.eu/sites/default/files/project/documents/20120406_001647_43476_TYROSAFEFINALSummaryReport.pdf (accessed on 7 September 2021).
57. Matamoros Cid, I. *Modelling Flexible Thrust Performance for Trajectory Prediction Applications in Air Traffic Management*; Universitat Politècnica de Catalunya: Barcelona, Spain, 2015; Available online: http://hdl.handle.net/2117/85938 (accessed on 7 September 2021).
58. Thacker, W.C. A brief review of techniques for generating irregular computational grids. *Int. J. Numer. Methods Eng.* **1980**, *15*, 1335–1341. [CrossRef]
59. Settari, A.; Aziz, K. Use of Irregular Grid in Reservoir Simulation. *Soc. Pet. Eng. J.* **1972**, *12*, 103–114. [CrossRef]
60. Standard Values of Atmospheric Absorption as a Function of Temperature and Humidity. *Soc. Automot. Eng. Comm. A-21* **1975**. [CrossRef]

61. Rickley, E.J.; Fleming, G.G.; Roof, C.J. Simplified procedure for computing the absorption of sound by the atmosphere. *Noise Control. Eng. J.* **2007**, *55*, 482. [CrossRef]
62. Report on Standard Method of Computing Noise Contours around Civil Airports. *Eur. Civ. Aviat. Conf.* **2016**, *3*. Available online: https://www.ecac-ceac.org/images/documents/ECAC-Doc_29_4th_edition_Dec_2016_Volume_3_Part_1.pdf (accessed on 7 September 2021).
63. European Union Aviation Safety Agency. Type Certificate Data Sheets (TCDS). 2021. Available online: https://www.easa.europa.eu/document-library/type-certificates (accessed on 7 September 2021).
64. International Standards and Recommended Practices, Environmental Protection, Annex 16. *Int. Civ. Aviat. Organ.* **2008**, *1*. Available online: https://www.caat.or.th/wp-content/uploads/2016/04/AN16_V1_cons.pdf (accessed on 7 September 2021).
65. HaoLei, L. *Research on the Evaluation Method of Aircraft Airworthiness Lateral Noise*; Civil Aviation University of China: Tianjin, China, 2018. [CrossRef]
66. Hong Kong Civil Aviation Department. Aircraft Noise Management. 2021. Available online: https://www.cad.gov.hk/english/ac_noise.html (accessed on 7 September 2021).
67. Albéri, M.; Baldoncini, M.; Bottardi, C.; Chiarelli, E.; Fiorentini, G.; Raptis, K.G.C.; Realini, E.; Reguzzoni, M.; Rossi, L.; Sampietro, D.; et al. Accuracy of Flight Altitude Measured with Low-Cost GNSS, Radar and Barometer Sensors: Implications for Airborne Radiometric Surveys. *Sensors* **2017**, *17*, 1889. [CrossRef] [PubMed]
68. Manolakis, D.; Lefas, C.; Rekkas, C. Computation of aircraft geometric height under radar surveillance. *IEEE Trans. Aerosp. Electron. Syst.* **1992**, *28*, 241–248. [CrossRef]
69. Poles, D.; Nuic, A.; Mouillet, V. Advanced Aircraft Performance Modeling for ATM: Analysis of BADA Model Capabilities. In Proceedings of the 29th Digital Avionics Systems Conference, Salt Lake City, UT, USA, 3–7 October 2010; p. 1-D. [CrossRef]
70. Sherry, L.; Neyshabouri, S. Estimating Takeoff Thrust from Surveillance Track Data. Transportation Research Board Annual Meeting. 2014. Available online: https://catsr.vse.gmu.edu/pubs/EstTakeoffThrustTrackData%5B4%5DTRB.pdf (accessed on 7 September 2021).
71. Nuic, A. User Manual for the Base of Aircraft Data (BADA) Revision 3.10. *Atmosphere* **2010**, *2010*, 1. Available online: https://www.eurocontrol.int/sites/default/files/library/022_BADA_User_Manual.pdf (accessed on 7 September 2021).
72. Poulain, K. *Numerical Propagation of Aircraft En Route Noise*; The Pennsylvania State University: University Park, PA, USA, 2011; Available online: https://etda.libraries.psu.edu/catalog/12491 (accessed on 7 September 2021).
73. Heath, S.; McAninch, G. Propagation Effects of Wind and Temperture on Acoustic Ground Contour Levels. In Proceedings of the 44th AIAA Aerospace Sciences Meeting and Exhibit, Reno, NV, USA, 9–12 January 2006; p. 411. [CrossRef]
74. Arntzen, M.; Hordijk, M.; Simons, D.G. Including Atmospheric Propagation Effects in Aircraft Take-off Noise Modeling. In Proceedings of the 43rd International Congress on Noise Control Engineering, Melbourne, Australia, 16–19 November 2014; Available online: https://acoustics.asn.au/conference_proceedings/INTERNOISE2014/papers/p307.pdf (accessed on 7 September 2021).
75. Dikshit, P.; Crossley, W. Airport Noise Model Suitable for Fleet-Level Studies. In Proceedings of the 9th AIAA Aviation Technology, Integration, and Operations Conference (ATIO) and Aircraft Noise and Emissions Reduction Symposium (ANERS), Hilton Head, SC, USA, 21–23 September 2009; p. 6937. [CrossRef]

aerospace

MDPI

Article

Application of Noise Certification Regulations within Conceptual Aircraft Design

Michel Nöding and Lothar Bertsch *

German Aerospace Center (DLR), Institute of Aerodynamics and Flow Technology, 37075 Göttingen, Germany; michel.noeding@dlr.de
* Correspondence: lothar.bertsch@dlr.de

Abstract: ICAO Annex 16 regulations are used to certify the acoustic performance of subsonic transport aircraft. Each aircraft is classified according to the measured EPNL levels at specific certification locations along the approach and departure. By simulating this certification process, it becomes possible to identify all relevant parameters and assess promising measures to reduce the noise certification levels in compliance with the underlying ICAO regulations, i.e., allowable operating conditions of the aircraft. Furthermore, simulation is the only way to enable an assessment of novel technology and non-existing vehicle concepts, which is the main motivation behind the presented research activities. Consequently, the ICAO Annex 16 regulations are integrated into an existing noise simulation framework at DLR, and the virtual noise certification of novel aircraft concepts is realized at the conceptual design phase. The predicted certification levels can be directly selected as design objectives in order to realize an advantageous ICAO noise category for a new aircraft design, i.e., simultaneously accounting for the design and the resulting flight performance. A detailed assessment and identification of operational limits and allowable flight procedures for each conceptual aircraft design under consideration is enabled. Sensitivity studies can be performed for the relevant input parameters that influence the predicted noise certification levels. Specific noise sources with a dominating impact on the certification noise levels can be identified, and promising additional low-noise measures can be applied within the conceptual design phase. The overall simulation process is applied to existing vehicles in order to assess the validity of the simulation resultsfcompared to published data. Thereafter, the process is applied to some DLR low-noise aircraft concepts to evaluate their noise certification levels. These results can then be compared to other standard noise metrics that are typically applied in order to describe aircraft noise, e.g., SEL isocontour areas. It can be demonstrated that certain technologies can significantly reduce the noise impact along most of an approach or departure flight track but have only a limited influence on the noise certification levels and vice versa. Finally, an outlook of the ongoing developments is provided, in order to apply the new simulation process to supersonic aircraft. Newly proposed regulations for such concepts are implemented into the process in order to evaluate these new regulations and enable direct comparison with existing regulations.

Keywords: aircraft noise simulation; conceptual aircraft design; noise certification; ICAO Annex 16; PANAM; RCE

Citation: Nöding, M.; Bertsch, L. Application of Noise Certification Regulations within Conceptual Aircraft Design. *Aerospace* **2021**, *8*, 210. https://doi.org/10.3390/aerospace8080210

Academic Editor: Michael Kokkolaras

Received: 16 July 2021
Accepted: 28 July 2021
Published: 3 August 2021

1. Introduction

In order to fully exploit novel technologies and specific measures to minimize aircraft noise, it is essential to incorporate noise prediction early within the conceptual aircraft design phase, e.g., see Reference [1]. Only at this early stage is the degree of freedom in available design and technology choices large enough to really enforce a low-noise aircraft design. Furthermore, the incorporation of all relevant disciplines within one simulation process enables all snowball effects that are caused by any modification to the aircraft, engine, or flight procedure to be accounted for, which affects all other related disciplines.

For example, any modification to a component design will not only affect a specific noise source on board but can also determine the flight performance of the final vehicle, e.g., as demonstrated in Reference [2]. According to the underlying flight performance along the approach and departure, the operating condition, and hence, the noise generation of each noise source, is directly and significantly affected. It becomes clear that any modification to a noise source on-board the aircraft, without consideration of its implications on the flight performance, will not result in realistic and physics-based results. These strong interdependencies have to be considered for each and every design or technology option under consideration; hence, a full approach-and-departure flight simulation is a prerequisite to any meaningful noise assessment of any new design or technology. Depending on the selected design objectives or noise metrics, the computational demands can be significantly increased, e.g., high computational costs can be expected if widespread noise contour areas are evaluated along the entire approach and departure flights for each design or technology option. The computational costs of such noise simulations for the overall aircraft typically scales linearly with the number of flight points along a simulated trajectory, multiplied by the number of observer locations on the ground, e.g., Reference [1]. The evaluation of a contour area (tens of thousands of grid points) along a realistic flight procedure (hundreds of flight points) for each and every vehicle option (hundreds of design variants, e.g., parameter variation of certain design parameters) will quickly scale up and can only be realized at an early design phase, when simple and fast simulation methods are still applicable. In addition to the above-mentioned determination of the noise immission along individual flights, e.g., assessment on the basis of footprints, the certification levels are of great importance, because the classification into different noise chapters also results in additional direct operating costs, e.g., through increased noise-related take-off and landing fees at certain airports, or severe disadvantages related to potential limitations of the number of flights or operating hours. Therefore, manufacturers as well as airlines are interested in an early and reliable prediction of such levels.

In addition, ICAO noise certification levels are simulated for a fraction of the computational costs compared to a full contour area assessment. In this case, only a few observer locations and shortened segments of the full approach-and-departure flight have to be simulated and assessed. Consequently, certification noise levels only capture a very limited and localized situation, and are not representative of the widespread noise exposure along the entire approach-and-departure flights. However, these certification levels can be used as initial noise indicators and are particularly suitable for a quick but coarse comparison of different aircraft designs and technologies due to their standardized form [3]. The use of simulated certification levels as design objectives in the context of low-noise aircraft design has been demonstrated by other researchers, e.g., NASA's N + 1 [4], N + 2 [5], and N + 3 [6,7] studies.

In general, a simulation of certification levels is essential, since any new vehicle would be subject to an ICAO noise certification procedure, and hence has to comply with the underlying regulations and limitations. Although computational demand is significantly reduced for an assessment of the certification situation, i.e., there are only three specific ground locations, simple and fast methods at the conceptual design phase still would be required when evaluating different technologies, aircraft designs, and flight procedures. According to Annex 16 [3], specific regulations and limitations are defined with respect to aircraft noise certification. Specific constraints, e.g., ambient weather conditions, and certain flight conditions are defined and have to be enforced when simulating the flight trajectories. To enable a feasible and physics-based assessment of the noise certification levels, vehicle design (source characteristics) and resulting flight performance (operating conditions) have to be accounted for simultaneously. Thereby, it is necessary to account for source characteristics and operating conditions along relevant parts of the approach-and-departure flight procedure, i.e., it can be demonstrated that at least the so-called 10 dB-down time has to be considered in order to adequately capture the certification levels [8,9].

Based on existing DLR software tools, and considering the ICAO regulations, a dedicated and fully automated simulation process has been assembled to realize a simulated noise certification [8,9]. The new and fully automated simulation process predicts certification levels within the conceptual aircraft design. The ICAO regulations can be assessed based on the available data within the simulation process. The underlying process of assessing the certification is complex and consists of (1) a detailed simulation of the flight path under the given constraints, (2) calculation of the sound emission based on the operating conditions and configurational parameters prevailing in proximity to the certification measurement points, and (3) the modeling of the sound propagation and ground attenuation effects. With this new simulation process, noise certification levels can now be predicted for both existing and novel aircraft concepts. The resulting cumulative noise levels can be directly processed as an optional design objective within the conceptual aircraft design. Different sensitivities of flightpath or design parameters on the certification noise can be investigated simulateously for the first time in one overall simulation process. Predictions for existing vehicles can be compared to published and available data in order to verify the simulation results. It should be noted that the presented activities at DLR do not aim to create a virtual replacement of the actual certification process. The goal of the presented activities is to provide reliable estimates of certification levels to enable a direct comparison of different technologies and designs. Furthermore, comparison to other simulation activities based on certification noise levels is enabled, e.g., research activities at NASA [4–7]. Furthermore, the certification assessment does enable quick comparative studies within the conceptual aircraft design, where significant variations in aircraft design, engine design, selected technology modifications, and flight performance are still realizable, and hence open up a huge solution space.

The prevailing certification regulations, according to the International Civil Aviation Organization (ICAO), are assessed and summarized in the next section of this article. Thereafter, the newly developed simulation process is described in more detail, i.e., Section 2. A verification of the simulation process and a detailed comparison of simulation results and official certification data are presented in Sections 2 and 4.1, respectively. The remaining differences between simulation results and official certification levels can be attributed to several effects that are discussed in Section 4.2. The novel process is finally applied to novel aircraft concepts, and the results of this investigation are presented in Section 5. Furthermore, suggested modifications to the existing regulations in order to enable an assessment of supersonic transport aircraft are presented in the application section. Finally, the findings of this study are summarized and discussed in Section 6. A brief outlook on ongoing and future activities in Section 7 completes this article.

2. ICAO Noise Certification

Noise certification of subsonic jet aircraft is defined by the International Civil Aviation Organization (ICAO) [10,11]. The ICAO defines different reference procedures and noise limits for different types of aircraft, i.e., mainly dependent on the aircraft maximum takeoff weight (MTOW) and the number of engines. In the following sections, the existing regulations for a subsonic jet aircraft are described. An initial discussion of an FAA proposal towards regulation of supersonic aircraft is addressed in this article. Certification regulations use the Effective Perceived Noise Level (EPNL) at three reference measurement points as a basis. Two of the reference measurement points are evaluated along take-off and one during landing. Different specific limits for the operation of subsonic aircraft are defined when assessing the EPNL at these measurement points. These regulations are finally implemented in the existing simulation process at DLR, e.g., as described in Reference [2], in order to assess the corresponding certification noise level for existing and novel aircraft designs within the conceptual design phase. Ultimately, the certification levels can be selected as design criteria, and hence can be subject to a low-noise optimization of the vehicle design.

Figure 1 shows these certification measurement points for departure relative to the runway, as specified by the ICAO. The flyover reference noise measurement point (K_1) is positioned 6500 m behind the break release point (Point A) on the extended runway centerline below the departure flight path.

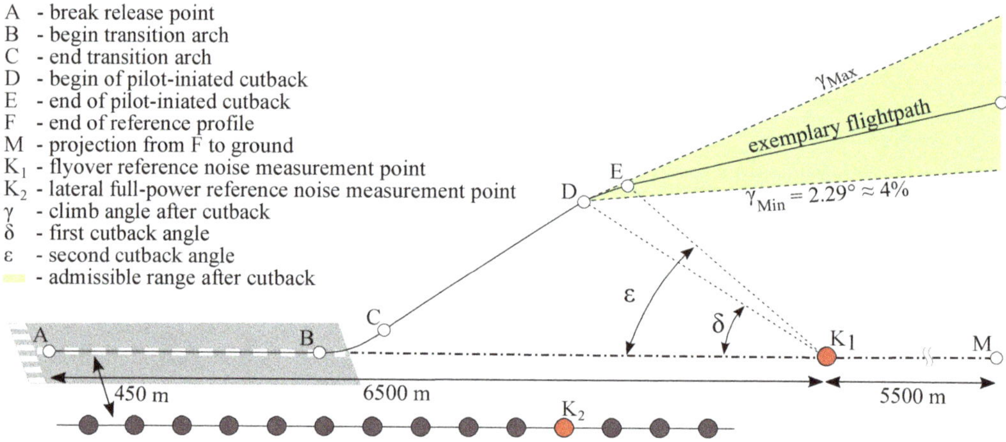

Figure 1. Take-Off Reference profile of subsonic aircraft for noise certification with flyover (K1) and lateral full-power reference measurement point (K2) and runway, adjusted from [3].

In addition, there is a lateral full-power reference noise measurement point (K_2), located on a line 450 m, parallel to the runway centerline. The position with the highest EPNL along this sideline is selected as the lateral full-power reference noise measurement point. For a valid certification flight, the maximum available thrust must be maintained until a cutback altitude (Point D), dependent on the number of engines, is reached. For a twin-engine aircraft, the cutback height is 300 m, for a three-engine 260 m and for a four-engine 210 m, above ground. This accounts for the excess thrust that an aircraft with fewer engines has over one with more engines due to the thrust requirements during an engine failure, i.e., One Engine Inoperative (OEI) case. After reaching the respective cutback altitude, the aircraft must climb at least with 4%, which corresponds to a climb angle of approximatly 2.29°. Since the reduction in the engine thrust is not instantaneous, ICAO defines a first and a second cutback angle (δ and ε related to point K_1) with the points D and E. Furthermore, the flight must be recorded up to 11 km (Point F) after the break release point. In addition, there are requirements for the calibrated airspeed at takeoff, defined relative to the aircraft-specific safe takeoff speed V2. The speed V2 + 10 knots must be reached as soon as possible after takeoff and the speed V2 + 20 knots must not be exceeded during the certification flight. Due to the requirement that the thrust setting be kept constant during calculation of EPNL, the 10 dB-down time must be completed for the lateral measurement point by the time point D is reached, in case a cutback is performed. For the flyover measurement point, the 10 dB-down time must be between points E and F for the same reason.

The noise during approach is evaluated at another measurement point during noise certification, which is shown in Figure 2, together with the specified approach path. The approach flightpath must be recorded from point G, which is 7400 m in front of the runway landing threshold. The projection of this point on the ground is called point P. This approach reference noise measurement point (K_3) is located 2000 m in front of the runway landing threshold (Point O) in the extended runway centerline. Taking into account the prescribed glide path of 3°, the aircraft flies over the measurement point at an altitude of 120 m (Point H), where the aircraft is in a stabilized state and the engine is not idling, as is the case in other phases of the approach. It should be noted that an altitude deviation of 20%, i.e., ±24 m, is allowed at Point H. The glide path ends at point I and the touchdown takes place at point J, about 300 m behind point O.

The EPNL metric is used for noise certification, and its calculation is explained below.

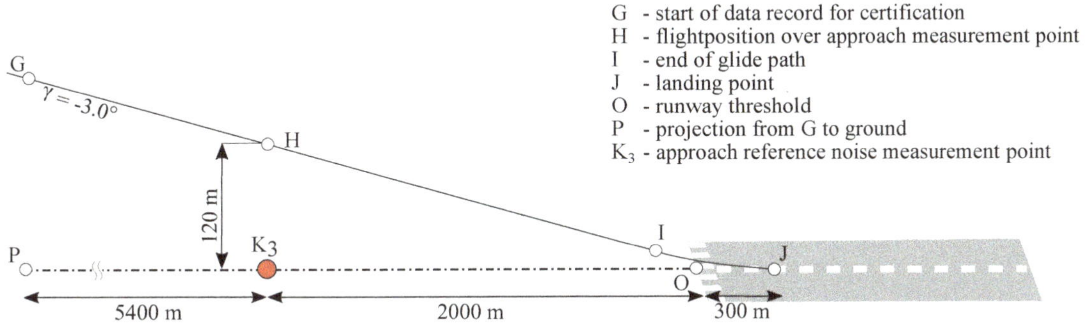

Figure 2. Reference profile of subsonic aircraft for noise certification with approach reference measurement point and runway.

In addition, a chart illustrating the calculation can be found in the Appendix A (Figure A1). At the three measurement points, the Sound Pressure Level (SPL) in the one-third octave band i is measured at intervals of 0.5 s in a microphone height of 1.2 m above ground for every timestep k. After frequency weighting of the measured SPL using noy tables into the Perceived Noisiness n(i,k) the individual one-third octave bands are summed over i and, to weight the highest band, the band SPL is multiplied by 0.85, while the other bands are multiplied by 0.15. A summation of the one-third octave band spectra gives the Total Perceived Noisiness N(k), which is then converted into the Perceived Noise Level PNL(k). In order to take the sensitivity of human hearing for discrete tones into account, a sound correction C(k) is applied, which adds a sound penalty to the PNL(k) depending on the relative differences between adjacent Sound Pressure Levels of the individual one-third octave bands, which leads to the Tone-Corrected Perceived Noise Level PNLT(k). There is only a penalty when Sound Pressure Level difference between the adjacent bands is above 1.5 dB. Afterwards, the EPNL is calculated by the time integration of all levels of PNLT(k), which are, at most, 10 dB below the maximum PNLTM. For better comparability, a reference flight procedure is defined under atmospheric reference conditions. The determined EPNL is corrected to this reference procedure by applying different correction terms. In these studies, it is assumed that both the trajectory and atmospheric conditions are simulated to be identical with the reference conditions and, therefore, no correction to the determined EPNL is necessary.

3. Simulation Environment

3.1. Aircraft Design

Aircraft and engine design was assessed with TU Braunschweig's PrADO (Preliminary Aircraft Design and Optimization) simulation tool [12]. PrADO is an aircraft design synthesis code comprised of individual software modules, which are each assigned to a certain task or discipline within the design process of aircraft and engines [12], e.g., dedicated modules will compute the aircraft mass breakdown based on the selected vehicle layout. PrADO not only accounts for all the individual disciplines but, most importantly, also takes the prevailing interaction effects into consideration, i.e., so-called snow-ball effects. The software modules are executed iteratively until preselected major design parameters reach convergence, e.g., the changes in aircraft component masses per process iteration stay below specified residuals.

After the convergence of all major design parameters, PrADO yields a valid and physics-based aircraft design that fulfills all the preselected design requirements, i.e., Top Level Aircraft Requirements. Thereby, the individual software modules can straightforwardly be replaced by other modules of higher fidelity or by external data, e.g., measured aircraft component weights. The PrADO tool has been upgraded to provide all the required input data for a subsequent system noise prediction, i.e., including aircraft and engine design parameters and a PrADO flight track [1]. With PrADO, existing aircraft can be simulated, as well as novel aircraft concepts. The process can then be applied to a system noise prediction of arbitrary tube-and-wing aircraft concepts [1].

External input data, such as engine performance maps, geometrical details, or the flight procedure, can be processed via dedicated interfaces in order to improve prediction accuracy if required and available [2]. Simulation results from corresponding software modules within PrADO are then directly replaced by external input data, and the PrADO modules will not be executed. For the noise assessment carried out within this study, such high-fidelity external input data for the engine were provided by the DLR Institute of Propulsion Technology, i.e., the simulation results of their tool GTlab [13]. The flight trajectory by PrADO was replaced by results from the DLR flight simulation tool Flightpath for Noise Analysis (FLIPNA), which is fully compatible with ECAC Doc. 29 regulations. At this point, the aircraft, the engine, and the resulting flight trajectories are completely described and the overall simulation process can be initiated.

3.2. Flight Simulation

Based on the complete aircraft description from the updated PrADO process, a detailed assessment of approach and departure procedures is enabled. For the purpose of the certification assessment, a newly developed flight simulation tool Flightpath for Noise Analysis (FLIPNA), e.g., Reference [14], replaces the corresponding PrADO simulation module because it can directly account for the ICAO regulations with respect to the certification rules. Thereby, FLIPNA directly processes the PrADO aircraft and engine output in order to simulate physics-based trajectories. The aircraft trajectory is defined as a series of flight points accounting for varying operational conditions, e.g., variations in engine operation or the usage of landing gear and a high-lift system. Certain control parameters for FLIPNA define the final flight trajectory, and can be varied and selected for optimization, e.g., glide slopes and velocities. At this point, detailed flight trajectories for arbitrary aircraft design can be generated from PrADO. The resulting description of the flight path incorporates all information, as required for a noise prediction with PANAM [14].

3.3. Noise Prediction

The overall aircraft noise can be described as the energetic sum of relevant components, i.e., so-called noise sources, and the influence of the most important interdependencies. The software PANAM [1] was developed to realize such a simulation approach, based on the available input data within the conceptual aircraft design process. PANAM is directly operated as a simulation module within PrADO, and will enable a system noise prediction within the design cycle, i.e., modeling the noise emission, transmission, and impact at arbitrary observer locations [1].

PANAM applies semi-empirical, parametric source models that are developed by DLR or adapted from models found in the open literature, i.e., airframe models are developed by DLR [1,15–19], whereas available external engine models have been adapted [20,21]. A list of the models used is given in Table 1. The application of these models yields a directional noise emission prediction in one-third octave bands for one operating condition. PANAM accounts for the individual sources separately. According to the underlying operating condition, the resulting noise source ranking is predicted, which varies significantly along an approach and can vary along a departure.

If required, noise shielding and refraction effects can be simulated with the DLR tool SHADOW [22]. The predicted noise level differences are then directly processed within PANAM and applied to the corresponding sources, e.g., to account for noise reflection of a landing gear mounted under a large wing surface.

Table 1. PANAM system noise assessment: source models selected for this study.

Noise Source/Element	Model
airframe noise models (airf)	
trailing edge	DLR [1,15–19]
leading edge	DLR [1,15–17]
main landing gear	DLR [1,15–17]
nose landing gear	DLR [1,15–17]
engine noise models (eng)	
fan broadband & tonal	modified Heidmann [20]
jet	modified Stone [21]
noise shielding effects (PAA)	
-	SHADOW [22]
sound propagation effects	
-	ISO 9613 [23]
ground attenuation effects	
-	SAE AIR 1751 [24]

Such an emission prediction is then repeated for each individual flight point along a simulated trajectory, e.g., as provided by FLIPNA. The emitted noise from each flight point is then further propagated through the atmosphere, accounting for the prevailing attenuation of the sound, i.e., geometric spreading [25] and atmospheric attenuation [23]. Optionally, ground attenuation effects can be accounted for, to reflect a specific observer situation, i.e., accounting for ground properties and observer details such as height-above-ground of the virtual observer. When reproducing simulating certification conditions, certain options become mandatory, as explained in the next section. The resulting noise from each flight step, as it is received on the ground, can finally be assembled to yield level time-histories and other common noise metrics.

The prediction capabilities of PANAM have been demonstrated for existing aircraft by comparison with experimental data from flyover campaigns, e.g., see Reference [1]. Furthermore, its applicability to novel technology and new aircraft concepts was assessed through a benchmark test of simulation tools from two other large research entities. The results of the benchmark test, as published in Reference [26], show a satisfying agreement and confirm confidence in the PANAM predictions. A direct quantification of the underlying prediction uncertainties is currently under investigation. An earlier study was published in 2019, see Reference [27], to identify and assess the various sources of uncertainty that mpact the overall prediction quality compared to experimental data. Ongoing activities focus on an automated uncertainty quantification within the conceptual aircraft design, i.e., applicable to existing and novel aircraft.

3.4. Virtual Noise Certification Simulation Process

It was decided to utilize DLR's distributed simulation environment, the Remote Component Environment (RCE [28]) to realize the virtual noise certification process. The reason for this was the simple applicability and direct availability of such an RCE process within the connected researchers. Aircraft designs and technologies from different groups that are connected to RCE can directly feed and apply the process toward a virtual certification of their aircraft concepts. As a mid-term goal, the process is not limited to PrADO designs but for example is also able to process alternative vehicle concepts from the Center of Excellence *Sustainable and Energy Efficient Aviation* (SE^2A) by TU Braunschweig. Ultimately, the new process can be connected to different aircraft design tools to enable certification noise-related decision making.

The new RCE-based process promises flexibility when updating the simulation with respect to any adjustment to the somewhat antique ANNEX 16 regulations that might be introduced to provide a more realistic representation of aircraft noise around airports. The developed process chain is shown in Figure 3. To investigate the influence of different parameters (e.g., design or flight path parameters), it is possible to specify a parameter variation. A manual modification of individual noise sources is also possible, which allows for an estimation of the influence of over- and underestimation of individual sources on the certification level. Additionally the effectiveness of targeted noise reduction measures on individual noise sources (e.g., the use of acoustic liners) can be investigated. Based on the definition of the engine and aircraft parameters, the process starts with a calculation of the flight trajectory using the FLIPNA software, on the basis of the previously selected settings. PANAM then uses this trajectory, in conjunction with other necessary input data, and calculates the noise immission at the respective certification measurement point.

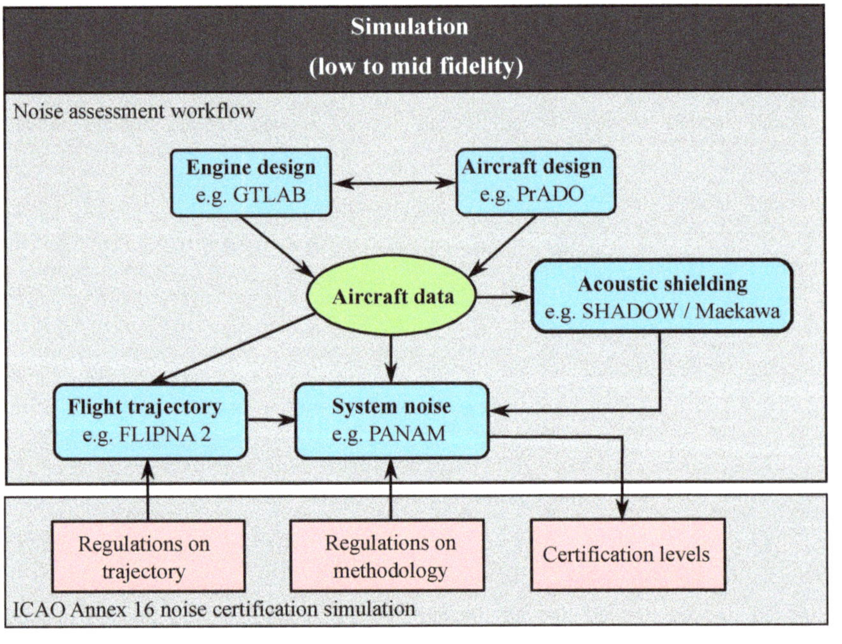

Figure 3. Schematic of the developed process for automated virtual noise certification.

The process automatically determines which parts of the trajectory are considered in EPNL calculation, i.e., where the 10 dB-down time starts and ends. This is supplemented by a consideration of the emission spectrum at the flight position that causes the highest PNLT (called PNLTM). A comparison of the spectra of the individual sources of emission and immission allows for a quick assessment of the influence of sound propagation effects, which have a significant impact on EPNL, e.g., at the flyover measurement point due to the relatively large distance between the source and observer. Finally, the virtual noise certification is performed in accordance with the implemented ICAO specifications (see Section 2). This status quo of the process chain can be used to automatically calculate the certification level of arbitrary aircraft within the framework of the preliminary aircraft design.

4. Verification

The new process was initially applied to two selected aircraft, i.e., a simulation model of an A319-100 powered by two CFM56-5A5 engines (net thrust per engine at sea level 104 kN) and a B747-400 powered by four CF6-80-C2-B5F engines (net thrust per engine at sea level 272 kN). An overview of the most important design parameters and the airspeeds used in the certification calculation are summarized in Table 2. The aircraft masses MTOM and MLM were adjusted to the values used and provided in the actual noise certification documents of this aircraft configuration.

Table 2. Data of two investigated aircraft.

Parameter	Unit	A319-100	B747-400
maximum takeoff mass	kg	68,000	396,893
maximum landing mass	kg	61,000	285,783
overall length	m	33.84	70.60
overall height	m	12.14	19.41
engine type	-	CFM56-5A5	CF6-80-C2-B5F
number of engines	-	2	4
static thrust per engine at sea level	N	104,043	272,530
bypass ratio	-	6	5
span	-	33.94	64.40
wing reference area	m^2	122.6	561.93
v_{Ref}	m/s	65.51	68.65
$v_{Landing}$	m/s	70.78	73.93
$v_{Flyover}$	m/s	75.13 to 80.13	87.90 to 92.89
$h_{Cutback}$	m	300	210

The simulation of the two aircraft types results in different cutback heights. A cutback height of 300 m is predicted for the A319-100, and 210 m is predicted for the B747-400. To allow for the minimization of certification levels at the flyover measurement point, the lowest allowed cutback height is not used. Instead, the cutback is performed just before the start of the 10 dB-down time of the flyover measurement point, which leads to an increase in cutback height. The later reduction in thrust leads to an increase in flight altitude and, thus, to a reduction in EPNL at the flyover measurement point due to stronger atmospheric propagation attenuation.

Altitude, thrust and calibrated airspeed profiles over the flightpath used to calculate certification levels are shown in Figure 4 for both aircraft. The authors highlight that there is no available information on official noise certification trajectories. Therefore, the trajectories shown and studied here are the result of an optimization to reduce EPNL at the individual measurement points, taking the certification regulations into account. As can be seen, the trajectories of the two aircraft studied differ greatly due to their significant differences in flight performance.

For both aircraft, the lowest certification level at the flyover measurement point was observed when the aircraft operated at the lowest allowed climb angle of 2.29°. Therefore, the relative engine speed could be reduced to N1 = 73% (A319-100) and N1 = 95% (B747-400). In this case, the effect of noise reduction on the ground due to a lower engine thrust outweighs the disadvantage of a reduced distance between source and observer. A variation of the engine thrust after cutback in the allowed range of the regulations showed a change in the certification level of 6.8 EPNdB (A319-100) and 4.9 EPNdB (B747-400) between the highest and lowest value. It is noticeable that the certification level for B747-400 increases abruptly between N1 = 95 and 96%, because this change triggers a significant increase in buzzsaw noise according to the implemented fan noise model, as the fan tip Mach number passes 1.

As previously mentioned, the lateral measurement point with the highest EPNL along the takeoff is identified and selected. This results in a distance to the break release point of 2700 m for the A319-100 where the aircraft has a flight altitude of 215 m, and for the B747-400 of 3600 m where the aircraft has a flight altitude of 225 m above ground. The position of the lateral measurement point is strongly dependent on whether the acceleration process on the runway is modeled correctly, so that inaccuracies resulting from aircraft preliminary design can have a large impact. The approach path, not shown here, has a glide path of 3° with an airspeed of 70.78 m/s (A319-100) and 73.93 m/s (B747-400). The engine thrust is constant along the relevant flight segment and amounts to 43.1 kN (A319-100) and 205.3 kN (B747-400). As specified for certification, the aircraft is simulated at precisely 120 m above the measurement location.

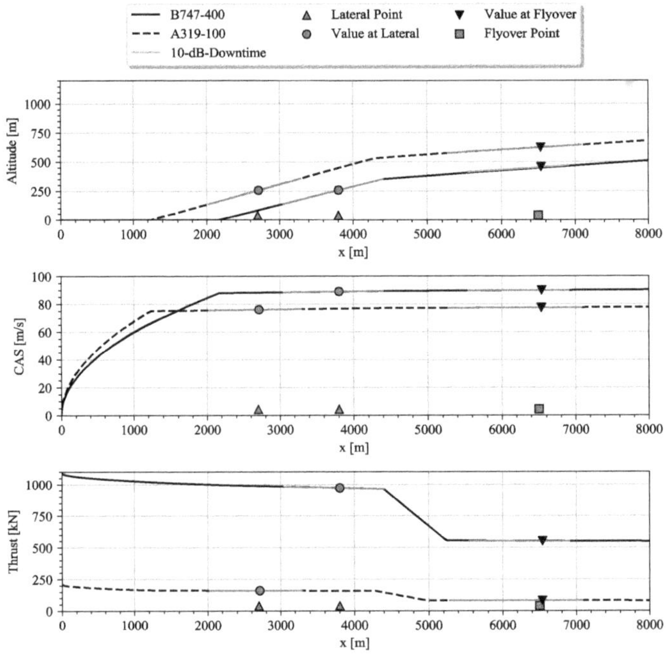

Figure 4. Departure Flightpath with lowest EPNL for virtual noise certification with flyover position and lateral reference meseaurement point.

4.1. Comparison of Predictions and Published Levels

Table 3 summarizes a comparison of the certification levels at the three measurement points for the two aircraft examined, with levels that are publicly accessible by European Union Aviation Safety Agency (EASA) , see Reference [29]. In contrast to the results presented earlier (DLRK, IB), the acoustic lining routines were adapted and the effects are incorporated into the results. Fan noise contribution was significantly reduced compared to earlier studies, due to the presence of an acoustic lining. There are discrepancies between the official certification levels and the results of this study. Deviations of +3.0 EPNdB (A319-100) and +2.8 EPNdB (B747-400) are observed at the flyover position, where certification regulations of flight trajectory are the least stringent. For the approach, reduced deviations to published levels are experienced, i.e., +1.8 EPNdB (A319-100) and +0.8 EPNdB (B747). The differences for the two investigated aircraft, observed at the lateral position, are −3.2 EPNdB (A319-100) and +0.8 EPNdB (B747-400). At this measurement point, a significant influence of the actual ground noise attenuation can be expected due to low elevation angles. This can lead to increasing uncertainties and inaccurate representation of the actual conditions of the real certification. The approach situation, according to the regulations, is much more regulated, resulting in a significantly smaller variability in acceptable flight conditions. It can be assumed that the simulated operational input for the noise prediction agrees more with the actual flight conditions of a real test flight for the approach situation, as compared to the departure situation. Other research studies show similar difficulties in predicting measured certification levels in aircraft preliminary design, e.g., Reference [30]. With the exception of the lateral full-power measurement point for A319-100, all simulated certification levels were overestimated relative to the official values. A good agreement is shown for the delta of the individual measurement points when comparing the two aircrafts. There are only significant deviations for the lateral measurement point, where the simplified modeling of the ground influence may play a role, as mentioned previously. In this context, different ground effect models were compared and it was observed that, according to the prevailing ground model, the levels can differ significantly from each other. Therefore, the authors would like to emphasize that the modeling of the ground attenuation effects is currently still associated with a large uncertainty and is part of ongoing research activities. More information on the actual condition, as experienced during the real certification, is not available but would be required for a fair comparison.

Table 3. Comparison of simulated and measured certification levels [29].

Observer Point	A/C	Simulation EPNL [EPNdB]	EASA EPNL [EPNdB]	Delta [EPNdB]
Flyover	A319-100	86.6	83.6	+3.0
Lateral	A319-100	90.7	93.9	−3.2
Approach	A319-100	96.3	94.5	+1.8
Flyover	B747-400	99.4	96.6	+2.8
Lateral	B747-400	101.0	100.2	+0.8
Approach	B747-400	104.1	103.3	+0.8

4.2. Discussion of Deviations

Simulation of the noise certification process and prediction of certification levels for existing aircraft is a very challenging task at the conceptual aircraft design phase. The direct comparison of simulation results to official noise certification data shows deviations in absolute levels which are also experienced by other researchers [30]. Based on the fidelity of the applied simulation tools and the lack of validation data for the flight simulation, this result is not surprising.

The semi-empirical noise source models listed in Table 1 and applied here are inherently associated with modeling uncertainties, e.g., see Reference [27]. Due to their empirical nature, a satisfying agreement between simulation results can be experienced when comparing the results to the measured data of existing aircraft. However, discrepancies are experienced when assessing certification noise levels. This can partially be attributed to the EPNL metric, which is especially sensitive to the spectral shape of the received signal. Obviously, the level of detail associated with the noise source models is not sufficient to precisely resolve the spectral shape for each noise source simulation under every operating condition. A much higher level of fidelity would be required for the simulation methods, which cannot be realized based on the selected models available in the conceptual aircraft design. The selected noise source models were specifically derived to capture standard metrics that are integrated over the frequency spectrum based on a third-octave band resolution. The experienced discrepancies between measured and predicted spectra at distant ground observers are further intensified due to the underlying and simplified sound propagation and ground attenuation models, as specified in Table 1. Ground attenuation is very uncertain, especially due to the microphone height of 1.2 m. A small overestimation of fan tones (1–2 dB) is possible in preliminary design, but has little to no impact on overall levels with conventional metrics such as OASPL; however, such inadequacies may be exacerbated by a greater weighting of the relevant frequency range of fan tones and tonal penalty when using metrics such as EPNL.

In addition to the described modeling challenges for EPNL at the conceptual design phase, the underlying input data introduces additional uncertainties since it originates from low-fidelity simulation at the conceptual aircraft design phase [27]. Especially, the flight trajectory as a direct input for the noise prediction is identified as a major source of uncertainty. In particular, the acceleration of the aircraft on the ground and the thrust setting after cutback, i.e., the actual takeoff trajectory, results in different flight altitudes above the flyover point. It has been demonstrated in various studies that a much better agreement between noise simulation and measured levels can be realized if the operating conditions are provided as an input from the actual flight test. A good agreement between simulation and measurements for metrics other than EPNL is documented for existing aircraft, e.g., Reference [1], if the noise prediction is fed with Flight Data Recorder (FDR) input. The importance of reliable input data is furthermore confirmed by a recent comparison of noise simulation tools between NASA, ONERA, and DLR [26]. It is demonstrated that a good level of agreement between the different tools can be achieved when the input data are clearly defined and kept identical or similar for all simulation tools. Not knowing the actual flight conditions during the certification process, especially for the flyover situation, does not support the direct comparison of absolute levels from simulation and official data. However, predicted level differences between various aircraft types show a much better agreement, supporting a comparative assessment. The reliability of the prediction results for a comparative assessment is furthermore confirmed by the aforementioned simulation tool comparison [26].

To further investigate the significant influence of the flight path on the certification levels, certification levels are predicted based on available flight data for an A319-100 aircraft [31]. The available approach-and-departure flight data from this measurement campaign at the Baltic Airport near Parchim (Germany) are used to calculate the respective certification levels. The available measured noise levels for A-weighted SPL were studied in great detail and show good agreement to the predictions; see Reference [1]. It should be noted that no EPNL measurements are available from this campaign. Due to the underlying motivation behind this flight campaign, no departure flight procedure fully complies with the allowable operating conditions according to the ICAO regulations [3]. The flight altitude above the approach measurement point is too high for approach paths 10 and 12 in Figure 5. This has to be kept in mind when comparing these predictions to the published certification levels. The other approach paths are certification-compliant and show very good agreement with the measured EASA levels. The minor differences between the procedures are due to the slightly different thrust settings and flight altitudes. When comparing with the simulated flight trajectories, it is noticeable that the approach speeds from the FDR data are lower, which lead to lower certification levels. This may indicate that the accuracy of the characteristic speeds in the preliminary design is limited. However, this comparison gives a good indication of the sensitivity of the flight procedures at the certification points, and also can confirm tendencies and trends. As shown in Figure 5, it can be demonstrated that all flight procedures result in higher noise levels at flyover measurement points in comparison with the noise levels published by EASA. Overestimation rates from 0.5 to 3 EPNdB are documented for the flyover situation. Obviously, the differences between the actual flight tracks of the certification and the simulated flights have a great impact. An underestimation of from 1.5 to 2.2 EPNdB at the sideline certification points is observed where the underestimated levels are also experienced when working with the simulated flight procedures. Comparing level differences between the individual flights confirms the expected trends and the applicability of simulation results for a comparative assessment, i.e., see Reference [32].

Another reason for the expected deviations can be identified in the ICAO documentation. Certain correction factors can be applied by the aircraft manufacturers to each of the actual noise levels at a certification point. These corrections are mentioned but not further disclosed in the ICAO documentation. The actual levels remain confidential and no further information has to be made available to the public. It is still unclear how these correction factors influence the final and official certification level. If the predicted levels are somewhat higher compared to the official levels, this is understood as a confirmation of the presented simulation results, since it is certain that any additional correction factor applied by an aircraft manufacturer will not further increase the certification levels. Ongoing research activities at DLR and its partners aim towards the quantification of simulation uncertainties for aircraft noise prediction, as applied here, e.g., see Reference [27]. Obviously, this is understood as an essential prerequisite when assessing differences between simulation results and available certification noise levels. However, this research is still ongoing and limited to an uncertainty prediction associated with less complex noise metrics or descriptors, i.e., the maximum A-weighted SPL. The direct application of available uncertainty quantification methods to a complex, frequency-dependent descriptor, including tonal corrections such as the EPNL, e.g., as applied by NASA [7], is not recommended at this moment. Uncertainty quantification associated with complex noise descriptors remains a challenging task for future research activities outside of the presented work.

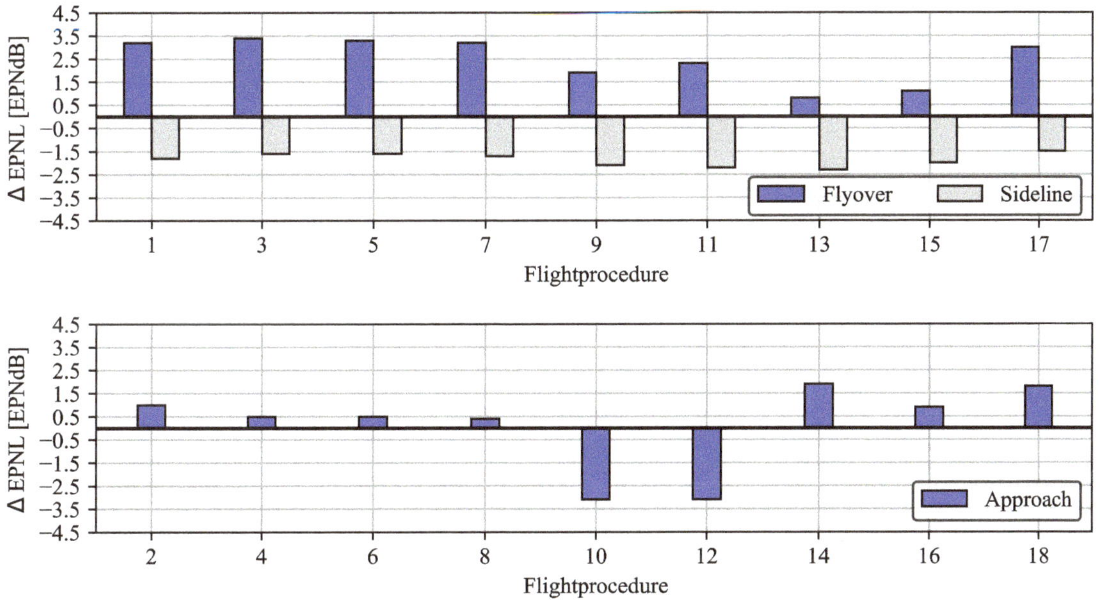

Figure 5. EPNL differences in simulated flight trajectories compared to published certification levels by the EASA.

4.3. Parameter Sensitivity Study: A319-100

In order to investigate the influence of the engine and airframe noise sources on the certification level, a sensitivity study of the certification level for A319-100 was performed for each of the three measurement points, starting from the reference case shown above. For this purpose, the predicted noise levels for engine and airframe components were artificially increased/reduced in the range from −9 to +9 dB. It was assumed that the level difference is equally applied over the entire frequency range.

The plots shown in Figure 6 represent the change in certification level relative to the reference case as a combination of the change in the Sound Pressure Level of the engine and airframe.

The blue lines show the change in certification level due to the changes applied to the engine, and the green lines show the change in certification level due to changes applied to the airframe contribution, with all other noise sources held constant.

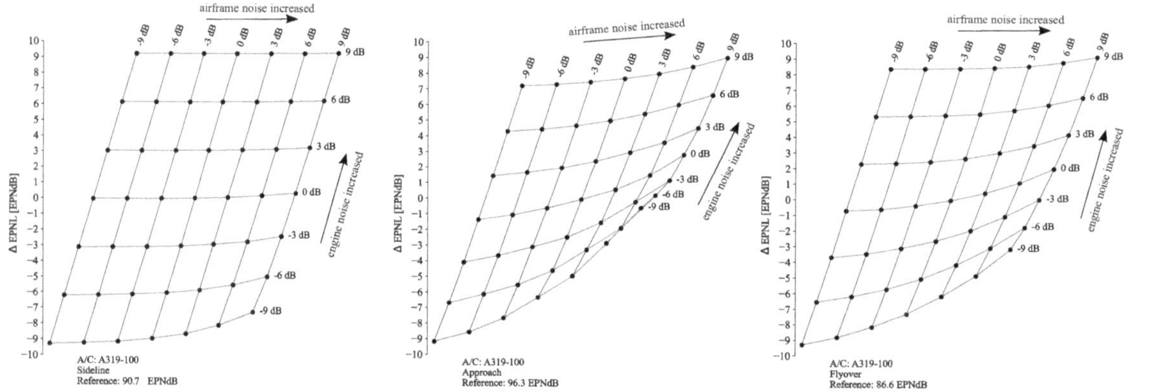

Figure 6. EPNL: sensitivity with respect to airframe and engine noise contribution for sideline (**left**), approach (**middle**) and flyover (**right**).

First, it can be shown that the noise sources of the engine dominate the immission at all three measurement points considered. For conventional tube-and-wing aircraft with turbofan engines, as considered here, airframe noise contribution is also insignificant at the approach certification noise measurement point. During its final approach, the flight segment increased-engine thrust settings are required to compensate the drag of deployed high-lift elements and landing gear while maintaining a prescribed flight velocity. Consequently, the engine noise clearly and constantly dominates the airframe noise contribution at the approach measurement point. Engine noise would have to be reduced drastically in order to see any impact of the airframe noise sources on the approach certification noise level, e.g., Figure 6. Consequently, for an efficient reduction in certification levels, noise abatement measures at the engine should be considered first; a quiet airframe would only after a drastic engine noise reduction. For the B747-400, the results are qualitatively very similar, but are not shown here for the sake of clarity. However, it is obvious that the influence of airframe noise on EPNL increases from sideline to flyover and from flyover to approach. This is due to the respective thrust setting, since this decreases in the aforementioned order. While the airframe has almost no influence on the sideline for all the considered modifications, there are changes in the certification level for flyover when the engine noise is significantly reduced. For the approach, this effect is even more pronounced, and smaller changes are also seen if the engine noise sources remain unchanged.

5. Application

5.1. Novel Aircraft Concepts

The developed process for the simulation of noise certification is used to evaluate different novel aircraft designs from a previous system noise assessment of a tube-and-wing aircraft with geared turbofan engines [2]. In this earlier study, the EPNL certification metric was evaluated on a 30 km × 20 km grid, but did not take ICAO noise certification regulations (e.g., flight path requirements) into account, which will be done in this study. The aircraft design denotations were adopted from Reference [2]. The aircraft design referred to as *zero* was selected as a reference aircraft for this study. The aircraft was initally designed in a former DLR project titled "Silent Leading Edge Design" (SLED) (see References [33,34]) and has a similar design to the A321 family.

As a reference engine, a two-spool unmixed high-bypass-ratio engine was used, similar to CFM International's CFM56-5A5 engine with a static thrust of 150 kN at sea level. The reference aircraft is compared to three aircrafts with different modifications to airframe and engine design. The first modification, referred to as *neo*, is the replacement of the engine by a modern GTF engine without changing the airframe (see Reference [2] for further details). The second modification, referred to as *neo (af)*, includes the GTF engine in conjunction with a low-noise airframe. In addition, another configuration (*fanex*) is compared, which is based on a different base aircraft (V2) (see also Reference [2]) and uses the GTF engines in addition to low-noise airframe technology. Here, the two engines are located above the aircraft, at the conjunction of fuselage and wing, in order to shield the engine noise. The aircraft design descriptions, maximum takeoff mass (MTOM) and maximum landing mass (MLM), as well as takeoff and landing speeds, are summarized in Table 4. Further details on the aircraft design studied here can be found in [2].

Table 4. Design aspects of applied aircrafts from previous system noise study [2].

A/C	Description	MTOM [t]	MLM [t]	Takeoff Speed [m/s]	Landing Speed [m/s]
zero	Reference aircraft with reference engines	83,548	72,082	67.86	63.17
neo	Reference aircraft with GTF engines	80,076	72,868	66.70	63.47
neo (af)	Reference aircraft with low noise airframe modifications and GTF engines	80,092	72,881	70.06	66.88
fanex	Variant of V2 (Reference [1]) with airframe adapted adapted to GTF	76,218	70,187	69.35	66.67

For better comparability of the concepts, the flight altitude at which the pilot-initiated cutback was performed was kept constant at 500 m above ground for all configurations. When optimizing the flight path to minimize EPNL at the flyover reference measurement point, the cutback altitude for each configuration was selected so that thrust was reduced shortly before the start of the 10 dB-downtime (integration time of EPNL). In addition, for all configurations, the thrust after cutback was selected so that the aircraft would subsequently continue to fly at a climb rate of 4%.

The noise certification simulation results are shown in Figure 7 for the three configurations *neo*, *neo (af)* and *fanex* relative to the reference aircraft *zero*. The corresponding lateral measurement position is adapted accordingly for each configuration. The distances from the lateral measurement point (highest EPNL along the takeoff flight path) to the break release point differ between the four configurations. For *zero* (2600 m) and *neo* or *neo (af)* (2500 m), there are only minor differences. For *fanex*, the lateral measurement position is significantly closer to the break release point (2000 m) as the forward-directed engine noise is shielded by the fuselage and wings. Reductions in certification levels are observed at all three measurement points, with *neo* and *neo (af)* showing very similar results. Small deviations may result from different characteristic airspeeds and slightly different aircraft masses. It can be demonstrated that the use of a low-noise airframe in the context of noise certification has only a small advantage. For *neo* and *neo (af)*, the 10 dB-downtimes are similar to those of the reference aircraft. Due to the fan's tonal contribution, the reference engine has an additional tonal penalty at the sideline measurement point, which does not occur with the GTF engine. This leads to a reduction in the PNLTM.

Due to the effective noise shielding, *fanex* shows significantly lower certification levels at all three measurement points compared to the *zero* and the two *neo* variants. The reductions differ significantly between the measurement points. The noise shielding has the most effective impact on the flyover measurement point, where a reduction of −16.2 EPNdB occurs. At the approach measurement point, there is a reduction of −10.9 EPNdB and at the lateral measurement point, there is only a reduction of −4.7 EPNdB. Due to the effective noise shielding of the forward-directed engine noise for the *fanex*, the 10 dB-downtime is significantly reduced, to 9.5 s (−51.6%), compared to the reference aircraft (18.4 s). This is in conjunction with the slightly reduced PNLTM, by 1.1 dB, to 94.5 dB.

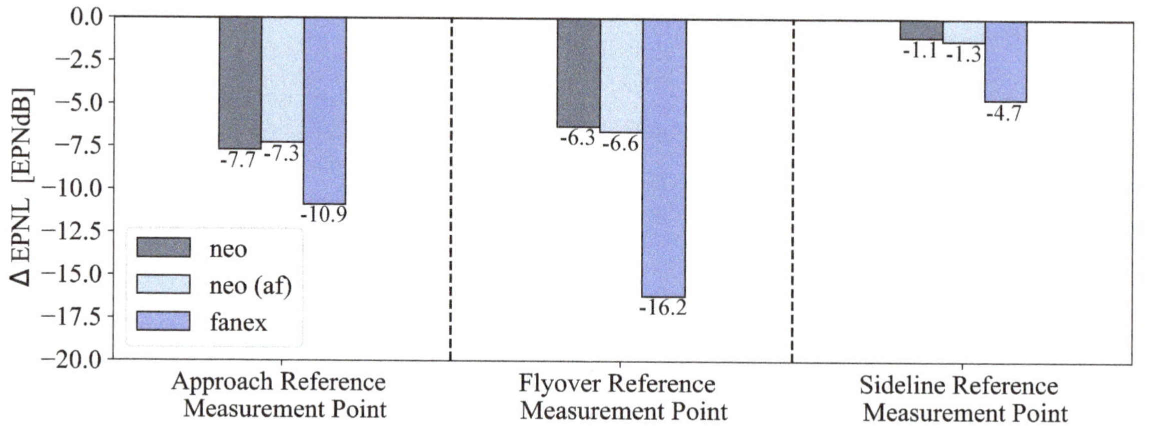

Figure 7. Reduction of certification levels for studied aircraft designs (*neo, neo (af), fanex*) relative to reference aircraft *zero* at three reference measurement points Approach, Flyover and Sideline.

For the four aircraft, the margins of the cumulative limits of Noise Chapter 4 are shown in Figure 8. The two neo-variants show similar results, with the a slightly larger margin for the *neo (af)*, with 18.2 EPNdB, compared to the *neo*, with 18.6 EPNdB. There is only a small margin for *zero* of 3.7 EPNdB, which means that this aircraft was the only one to not achieve a Chapter 14 certification, since, for this, the cumulative limit must be reduced by 7 EPNdB for Chapter 4 to Chapter 14.

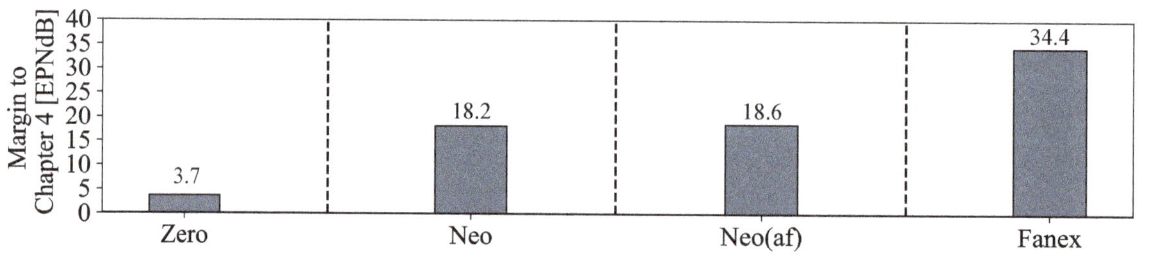

Figure 8. Margin to cumulative Limit of chapter 4 for studied aircraft designs (*zero, neo, neo (af), fanex*).

It is demonstrated that noise abatement measures for the individual aircraft have different effects on the certification levels. Certification represents only one aspect of the noise assessment of an aircraft during takeoff and landing. The noise-reducing effect of individual measures on the aircraft geometry and engine is only identified at individual measurement points, very close to the runway, and is not considered over a large area. This leads to deviations in the evaluation of individual measures, such as for the low-noise airframe, which shows no significant effect on the certification level in noise certification, although a significant advantage can be demonstrated for large-area contours.

5.2. *Towards Certification Regulations for Supersonic Transport Aircraft*

Recently, noise certification rules for future supersonic aircraft have been presented by the Federal Aviation Agency (FAA) and the Department of Transportation (DOT) [35] as a consequence of recent industry activities, e.g., product announcements of supersonic business jet developers such as Boom and Aerion. Again, specific limits for the operation are defined and the EPNL is assessed at the three measurement points. The proposed regulations solely focus on the take-off situation and can be directly implemented into the existing simulation framework to enable a virtual certification process, i.e., under the assumption of constant approach definitions for subsonic and supersonic vehicles. Due to the high thrust requirements for supersonic cruise flight, these SST vehicles use engines with a high specific thrust, tailored to cruise conditions. This results in large excess thrust levels during takeoff, i.e., high jet exit velocities and jet noise generation, as demonstrated by various researchers, e.g., References [36,37]. As a consequence, it can be expected that the direct application of current regulations for subsonic aircraft to supersonic vehicles would result in significantly increased take-off noise levels compared to subsonic transport vehicles, particularly at the lateral measurement point. Therefore, new take-off regulations for supersonic aircraft were proposed by the FAA in order to specifically enable further reductions in take-off thrust settings in proximity to the measurement points and allow for different flight velocities according to the specific SST aerodynamics, e.g., as discussed in Reference [36]. The overall goal of the modifications is to achieve noise certification levels for supersonic aircraft that remain below the published noise limits, applicable to subsonic aircraft of a certain noise category. The discussed rule change regarding take-off, and its influence on certification flight altitude, total thrust and airspeed, are depicted in Figure 9. At the top of this figure, the altitude profiles are shown, commparing subsonic and supersonic aircraft. The first proposed modification focuses on the take-off speed regulations, which are defined relative to the aircraft-specific safe takeoff speed *V2*. The default speed range is removed, and only the default maximum airspeed of 250 kts is maintained for the SST, to enable a high-speed climb, which can be associated with reduced ground noise impact along the take-off trajectory. According to Reference [36], the higher airspeed reduces the required engine thrust, so that, for the flyover measurement point, the lower thrust should lead to a reduced certification level. This is achieved by a delayed rotation takeoff, i.e., longer acceleration segment on the ground, which can also be seen in the calibrated airspeed profile at the bottom of the figure (from Point A to Point F′). This should enable higher lift-to-drag ratios, and thus optimize flight performance. Furthermore, an additional thrust reduction of up to 10% before the typical pilot-initiated cutback is being considered, which is not allowed for subsonic aircraft. This modified procedure is called *Programmed Lapse Rate* (PLR), and was developed in the 1990s [38,39] and applied to SST aircraft by NASA [36,40]. This thrust reduction takes place after the obstacle height is reached, and can be seen in the thrust profile in the middle of the figure, between Point C and D′.

No modifications to the subsonic regulations have been proposed with respect to the approach situation to date. However, it can be expected that novel SST vehicles do not have distinct slow flight characteristics, and any installation of a classical high-lift system on-board slim SST wings will be challenging. High-approach velocities can directly translate to increased aerodynamic and fan noise contribution. It is expected that approach noise could become a big challenge for SST vehicles. Only advanced flight procedures, enabled by relaxed regulations and additional low-noise technologies for major noise sources, can relax the noise issue associated with the landing and take-off operation of these vehicles.

The initial implementation of the proposed regulation modifications is intended as a starting point for the evaluation and estimation of certification levels for the new generation of supersonic business jets, which are currently under development. The proposed rule changes, as briefly mentioned here, will then be compared to the current regulations. The impact of these measures on the SST certification levels can then be investigated and discussed. This update will ultimately contribute to the studies that have already been conducted on this research area, to support decision-making in the context of SST certification legislation.

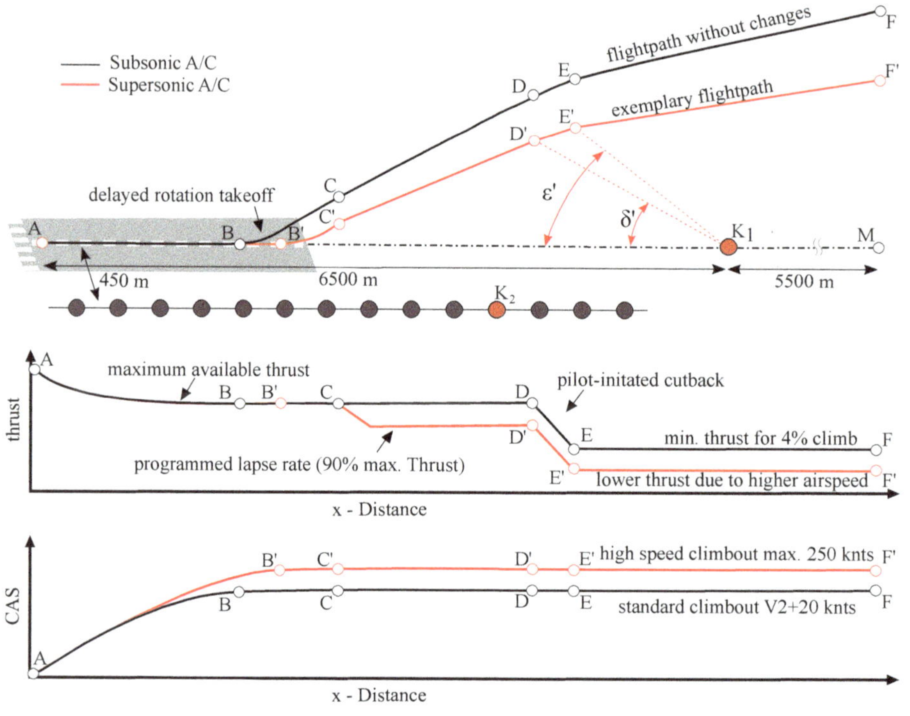

Figure 9. Take-Off reference flight path profile of subsonic and supersonic aircraft for noise certification (**top**). Changes of thrust (**middle**) and calibrated airspeed (**bottom**) profiles for discussed new supersonic noise certification regulations due to high-speed climbout and Programmed Lapse Rate.

6. Discussion and Conclusions

This article describes the initial work toward a virtual noise certification process within conceptual aircraft design. An overview of the regulations of ICAO is provided, and a corresponding simulation process is established. It is demonstrated that all the required input information can be made available at the preliminary design phase in order to estimate certification levels and assess the influencing variables. The application of this process to existing aircraft and a comparison of the predicted and published certification noise levels shows discrepancies. A detailed investigation of the influence of all underlying simulation steps is performed and results are presented, including aircraft and engine design, simulation of the flight tracks, and the overall noise prediction methodology. It can be demonstrated that the complexity of the EPNL metric provides difficulties at the conceptual aircraft design phase, and may play a role in the differences. At this phase, only simplified definitions of the actual aircraft design and simplified and semi-empirical noise source models are available. The applied noise source models were specifically developed to enable a comparative assessment of different technologies and aircraft, instead of the prediction of precise absolute levels in complex metrics such as the EPNL. Based on the identified challenges of simulating the certification situation and the prevailing discrepancies with official data, it is recommended that the process only be applied for comparative assessment. The process will predict the noise level differences for technologies under investigation, i.e., aircraft and flight procedure. The impact of novel technologies on the certification levels can be predicted and used to support the decision-making process.

Available simple propagation models furthermore contribute that prediction results are not adequate for a detailed assessment of the spectral information as received on the ground. Comparing prediction results to experimental data (measurements) using different and less complex metrics, e.g., A-weighted SPL or SEL, can result in significantly smaller discrepancies, as demonstrated in previous studies, e.g., References [1,41].

However, a direct comparison of prediction results from different simulation tools confirms the reliability of the underlying simulation methods, even for EPNL predictions. Running such different simulation tools based on similar input data results in good agreement between the simulation results [26]. This is a strong indication that much of the experienced discrepancies in the published certification levels is caused by the actual input data, particularly the underlying simulated flight.

In conclusion, at present, it is recommended to only apply the new simulation process to evaluate level differences, instead of absolute levels. Too many uncertainties prevail when comparing the predicted absolute levels with the available certification data, as also experienced by other researchers, e.g., Reference [30]. Due to the lack of information on the precise certification process, i.e., actual flight procedures and availability of all additional corrections to the final results, it is difficult to assess the absolute levels at the certification locations and directly compare these with the available certification data. The process application is, therefore, limited to a comparative assessment of different technologies or aircraft concepts, which was the main goal of these activities. The investigation of different low-noise technologies (e.g., low-noise airframes) reveals that a comprehensive assessment should not only be based on the certification level and metric EPNL at the three representive reference noise measurement points alone. Instead, the assessment should be a holistic evaluation consisting of a large-scale assessment of contours in the vicinity of the airport and multiple noise metrics, e.g., SEL and certification levels, in order to adequately assess the noise issue associated with novel vehicles in conceptual aircraft design.

7. Outlook

A revisiting and improvement of the available input data situation and the simulation capabilities is planned in the near future. A better validation of the simulations would result from a comparison with the experimental trajectory data of a valid official noise certification. Typically, the available data from flyover noise measurement campaigns are very limited. Only a few aircraft have been subject to such a detailed and costly assessment, e.g., as described in Reference [17]. In general, the focus of such campaigns does not lie in certification assessment; the goal is to capture the widespread effects of a novel technology on-board the aircraft and/or the effects of novel flight procedures, e.g., Reference [41]. Therefore, the underlying flight procedures usually do not comply with the certification regulations, and, accordingly noise is not measured, i.e., the certification points are not captured or the defined 1.2 m microphones are not used. Other campaigns capture the defined locations and use the required microphones but do not consider the regulations with respect to the flight procedures, since the research is focused on other aspects, e.g., Reference [31]. In conclusion, such data are most valuable for scientific studies but are not available at present. The assumptions and simplifications regarding the presented simulation of a certification flight can only be verified by a comparison with actual flight data, recorded during a real certification procedure. At this point, it is not possible to fully validate the most important input for the certification simulation, i.e., the actual flight track, which explains the prevailing operating conditions.

The insufficient input data situation can be justified by the assumption of increasing simulation uncertainties. Ongoing research activities toward uncertainty quantification in the context of aircraft noise simulation, e.g., Reference [27], will focus on the certification situation. At the same time, the simulation uncertainties might be reduced by the integration of high-fidelity simulation results into the overall process. Although such high-fidelity methods are typically not available for such an assessment, due to their inherent simulation costs, a mixed-fidelity or multi-fidelty approach and the combination of results is most promising. High-fidelity tools could be applied to confirm the results of the simplified tools, or provide these tools with more reliable input data, and hence improve the overall quality of results.

Author Contributions: Conceptualization, M.N. and L.B.; methodology, M.N. and L.B.; software, M.N.; formal analysis, M.N. and L.B.; validation, M.N.; writing—original draft preparation, M.N. and L.B.; writing—review and editing, L.B.; visualization, M.N.; supervision, L.B.; project administration, L.B.; funding acquisition, L.B. All authors have read and agreed to the published version of the manuscript.

Funding: The authors gratefully acknowledge a partial funding by the Deutsche Forschungsgemeinschaft (DFG, German Research Foundation) under Germany's Excellence Strategy—EXC 2163/1—Sustainable and Energy Efficient Aviation—Project-ID 390881007. All other funding is associated with DLR internal research activities.

Data Availability Statement: Not applicable.

Acknowledgments: The consulting and support of Jason Blinstrub (DLR) with respect to the flight simulation is greatly appreciated. Furthermore we would like to thank Marc Koch for his assistance with respect to the B747-400 assessment.

Conflicts of Interest: The authors declare no conflict of interest.

Abbreviations

The following abbreviations are used in this manuscript:

C	sound correction, penalty
CAS	calibrated airspeed
EPNL	Effective Perceived Noise level
FDR	Flight Data Recorder
FLIPNA	Flightpath for Noise Analysis
ICAO	International Civil Aviation Organization
MLM	Maximum landing mass
MTOM	Maximum takeoff mass
n	Perceived Noisiness as a function of time and frequency
N	Total Perceived Noisiness as a function of time
PANAM	Parametric Aircraft Noise Analysis Module, DLR software
PNL	Perceived Noise Level as a function of time
PNLT	Tone Corrected Perceived Noise Level as a function of time
PNLTM	maximum of Tone Corrected Perceived Noise Level
PrADO	Preliminary Aircraft Design and Optimization Program
SAE	Society of Automotive Engineers
SPL	Sound Pressure Level as a function of time
SST	Supersonic Transport
Variables:	
A	break release point
B	begin transition arch
B′	begin transition arch (SST regulations)
C	end transition arch
C′	end transition arch (SST regulations)
D	begin of pilot-iniated cutback
D′	begin of pilot-iniated cutback (SST regulations)
E	end of pilot-iniated cutback
F	end of reference profile
G	start of data record for certification
H	flightposition over approach measurement point
I	end of glide path
J	landing point
K_1	flyover reference noise measurement point
K_2	lateral full-power reference noise measurement point
K_3	approach reference noise measurement point
M	projection from F to ground
O	runway threshold

P	projection from G to ground
γ	climb angle
δ	first cutback angle
δ'	modified first cutback angle (SST regulations)
ε	second cutback angle
ε'	modified second cutback angle (SST regulations)

Appendix A

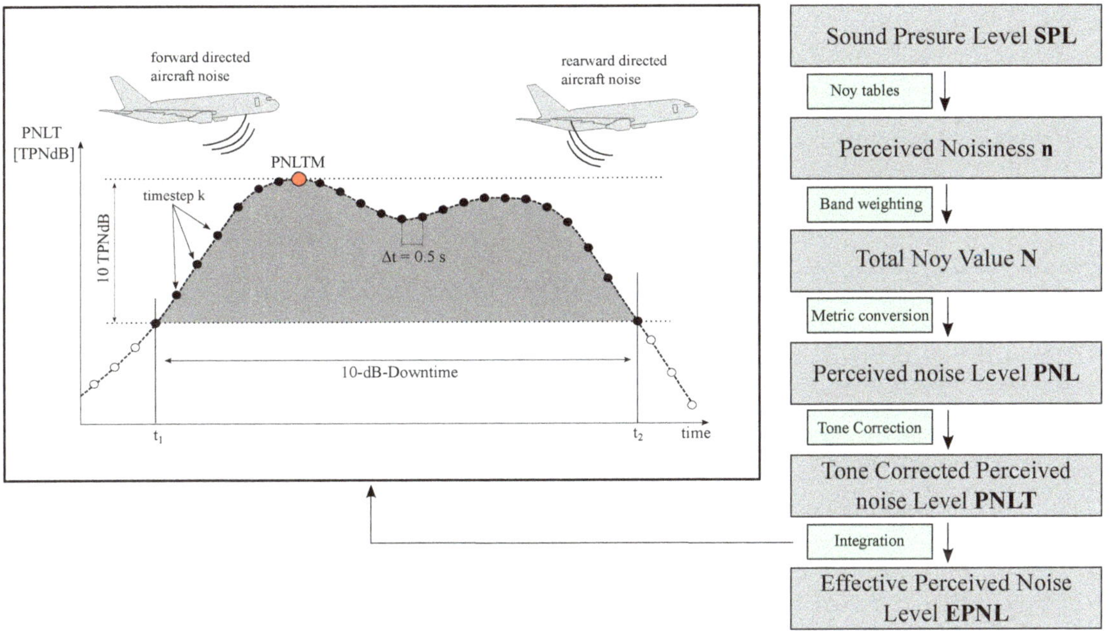

Figure A1. Visualization of Effective Perceived Noise Level Calculation from measured Sound pressure Level.

References

1. Bertsch, L. *Noise Prediction Within Conceptual Aircraft Design*; Technical Report DLR-FB-2013-20; Deutsches Zentrum für Luft- und Raumfahrt (DLR): Göttingen, Germany, 2013. [CrossRef]
2. Bertsch, L.; Wolters, F.; Heinze, W.; Pott-Pollenske, M.; Blinstrub, J. System Noise Assessment of a Tube-and-Wing Aircraft with Geared Turbofan Engines. *AIAA J. Aircr.* **2019**, *56*, 1577–1596. [CrossRef]
3. International Civil Aviation Organization (ICAO). *Environmental Protection. Annex 16 to the Convention on International Civil Aviation. Vol. I, Aircraft Noise*, 5th ed.; Technical Report Annex 16; ICAO: Montreal, QC, Canada, 2008; Volume I.
4. Berton, J.; Envia, E.; Burley, C. An analytical assessment of NASA's N+ 1 subsonic fixed wing project noise goal. In Proceedings of the 15th AIAA/CEAS Aeroacoustics Conference (30th AIAA Aeroacoustics Conference), Miami, FL, USA, 11–13 May 2010; p. 3144.
5. Rizzi, S.A.; Burley, C.L.; Thomas, R.H. Auralization of NASA N+ 2 aircraft concepts from system noise predictions. In Proceedings of the 22nd AIAA/CEAS Aeroacoustics Conference, Lyon, France, 30 May–1 June 2016; p. 2906.
6. Guo, Y.; Thomas, R.H.; Clark, I.A.; June, J.C. Far-Term Noise Reduction Roadmap for the Midfuselage Nacelle Subsonic Transport. *AIAA J. Aircr.* **2019**, *56*. [CrossRef]
7. June, J.; Thomas, R.; Guo, Y. System Noise Prediction Uncertainty Quantification for a Hybrid Wing–Body Transport Concept. *AIAA J. Aircr.* **2020**, *58*, 1157–1170. [CrossRef]
8. Nöding, M. Virtuelle Lärmzertifizierung im Flugzeugvorentwurf. Master's Thesis, Deutsches Zentrum für Luft- und Raumfahrt (DLR), Göttingen, Germany, 2020.
9. Nöding, M.; Bertsch, L. Beitrag zu einer virtuellen Lärmzertifizierung im Flugzeugvorentwurf. In *Deutscher Luft- und Raumfahrtkongress*; DLRK: Virtual conference, 2020.
10. *ICAO Annex 16: Environmental Protection, Volume I—Aircraft Noise*; International Civil Aviation Organization (ICAO): Montreal, QC, Canada, 2011.

11. *ICAO: Environmental Technical Manual, Volume I-Procedures for the Noise Certification of Aircraft*; International Civil Aviation Organization (ICAO): Montreal, QC, Canada, 2018.
12. Heinze, W. *Beitrag zur Quantitativen Analyse der Technischen und Wirtschaftlichen Auslegungsgrenzen Verschiedener Flugzeugkonzepte für den Transport Grosser Nutzlasten*; Technical Report ZLR-Forschungsbericht; ZLR: Braunschweig, Germany, 1994; ISBN 3-928628-14-3.
13. Deidewig, F. *Ermittlung der Schadstoffemissionen im Unter- und Ueberschallflug*; Technical Report; DLR: Cologne, Germany, 1998. [CrossRef]
14. Blinstrub, J. *Immission-Based Noise Reduction within Conceptual Aircraft Design*; Technical Report DLR-FB-2019-12; Deutsches Zentrum für Luft- und Raumfahrt (DLR): Göttingen, Germany, 2019. [CrossRef]
15. Dobrzynski, W.; Chow, L.; Guion, P.; Shiells, D. A European Study on Landing Gear Airframe Noise Sources. In Proceedings of the 6th AIAA/CEAS Aeroacoustics Conference, Lahaina, HI, USA, 12–14 June 2000. [CrossRef]
16. Dobrzynski, W.; Pott-Pollenske, M. Slat Noise Source Studies for Farfield Noise Prediction. In Proceedings of the 7th AIAA/CEAS Aeroacoustics Conference, Maastricht, The Netherlands, 28–30 May 2001. [CrossRef]
17. Pott-Pollenske, M.; Dobrzynski, W.; Buchholz, H.; Gehlhar, B.; Walle, F. Validation of Semiempirical Airframe Noise prediction Method through Dedicated A319 Flyover Noise Measurements. In Proceedings of the 8th AIAA/CEAS Aeroacoustics Conference, Breckenridge, CO, USA, 17–19 June 2002. [CrossRef]
18. Rossignol, K.S. Development of an empirical prediction model for flap side-edge noise. In Proceedings of the 16th AIAA/CEAS Aeroacoustics Conference, Stockholm, Sweden, 7–9 June 2010. [CrossRef]
19. Rossignol, K.S. Empirical Prediction of Airfoil Tip Noise. In Proceedings of the 17th AIAA/CEAS Aeroacoustics Conference, Portland, OR, USA, 5–8 June 2011. [CrossRef]
20. Heidmann, M. *Interim Prediction Method for Fan and Compressor Source Noise*; Technical Report NASA TMX-71763; NASA Langley Research Center: Langley, VA, USA, 1979. [CrossRef]
21. Stone, J.; Groesbeck, D.; Zola, C. Conventional profile coaxial jet noise prediction. *AIAA J.* **1983**, *21*, 336–342.
22. Lummer, M. Maggi-Rubinowicz Diffraction Correction for Ray-Tracing Calculations of Engine Noise Shielding. In Proceedings of the 14th AIAA/CEAS Aeroacoustics Conference, Vancouver, BC, Canada, 5–7 May 2008. [CrossRef]
23. International Organization for Standardization (ISO). *Acoustics—Attenuation of Sound during Propagation Outdoors. Part 1: Calculation of the Absorption of Sound by the Atmosphere*; Technical Report ISO 9613-1:1993; ISO: Geneva, Switzerland, 1993.
24. Society of Automotive Engineers. *Prediction Method for Lateral Attenuation of Airplane Noise During Takeoff and Landing*; Society of Automotive Engineers: Warrendale, PA, USA, 1981. [CrossRef]
25. Smith, M. *Aircraft Noise*; Cambridge Aerospace Series; Cambridge University Press: Cambridge, UK, 2004; ISBN 0-521-61699-9. [CrossRef]
26. Bertsch, L.; Sanders, L.; Thomas, R.H.; LeGriffon, I.; June, J.C.; Clark, I.A.; Lorteau, M. Comparative Assessment of Aircraft System Noise Simulation. *AIAA J. Aircr.* **2021**, 1–18. [CrossRef]
27. Bertsch, L.; Schäfer, B.; Guerin, S. Uncertainty analysis for parametric aircraft system noise prediction. *AIAA J. Aircr.* **2019**, *56*, 529–544. [CrossRef]
28. Boden, B.; Flink, J.; Mischke, R.; Schaffert, K.; Weinert, A.; Wohlan, A.; Schreiber, A. RCE: An Integration Environment for Engineering and Science. *SoftwareX* **2021**, *15*, 100759. [CrossRef]
29. EASA Jet Aeroplane Noise Database for Certification. Available online: https://www.easa.europa.eu/domains/environment/easa-certification-noise-levels (accessed on 17 April 2021). [CrossRef]
30. Thomas, J.L.; Hansman, R.J. Framework for Analyzing Aircraft Community Noise Impacts of Advanced Operational Flight Procedures. *AIAA J. Aircr.* **2019**, *56*, 1407–1417. [CrossRef]
31. Pott-Pollenske, M.; Dobrzynski, W.; Buchholz, H.; Guérin, S.; Saueressig, G.; Finke, U. Airframe Noise Characteristics from Flyover Measurements and Prediction. In Proceedings of the 12th AIAA/CEAS Aeroacoustics Conference, Cambridge, MA, USA, 8–10 May 2006. [CrossRef]
32. Deutsches Zentrum für Luft- und Raumfahrt e.V. (DLR). *Leiser Flugverkehr II—Abschlussbericht*; Resreport; Deutsches Zentrum für Luft- und Raumfahrt e.V.: Göttingen, Germany, 2007.
33. Pott-Pollenske, M.; Wild, J.; Bertsch, L. Aerodynamic and Acoustic Design of Silent Leading Edge Devices. In Proceedings of the 20th AIAA/CEAS Aeroacoustics Conference, Atlanta, GA, USA, 16–20 June 2014. [CrossRef]
34. Wicke, K.; Bertsch, L. *SLED—Silent Leading Edge Devices—Abschlussbericht zum Flugzeugentwurf und der Systemanalyse*; DLR Institutsbericht IB-328-2013-30; Deutsches Zentrum fuer Luft- und Raumfahrt: Hamburg, Germany, 2014.
35. Department of Transportation and Federal Aviation Administration. *Noise Certification of Supersonic Airplanes*; Technical Report; Federal Register, Notice of Proposed Rulemaking (NPRM): Washington, DC, USA, 2021; Volume 85, pp. 20431–20447. Available online: https://www.govinfo.gov/content/pkg/FR-2020-04-13/pdf/2020-07039.pdf (accessed on 20 November 2020).
36. Berton, J.J.; Huff, D.L.; Geiselhart, K.; Seidel, J. Supersonic Technology Concept Aeroplanes for Environmental Studies. In Proceedings of the AIAA Scitech 2020 Forum, Orlando, FL, USA, 6–10 January 2020; p. 0263.
37. Rizzi, S.A.; Berton, J.J.; Tuttle, B.C. Auralization of a Supersonic Business Jet Using Advanced Takeoff Procedures. In Proceedings of the AIAA Scitech 2020 Forum, Orlando, FL, USA, 6–10 January 2020; p. 0266.
38. Shaw, R.J.; Hines, R.; Gilkey, S. Engine Technology Challenges for a 21st Century High-Speed Civil Transport. In the Proceedings of the International Symposium on Air Breathing Engines, Tokyo, Japan, 20–24 September 1993.

39. Jackson, E.B. *Piloted Simulation Assessment of a High-Speed Civil Transport Configuration*; NASA Langley Research Center: Langley, VA, USA, 2002. [CrossRef]
40. Huff, D.L.; Henderson, B.S.; Berton, J.J.; Seidel, J.A. Perceived noise analysis for offset jets applied to commercial supersonic aircraft. In Proceedings of the 54th AIAA Aerospace Sciences Meeting, San Diego, CA, USA, 4–8 January 2016; p. 1635.
41. Bertsch, L.; Looye, G.; Anton, E.; Schwanke, S. Flyover Noise Measurements of a Spiraling Noise Abatement Approach Procedure. *AIAA J. Aircr.* **2011**, *48*, 436–448. [CrossRef]

aerospace

MDPI

Article

Sound-Quality-Based Decision Making in Multiobjective Optimisation of Operations for Sustainable Airport Scenarios

Umberto Iemma * and **Francesco Centracchio**

Departmentof Engineering, Roma Tre University, Via Vito Volterra, 62, 00146 Roma, Italy; francesco.centracchio@uniroma3.it

* Correspondence: umberto.iemma@uniroma3.it

Abstract: The paper deals with a community-oriented approach to the multiobjective optimisation of sustainable takeoff and landing procedures of commercial aircraft. The objective functions to be minimised are defined as the measure of area surrounding the airport where the Sound Exposure Level (SEL) is higher than 60 dBA, and the amount of fuel burned during the procedure. The first merit factor is a measure of the number of citizens affected by a potentially harmful noise level, whereas the second is proportional to the chemical emissions. The novelty of the present approach is the use of a criterion based on sound quality for the selection of the optimal procedure from the Pareto front set. The spectrum of the noise produced by each non-dominated solution is compared to a reference spectrum, the target sound. This is synthesised to meet the acceptance requirements that emerged by a campaign of psychometric tests. The rationale underlying the research is tightly linked to the expected transformation of civil aviation, with the advent of new air transport solutions in urban and suburban environments. The breakthrough nature of the emerging scenarios requires a drastic renewal of the approaches used in the management of operations, and the present work represents a contribution to this evolution. The optimisation is attained adopting a global, deterministic method, and numerical results are obtained for single- and twin-aisle aircraft.

Keywords: multiobjective optimisation; community noise; air quality; sound quality; noise abatement procedures; spectral matching

Citation: Iemma, U.; Centracchio, F. Sound-Quality-Based Decision Making in Multiobjective Optimisation of Operations for Sustainable Airport Scenarios. *Aerospace* **2022**, *9*, 310. https://doi.org/10.3390/aerospace9060310

Academic Editors: Lothar Bertsch and Adrian Sescu

Received: 27 February 2022
Accepted: 31 May 2022
Published: 8 June 2022

1. Introduction

The evolution of the air transportation system is nowadays experiencing a turning point that will lead to the advent of a completely new scenario in the next decade. Electrification of propulsion, Urban Air Mobility (UAM), suburban air shuttles, and alternative fuels are some of the factors that are inducing a change of paradigm for civil aviation. The effect on society of these ground-breaking solutions is amplified by the enduring expansion of urban areas around airports and the constant increase in air traffic in terms of daily movements and passengers (for additional information and details see Knobloch et al. [1]). In such a rapidly evolving context, sustainability can be guaranteed only introducing new approaches to the management of the air transportation system, capable of satisfying market requests and, at the same time, preserving the health and the quality of life of citizens. The European community has sponsored several research projects over the last twenty years to develop disruptive technologies and operational procedures to lower the impact on the community of the chemical and acoustic emissions related to civil aviation.

A standard trajectory optimisation is aimed at calculating the best flight path through the minimisation (or maximisation) of a performance index that can deal with noise reduction (in a specific point or in a certain area), fuel consumption or the total cost [2]. The growth of computational resources, together with the development of sophisticated multiobjective optimisation algorithms have made it possible to deal with the design of the trajectories as a vector minimisation problem (a valuable review of Multi-Objective

Trajectory Optimisation techniques for transport aircraft flight operations can be fund in Gardi et al. [3]), so that multiple environmental factors can be minimised at the same time [4]. When the main focus is on noise reduction, the resulting flight path can be named a noise abatement procedure (NAP). NAPs can involve, in order to avoid overflight of populated areas, preferential routings [5] to cope with the well-being of populations living near airports. The NAP design for airport noise mitigation is usually carried out by airlines [6] under the guidance of the International Civil Aviation Organization (ICAO) and regulatory entities such as the Federal Aviation Administration (FAA). Notwithstanding, over the past decades, universities and research centres have also been involved in the design of environmentally friendly airport procedures: indeed, this activity provides the possibility of developing sophisticated flight simulation tools involving flight dynamics and control, multilevel computational aeroacoustics techniques together with numerical optimisation strategies [7–11]. To date, one of the most challenging research topics related to the design of low environmental impact trajectories is the adaptation of the NAP design techniques to the new aircraft concepts, such as the Blended Wing Body (BWB) configuration: for such disruptive layouts the designer cannot rely on past experience; thus, a viable strategy appears to be metamodelling the BWB shielding effect [12,13], with the aim of exploiting the existing well-assessed noise estimation tools for the prediction of future airport scenarios.

The work presented in this paper is different from traditional trajectory optimisations, and it can be considered as an evolution of the approach first introduced during the SEFA projects (Sound Engineering For Aircraft, FP6, 2004–2007 [14–16]) to assess the perceptive qualities of the noise produced by a civil aircraft, and subsequently evolved to an airport scenario (i.e., a prescribed sequence of takeoff and landing procedures) during COSMA (community noise solutions to minimise aircraft noise annoyance, FP7, 2009–2012 [17,18]). Here, the sound quality assessment is applied in the multiobjective context adopted in the ANIMA project (Aviation Noise Impact Management through Novel Approaches, H2020, 2017–2021, [1,19]) as a criterion to select the least annoying procedure from the Pareto set of a two-objective optimisation. The objective functions used are indirect measures of the number of citizens affected by a potentially harmful noise level, and of the amount of greenhouse gases injected in the atmosphere.

Specifically, this work aims at exploring multiple noise descriptors for the design of low-noise flight paths. Indeed, the purpose of the analysis presented here is to simultaneously minimise both the chemical and the acoustic emissions: the case studies concern the takeoff and the approach procedures of single- and twin-aisle aircraft, representative of aeroplane classes used for most of the international commercial routes. Since the Multi Objective Optimisation (MOP) solution is composed of a set of non-dominated points, the sound quality assessment is applied to select the optimal trajectories. The aircraft operations have been optimised within the in-house Multidisciplinary Conceptual Robust Design Optimisation (MCRDO) framework FRIDA (Framework for Innovative Design in Aeronautics [12,15,16,18,20–22]). The two objectives to be minimised are the Sound Exposure Level (SEL) 60 dBA contour area and the amount of fuel burnt during the entire airport operation. The decision making criterion makes use of a third objective function, built based on the sound-quality assessment: the latter is formalised as the L^p norm in the vector space defined by the difference between the noise produced by the aircraft and a weakly annoying target sound (the interested reader can find useful information on the target sound definition in [18]). It is worth highlighting that L^p norms of different orders can be used to build objectives that focus on local and distributed differences: this behaviour has been investigated concerning benchmark problems [16], and it can be demonstrated that it can be exploited when the effect of tonal components is explicitly available. The constrained optimisation problem is solved within FRIDA by using the Deterministic Particle Swarm Optimisation (DPSO) method, a gradient-free global technique introduced by Kennedy and Russel [23], in its original deterministic implementation [24,25]. All the numerical results are presented in terms of approximated Pareto frontiers for takeoff and landing conditions for both the single- and twin-aisle aircraft.

The paper is organised as follows. Section 2 provides details on the optimisation problem setup, including the definition of the two objectives functions. The main characteristics of both the single- and twin-aisle aircraft are reported in Section 3. In Section 4, the optimisation results are presented, and the criteria for the sound-quality-based decision making are reported in Section 5 with the final designer choices. Finally, Section 6 presents some concluding remarks.

2. Optimisation Problem Setup

A Generic Constrained Multiobjective Optimisation Problem (MOP) is formalised as

$$\begin{aligned} \min/\max[J_k(\mathbf{x})], &\quad k = 1,\ldots,K \text{ and } \mathbf{x} \in \mathcal{D} \\ \text{with bounds } x_n^L \leq x_n \leq x_n^U, &\quad n = 1,\ldots,N \\ \text{subject to } g_i(\mathbf{x}) \leq 0, &\quad i = 1,\ldots,I \\ \text{and } h_j(\mathbf{x}) = 0, &\quad j = 1,\ldots,J \end{aligned} \tag{1}$$

where $J_k(\mathbf{x})$ is the k-th objective function (with $\mathbf{x} \in \mathcal{D}$ being the vector containing the N design variables), and $g_i(\mathbf{x})$ and $h_j(\mathbf{x})$ are the inequality constraints and the equality constraints, respectively. The solution of the above problem consists of the set of non-dominated solutions that forms the Pareto frontier.

In the present multiobjective approach, the two merit functions are related to acoustic and chemical emissions. The acoustic descriptor used here is the Sound Exposure Level (SEL) which represents the most suitable strategy for comparing airport procedures of different durations. Here, the objective function to be minimised is chosen to be a measure of the number of citizens affected by a noise level above a certain threshold. More specifically, $J_1(\mathbf{x})$ is chosen as the area bounded by the SEL 60 dBA contour, formalised as

$$J_1(\mathbf{x}) = \iint_{A_{60}} dA \quad \{A_{60} : \forall(\xi,\eta) \in A_{60} \Rightarrow \text{SEL}(\mathbf{x},\xi,\eta) \geq 60 \text{ dBA}\} \tag{2}$$

where A_{60} is the area surrounding the airport where SEL is greater or equal to 60 dBA, with ξ and η being the reference spatial coordinates of the noise map. The objective function related to the chemical emission is simply the amount of fuel burnt during the manoeuvre,

$$J_2(\mathbf{x}) = \int_0^{t_{10k}} cTdt \tag{3}$$

where t_{10k} is the time that the aircraft takes to reach an altitude of 10,000 ft, T is the required thrust and c is the specific fuel consumption, here modelled as a function of the true air speed (TAS).

With the aim of ensuring the simulation of a realistic operation, suitable constraints were imposed. First of all, the high-lift devices deployment combined with the angle of attack must ensure a vertical equilibrium, preventing the stall at the p-th trajectory sample. Thus, the following constraint is imposed

$$g_1(\mathbf{x}) = \sum_{p=1}^{P} \max\left(0, \frac{\alpha_p}{\alpha_{ST}} - 1\right) \tag{4}$$

where α_{ST} is the stall angle. Furthermore, since N1 (the rotational velocity of the low-pressure spool) must not exceed the overspeed in takeoff and must not fall below the idle setting in the approach (T and A indicate the takeoff and the approach procedures), the following constraints are also used

$$g_2(\mathbf{x})|_T = \sum_{p=1}^{P} \max\left(0, \frac{N1_p}{N1_{OS}} - 1\right)$$
$$g_2(\mathbf{x})|_A = \sum_{p=1}^{P} \max\left(0, 1 - \frac{N1_p}{N1_{ID}}\right) \tag{5}$$

where the subscripts OS and ID stand for overspeed and idle, respectively. To avoid structural failures, the following constraints, which account for the normal load factor n variation, are imposed

$$g_3(\mathbf{x}) = \sum_{p=1}^{P} \max\left(0, \frac{n_p}{n_L^+} - 1\right) \tag{6}$$

$$g_4(\mathbf{x}) = \sum_{p=1}^{P} \max\left(0, 1 - \frac{n_p}{n_L^-}\right) \tag{7}$$

where n_L^+ and n_L^- are the positive and negative limit normal load factor. Equations (4)–(7) are calculated at each p-th sample of the trajectory. In addition, considering the q-th trajectory segment the following constraint was introduced

$$g_5(\mathbf{x}) = \sum_{q=1}^{Q} \max\left(0, \frac{|\gamma_q|}{\gamma_{MAX}} - 1\right) \tag{8}$$

to account for the maximum absolute value of the ramp angle over the Q trajectory segments. Finally, the maximum change in slope $\Delta\gamma_{MAX}$ between two consecutive trajectory segments was imposed as

$$g_6(\mathbf{x}) = \sum_{q=1}^{Q} \max\left(0, \frac{|\gamma_q - \gamma_{n-1}|}{\Delta\gamma_{MAX}} - 1\right) \tag{9}$$

where $2 < q < Q$, to ensure passenger comfort.

It is important to underline that the correct estimation of $J_1(\mathbf{x})$ requires the use of suitable aeroacoustic models: indeed, the prediction of the noise at a prescribed location requires both the modelling of the relevant physical phenomena involved in the sound generation and propagation mechanisms. Within the optimisation context, the identification of a proper trade-off between accuracy and computational cost is a crucial aspect: thus, well-assessed semi-empirical models were used here, capable of predicting the aircraft noise spectra at specified location (for useful details, see Appendix A). In addition, it is worth noting that the deep interplay between aeroacoustics and flight mechanics makes a correct description of the aircraft dynamics and its operating conditions mandatory. Specifically:

- The airframe noise during the airport operations is also linked to the deployment of the high lift devices;
- Due to the strong directivity of some acoustic sources, the attitude of the aircraft influences the spectrum that reaches a listener;
- The noise emitted by the main sources related to the propulsion system (fan, compressor, turbine, buzz-saw and jet) depends on the engine's settings in terms of N1 and N2, which depends on the required thrust;
- The relative speed and the distance between the aircraft and the listener influence the spectrum of the received signal through the Doppler effect and the atmospheric absorption.

Therefore, a detailed description of the aircraft flight mechanics makes it possible to provide an adequate estimate of the noise emissions. Interestingly, since $J_2(\mathbf{x})$ also depends on the required thrust, the two objective functions are intrinsically connected.

3. Case Studies

The case studies involve both a single- and a twin-aisle aircraft, representative of aeroplane classes used for most of the international commercial routes: the first one for short- and medium-haul flights, and the second for medium- to long-haul ones. The aircraft characteristics are reported in Table 1.

Table 1. Main characteristics of the single- and twin-aisle aircraft.

A/C Charachteristics	Single-Aisle	Twin-Aisle
Number of seats	164	406
Cruise Mach number	0.78	0.82
Cruise altitude (ft)	42,000	42,000
Range (nmi)	3250	5500
MTOW (ton)	78	230
OEW (ton)	43	122
Number of engines	2	2
Maximum thrust per engine (kN)	111.2	316.4
Bypass ratio	6.5	9.6
Engine placement	under the wing	under the wing

The framework FRIDA builds the aircraft geometrical model, performing a complete analysis for the estimate of aerodynamic and the structural properties. In Figure 1 are presented the pictorial representations of the models built within FRIDA related to the single- and the twin-aisle aircraft.

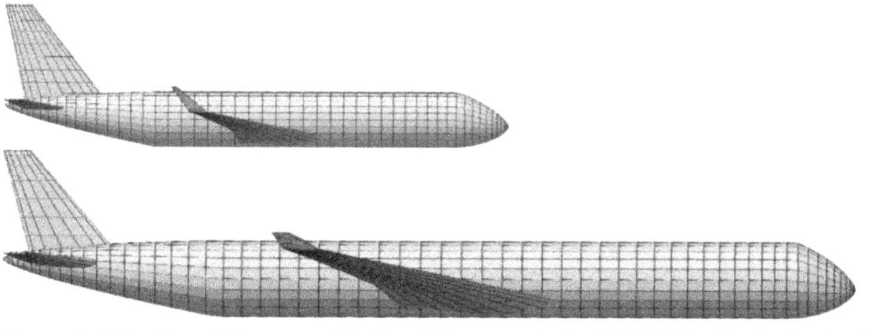

Figure 1. Side view of the geometrical models generated by FRIDA for the single- (**top**) and the twin-aisle (**bottom**) aircraft.

All the data coming from the analysis modules are collected and passed to the FRIDA flight simulation environment. With the aim of reducing the computational effort, suitable corrections are applied to the aerodynamic and inertial data to account for the specific flight conditions: by doing this, the airport manoeuvres can be calculated with the expense of a single aircraft analysis, maintaining an accuracy more than acceptable.

In this work, both takeoff and landing procedures are optimised. Specifically, the reference takeoff manoeuvres are modelled starting from the ICAO procedures for aircraft compatible with the test cases analysed. The trajectories consist of five segments (six input nodes), starting from the brake release up to a distance related to an altitude equal to 10.000 ft (about 35 km from the runway for both the aircraft). The trajectory description is built starting from nodal variables (geometric and kinematic), reported in Table 2.

Table 2. Reference takeoff manoeuvre for the single- and twin-aisle aircraft: geometric and kinematic variables.

	Single-Aisle			Twin-Aisle		
Op. #	**x (m)**	**z (m)**	**v (m/s)**	**x (m)**	**z (m)**	**v (m/s)**
1	0.0	0.0	0.0	0.0	0.0	0.0
2	1698.6	0.0	83.7	1908.1	0.0	87.9
3	4500.4	457.2	85.6	4772.2	457.2	89.9
4	8648.7	914.4	87.5	9571.5	914.4	91.9
5	15,852.8	1179.4	136.2	16,285.3	1132.4	135.9
6	35,441.3	3048.0	149.7	35,833.9	3048.0	149.7

The initial aircraft masses for the takeoff phase were set approximately equal to 77 tons for the single-aisle aircraft and 217 tons for the twin-aisle, corresponding to 98% and 95% of the MTOW, respectively. In Figure 2 are depicted the reference takeoff trajectories for the single- and the twin-aisle aircraft with the 60 dBA isolevel.

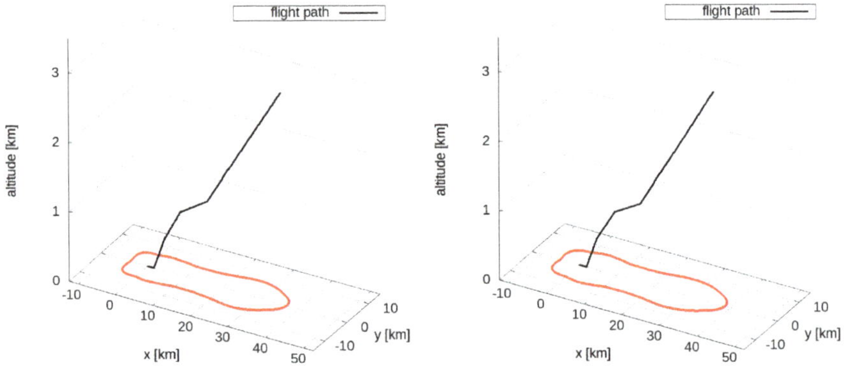

Figure 2. Flight path and 60 dBA isolevel for the reference takeoff manoeuvre. (**a**) single–aisle aircraft and (**b**) twin–aisle aircraft.

The reference approach phases are modelled based on the standard ICAO procedures for two aircrafts compatible with the test cases chosen. The trajectory consists of six segments (seven input nodes), starting at about 40 km from the airport and ending at the touchdown point on the runway. The geometric and the kinematic variables of the nodes are reported in Table 3.

Table 3. Reference approach manoeuvre for the single- and twin-aisle aircraft: geometric and kinematic variables.

	Single-Aisle			Twin-Aisle		
Op. #	**x (m)**	**z (m)**	**v (m/s)**	**x (m)**	**z (m)**	**v (m/s)**
1	−39,213.1	1828.8	140.6	−43,346.7	1828.8	140.6
2	−24,262.8	914.4	134.4	−26,462.8	914.4	134.4
3	−17,447.8	914.4	98.0	−17,447.8	914.4	89.3
4	−15,202.8	796.7	92.9	−14,813.2	776.3	82.3
5	−10,297.6	539.7	70.0	−12,164.6	637.5	69.4
6	0.0	0.0	68.2	0.0	0.0	68.4
7	925.1	0.0	15.4	641.4	0.0	15.4

The initial masses for the approach phase for the single- and twin-aisle aircraft are about 58 tons and 157 tons, respectively, corresponding to 135% of the OEW for the single-aisle aircraft and 129% of the OEW for the twin-aisle. The reference approach trajectories for the single- and the twin-aisle aircraft with the 60 dBA isolevel are depicted in Figure 3.

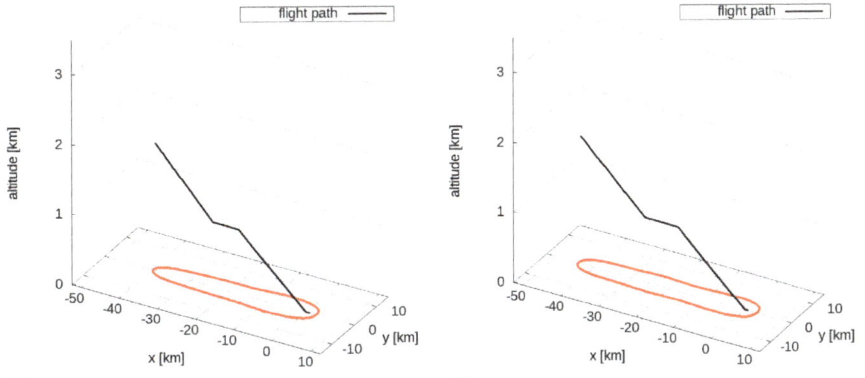

(a) (b)

Figure 3. Flight path and 60 dBA isolevel for the reference approach manoeuvre. (**a**) single–aisle aircraft and (**b**) twin–aisle aircraft.

4. Optimisation Results and Discussion

The optimisations were carried out within FRIDA, using the Deterministic Particle Swarm Optimisation (DPSO) algorithm, an original implementation of the Particle Swarm Optimisation (PSO) method, making use of $10n_{DV}$ particles (with n_{DV} being the number of design variables) and 250 PSO iterations. Instead of solving an unconstrained minimisation problem, pseudo-objective functions are used to account for the constraints described by Equations (4)–(9): the pseudo-objective $\hat{J}_k$ is defined using the external quadratic penalty function as follows

$$\hat{J}_k(\mathbf{x}) = J_k(\mathbf{x}) + \frac{1}{\varepsilon}\sum_i \max[0, g_i(\mathbf{x})]^2 \tag{10}$$

where ε is the penalty coefficient. The optimisation process for the takeoff is aimed at finding the optimal path in terms of spatial and kinematic variables of the third, the fourth and the fifth trajectory nodes, whereas the landing optimisation involves the second, the third and the fourth trajectory nodes (see Tables 2 and 3). The results are reported below in terms of DPSO solutions with the approximated Pareto frontiers: for both the single- and the twin-aisle aircraft under consideration, the optimal solutions related to the minimum noise and the minimum fuel are presented and discussed.

4.1. *Single-Aisle Aircraft*

4.1.1. Takeoff

The optimisation process for the takeoff is aimed at finding the optimal path in terms of spatial and kinematic variables of three nodes (the third, the fourth and the fifth trajectory node). The design variables with the upper and lower bounds are reported in Table 4.

Figure 4 shows the optimisation results for the single-aisle aircraft takeoff operation in terms of DPSO solutions (normalised with respect to the reference solution) and the approximated Pareto front.

Table 4. Takeoff manoeuvre optimisation variables for the single-aisle aircraft: reference values with upper and lower bounds.

Variable	Lower Bound	Reference	Upper Bound
x_3 (m)	4000.0	4500.4	5000.0
z_3 (m)	400.0	457.2	500.0
v_3 (m/s)	85.0	85.6	86.0
x_4 (m)	7000.0	8648.7	10,000.0
z_4 (m)	800.0	914.4	1000.0
v_4 (m/s)	87.5	87.5	110.0
x_5 (m)	13,000.0	15,852.8	17,000.0
z_5 (m)	1000.0	1179.4	1300.0
v_5 (m/s)	110.0	136.2	140.0

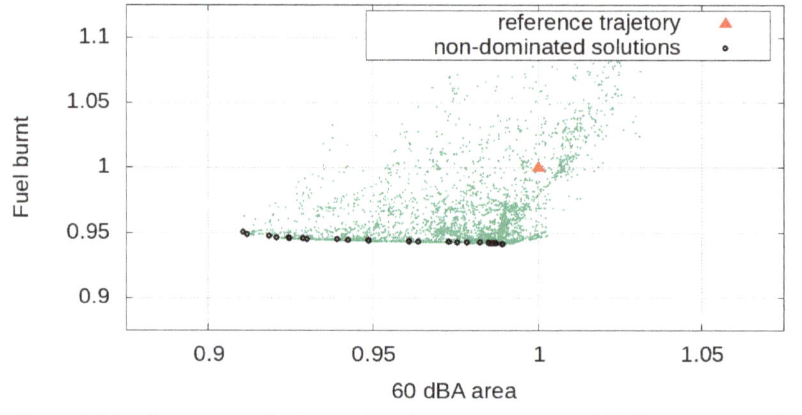

Figure 4. Takeoff manoeuvre for the single-aisle aircraft: normalised DPSO solutions with reference trajectory and approximated Pareto front.

The analysis of the non-dominated solutions in Figure 4 highlights that there is the possibility of decreasing the 60 dBA area by about 9% starting from the minimum fuel solution moving along the Pareto front. In addition, it is worth noting that the entire approximated Pareto frontier is composed of solutions that improve both the acoustical and chemical emissions with respect to the reference takeoff trajectory. The optimal solutions related to the minimum noise and the minimum fuel consumption are presented in Figure 5 with 60 dBA contours.

It is easy to note that, as shown in Figure 5a, the flight paths related to the minimum noise are remarkably different with respect to the minimum fuel solution. Such solutions are also distant from each other in the objectives space: this is highlighted in Figure 5b, which shows substantial differences between the 60 dBA areas. The minimum noise solution consists of an initial steep climb with a maximum ramp angle compatible with the engine overspeed: the steep climb is followed by a low-ramp path. this is reflected in the 60 dBA area enlargement downstream from the runway (at x = 10 km), compensated by its shortening at x = 34 km. It is interesting to note that the minimum fuel solution 60 dBA contour is similar to that related to the reference takeoff (see Figure 5b), which could prove that the reference flight path is designed to reduce chemical emissions.

To better understand the reasons that lead to the modification of the 60 dBA contour shape, let us observe the procedural variables related to the takeoff optimal solutions, reported in Figure 6.

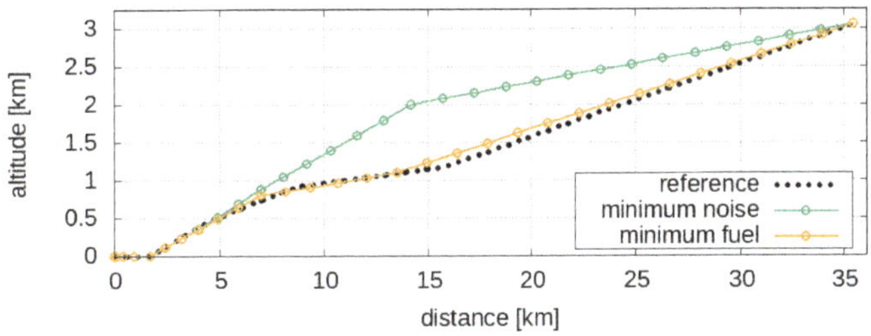

(**a**)

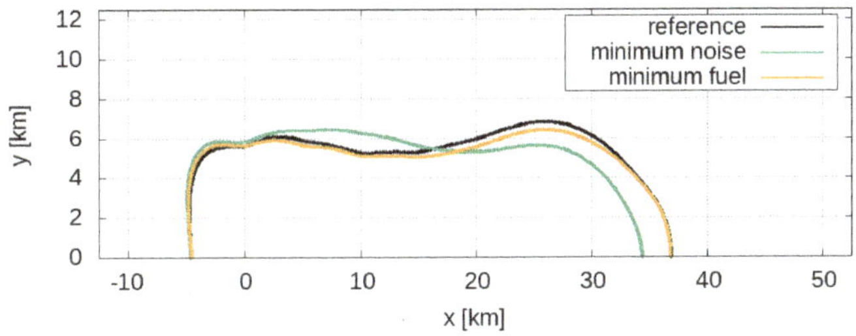

(**b**)

Figure 5. Takeoff manoeuvre for the single-aisle aircraft: flight paths and 60 dBA contours for the reference trajectory and the optimal solutions related to the minimum noise and the minimum fuel consumption: (**a**) flight path, (**b**) 60 dBA contours.

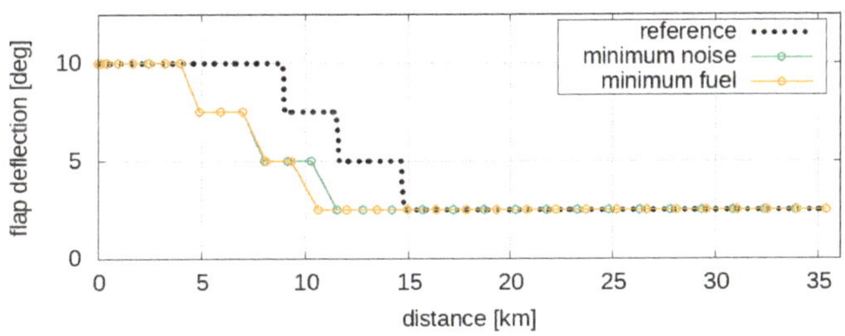

(**a**)

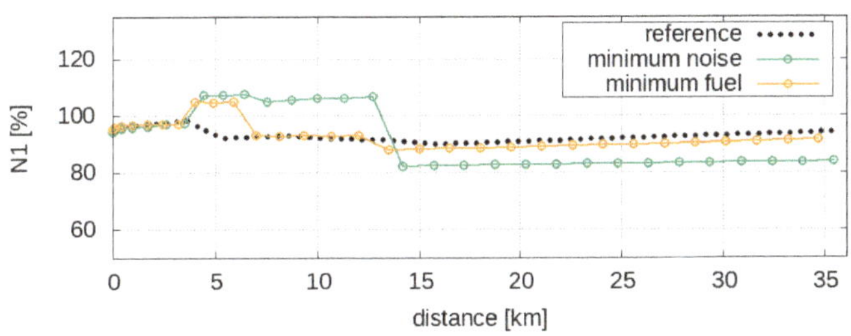

(**b**)

Figure 6. Takeoff manoeuvre for the single-aisle aircraft. Procedural variables. (**a**) flap deflection, (**b**) N1.

Figure 6a shows the time history of the flap deflection: for both optimal solutions, the flap retraction is anticipated, producing less noise with respect to the reference manoeuvre. It is worth noting that, as shown in Figure 6b, the rotational speed N1 for the low-noise optimal solution exceeds 100%, reaching the overspeed in just under 10 min: this generates more propulsive noise, but the steep climb, as mentioned above, shortens the 60 dBA contour from x = 37 km to x = 34 km at y = 0 km.

4.1.2. Approach

The second, the third and the fourth trajectory nodes are optimised for the problem related to the approach trajectory: the optimisation variables are reported in Table 5 with the upper and lower bounds.

Table 5. Approach manoeuvre optimisation variables for the single-aisle aircraft: reference values with upper and lower bounds.

Variable	Lower Bound	Reference	Upper Bound
x_2 (m)	−27,500.0	−24,262.8	−22,500.0
z_2 (m)	914.4	914.4	1100.0
v_2 (m/s)	115.0	134.4	140.6
x_3 (m)	−19,000.0	−17,447.8	−16,500.0
z_3 (m)	850.0	914.4	914.4
v_3 (m/s)	95.0	98.0	115.0
x_4 (m)	−15,500.0	−15,202.8	−13,000.0
z_4 (m)	600.0	796.7	850.0
v_4 (m/s)	75.0	539.7	95.0

The DPSO solutions (normalised with respect to the reference solution) for the approach operation related to the single-aisle aircraft are presented in Figure 7 with the approximated Pareto front.

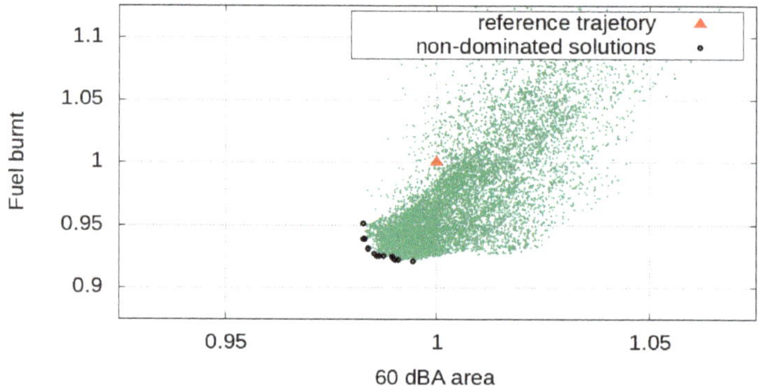

Figure 7. Approach manoeuvre for the single-aisle aircraft: normalised DPSO solutions with reference trajectory and approximated Pareto front.

Figure 7 shows that, also for the approach operation case, as already highlighted for takeoff, the non-dominated solutions are better than the reference solution in terms of both the objective functions. Notwithstanding, in this case the variation of the 60 dBA area is just above 2 km^2 comparing the minimum fuel and the minimum noise optima (which correspond to a relative variation of about 1.5%). In Figure 8, the reference trajectory and the optimal solutions related to the minimum noise and the minimum fuel consumption are presented.

The analysis of Figure 8b shows that the contours of the 60 dBA areas almost overlap, but both are different from the 60 dBA area contour of the reference approach flight path. Indeed, as can be seen in Figure 8a, the minimum noise and the minimum fuel trajectories are almost identical, both tending to the continuous descent approach (CDA) with a $-3°$ slope.

The analysis of the procedural variables related to the approach optimal solutions, reported in Figure 9, also shows that the optimal operations are very similar.

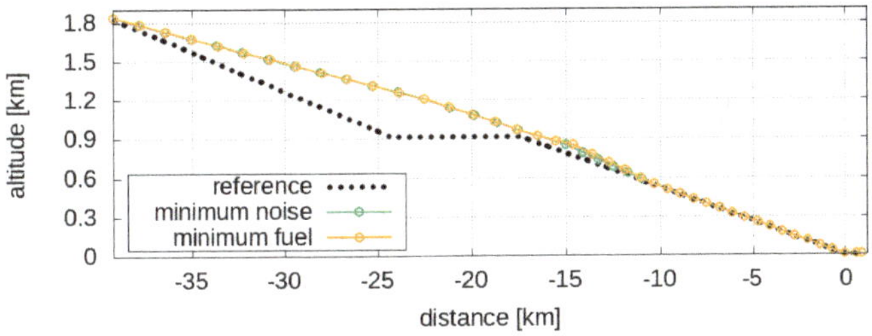

(a)

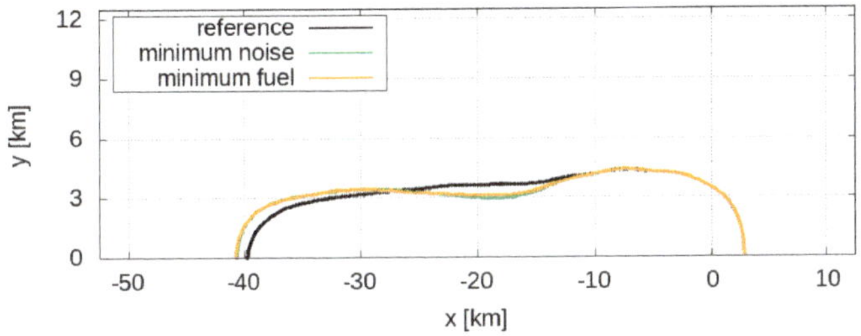

(b)

Figure 8. Approach manoeuvre for the single-aisle aircraft: flight paths and 60 dBA contours for the reference trajectory and the optimal solutions related to the minimum noise and the minimum fuel consumption: (**a**) flight path, (**b**) 60 dBA contours.

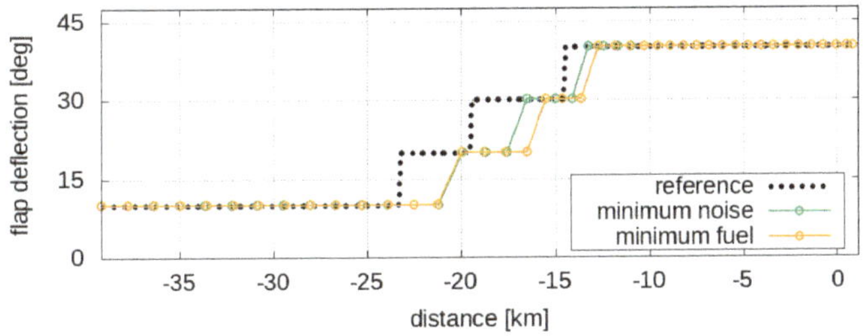

(a)

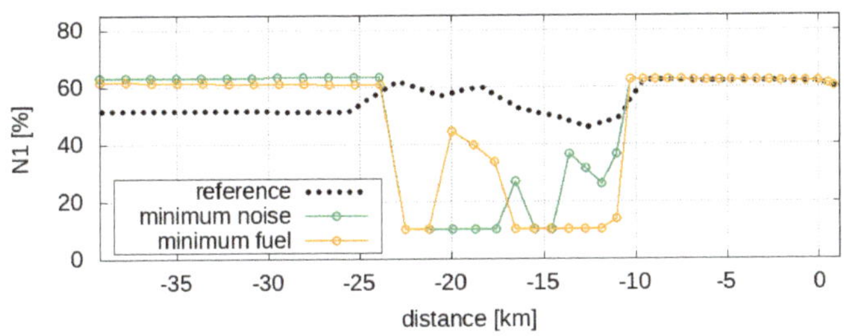

(b)

Figure 9. Approach manoeuvre for the single-aisle aircraft. Procedural variables. (**a**) flap deflection, (**b**) N1.

The flap deployment of both the optimal operations, reported in Figure 9a, is delayed if compared to the reference manoeuvre, whereas N1, shown in Figure 9b, tends to the idle condition in the flight path segment from x = −25 km to x = −10 km. The different engine operating point, combined with a greater altitude, implies a reduction in the 60 dBA contour at x = −20 km related to the optimal solution with respect to the reference one, as depicted in Figure 8b.

4.2. *Twin-Aisle Aircraft*

4.2.1. Takeoff

The takeoff optimisation, as for the single-aisle aircraft, aims at finding the optimal flight path in terms of three nodal variables (the third, the fourth and the fifth trajectory nodes). The design variables with the upper and lower bounds are reported in Table 6 with the upper and lower bounds.

Table 6. Takeoff manoeuvre optimisation variables for the twin-aisle aircraft: reference values with upper and lower bounds.

Variable	Lower Bound	Reference	Upper Bound
x_3 (m)	4000.0	4772.2	5500.0
z_3 (m)	350.0	457.2	700.0
v_3 (m/s)	87.9	85.6	90.0
x_4 (m)	7000.0	9571.5	11,000.0
z_4 (m)	700.0	914.4	1100.0
v_4 (m/s)	90.0	87.5	115.0
x_5 (m)	13,500.0	16,285.3	20,000.0
z_5 (m)	1100.0	1179.4	2000.0
v_5 (m/s)	115.0	136.2	149.7

The DPSO solutions (normalised with respect to the reference solution) and the approximated Pareto front are depicted in Figure 10 for the takeoff optimisation related to the twin-aisle aircraft.

The non-dominated solutions of Figure 10 show that it is possible to decrease the 60 dBA area by about 10% starting from the minimum fuel solution moving along the Pareto front (as already discussed for the single-aisle takeoff optimisation shown in Figure 4). Once again, the approximated Pareto frontier is composed of solutions that improve both objectives if compared with the reference takeoff flight path. Figure 5 depicts the the optimal solutions related to the minimum noise and the minimum fuel consumption with the 60 dBA contours.

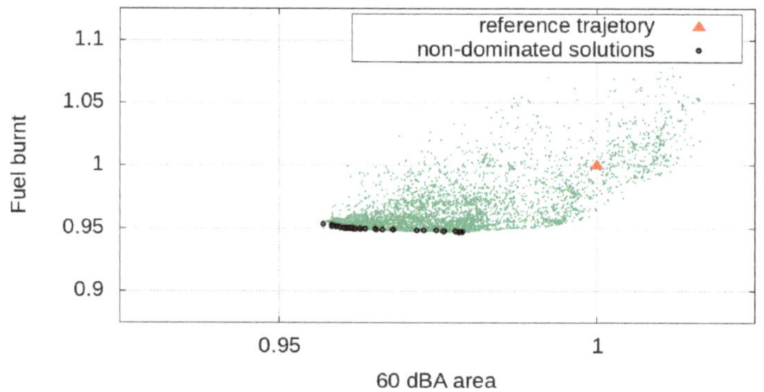

Figure 10. Takeoff manoeuvre for the twin-aisle aircraft: normalised DPSO solutions with reference trajectory and approximated Pareto front.

Figure 11a shows that, as already noted in Figure 5a for the single-aisle aircraft, the minimum fuel solution and the minimum noise one are remarkably different. The behaviour highlighted above for the minimum noise optimal solution is here confirmed: the twin-aisle minimum noise takeoff also corresponds to an initial steep climb followed by a low-ramp flight path. The 60 dBA contour distortion with respect to the reference trajectory is similar to the previous case, but is less evident because the aircraft is larger and the engines need to provide more thrust, causing a more intense propulsive noise. The minimum fuel solution flight path is quite different from the reference trajectory, but their 60 dBA contours (in Figure 5b) manifest many similarities.

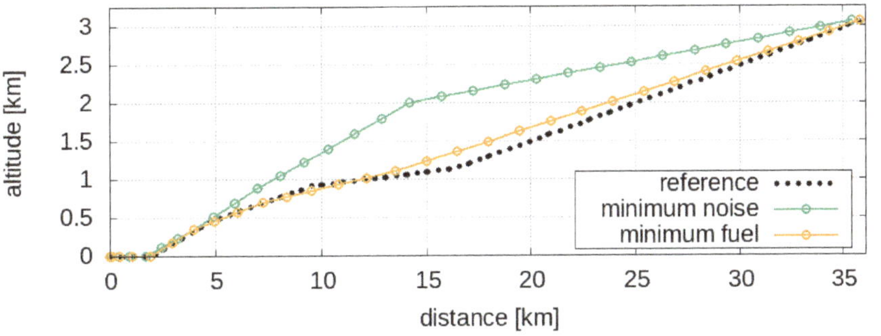

(**a**) Flight paths

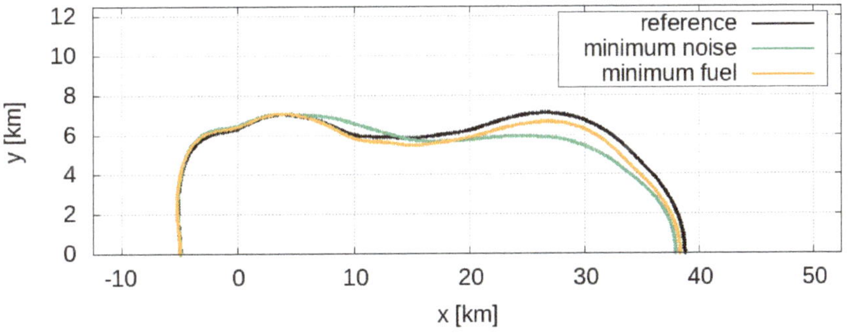

(**b**) 60 dBA contours

Figure 11. Takeoff manoeuvre for the twin-aisle aircraft: flight paths and 60 dBA contours for the reference trajectory and the optimal solutions related to the minimum noise and the minimum fuel consumption: (**a**) flight path, (**b**) 60 dBA contours.

As already carried out for the single-aisle aircraft, let us analyse the procedural variables (in terms of flap and N1 time history) related to the takeoff optimal solutions, reported in Figure 12.

Again this time, the flap deployment anticipates both the solutions, as can be observed in Figure 6b, as a result of higher aircraft acceleration. However, as regards N1, only the solution related to the minimum noise is characterised by the engine overspeed in the first phase of the climb, as shown in Figure 6b: instead, the engine operating points related to the minimum fuel solution are comparable to those of the reference solution, as well as the flight path, resulting in a similar 60 dBA contour (see Figure 11b).

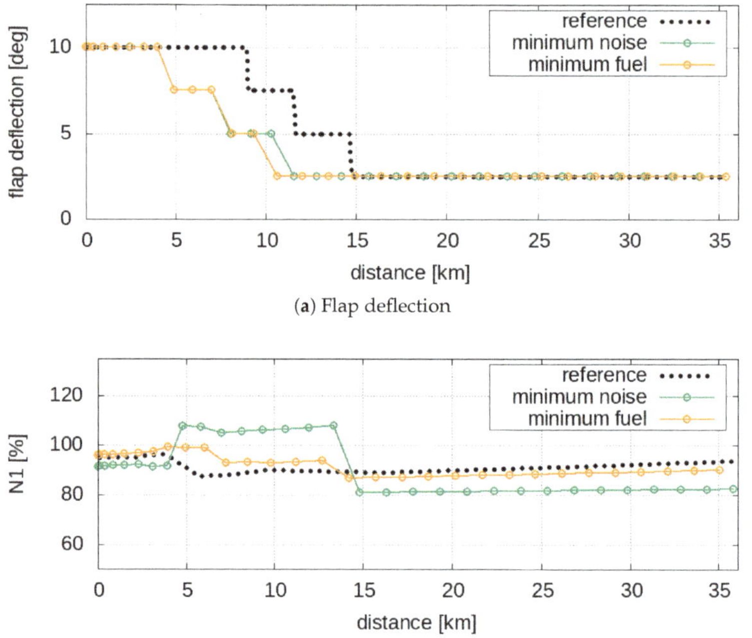

(**a**) Flap deflection

(**b**) N1

Figure 12. Takeoff manoeuvre for the twin-aisle aircraft. Procedural variables. (**a**) flap deflection, (**b**) N1.

4.2.2. Approach

The optimisation process for the landing is aimed at finding the optimal path in terms of spatial and kinematic variables of three nodes (the second, the third and the fourth trajectory nodes): the optimisation variables, with the upper and lower bounds, are reported in Table 7.

Table 7. Approach manoeuvre optimisation variables for the twin-aisle aircraft: reference values with upper and lower bounds.

Variable	Lower Bound	Reference	Upper Bound
x_2 (m)	−30,000.0	−24,262.8	−22,500.0
z_2 (m)	914.4	914.4	1200.0
v_2 (m/s)	115.0	134.4	140.6
x_3 (m)	−20,000.0	−17,447.8	−16,500.0
z_3 (m)	845.0	914.4	914.4
v_3 (m/s)	85.0	89.3	115.0
x_4 (m)	−15,500.0	−14,813.2	−13,500.0
z_4 (m)	637.5	796.7	845.0
v_4 (m/s)	69.4	82.3	85.0

Figure 13 presents the DPSO solutions (normalised with respect to the reference solution) with the reference trajectory and the approximated Pareto front related to the approach operation optimisation for the twin-aisle aircraft.

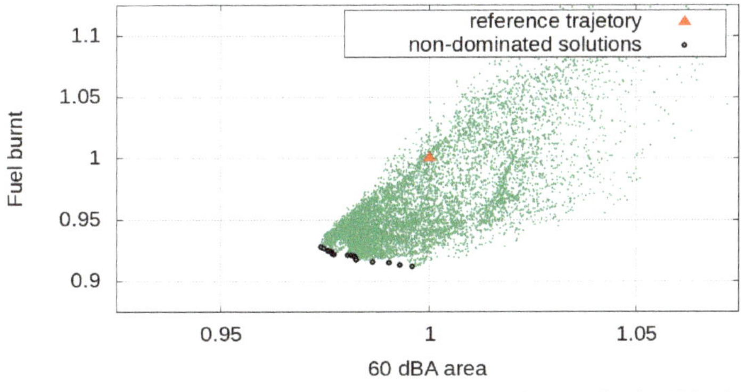

Figure 13. Approach manoeuvre for the twin-aisle aircraft: normalised DPSO solutions with reference trajectory and approximated Pareto front.

All the non-dominated solutions of Figure 13 improve the reference solution in terms of both the noise emission and the fuel consumption (as already noted for the cases of Figures 4, 7 and 10). The 60 dBA area relative variation of the minimum noise solution with respect to the minimum fuel one is 2%, corresponding to about 7 km^2. In Figure 14 are depicted the optimal solutions related to the minimum noise and the minimum fuel consumption with the reference trajectory.

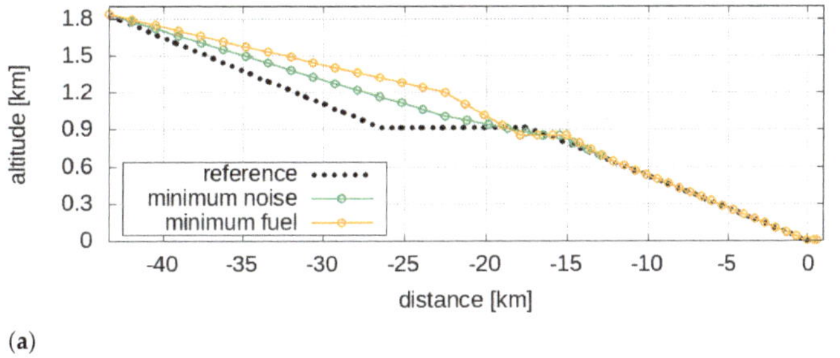

(**a**)

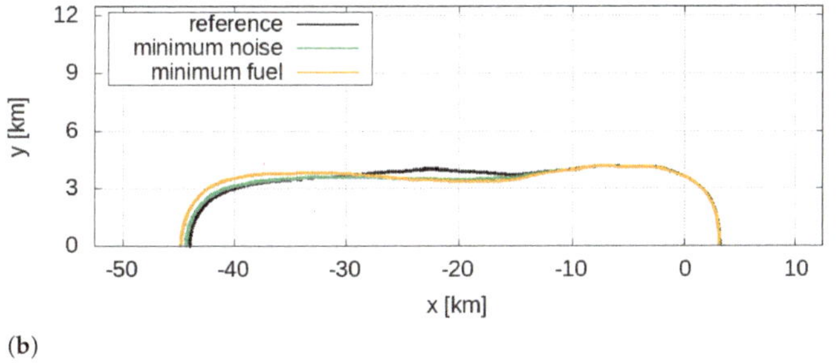

(**b**)

Figure 14. Approach manoeuvre for the twin-aisle aircraft: flight paths and 60 dBA contours for the reference trajectory and the optimal solutions related to the minimum noise and the minimum fuel consumption: (**a**) flight path, (**b**) 60 dBA contours.

Despite the similar 60 dBA areas, Figure 14b shows that the contours related to the minimum noise and the minimum fuel solutions do not overlap with that related to the reference trajectory, especially at x = −45 km and x = −20 km. In fact, the two analysed optimal solutions correspond to two rather different paths: the minimum noise trajectory tends to the CDA (as for the single-aisle case of Figure 8a), whereas the minimum fuel path shows two different slope angles up to x = −18 km and preserves a zero-slope path for $18 < x < -15$ km.

Figure 15 shows the procedural variables related to the approach's optimal solutions for the approach's optimal operations.

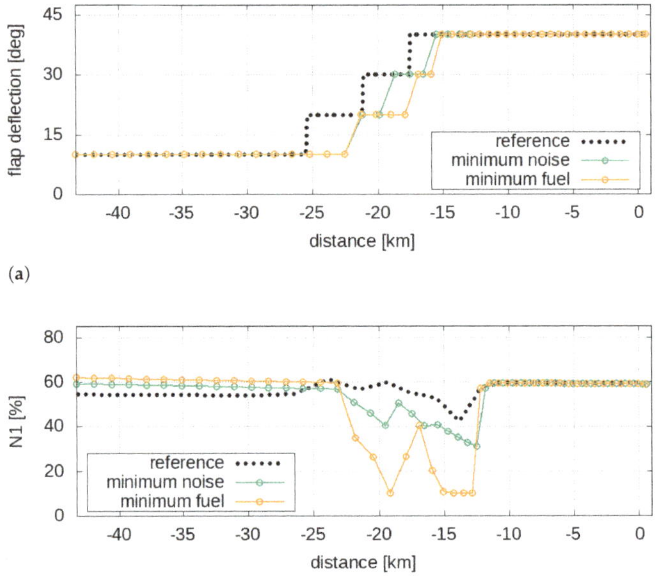

(b)

Figure 15. Approach manoeuvre for the twin-aisle aircraft. Procedural variables. (**a**) flap deflection, (**b**) N1.

The same considerations made for the single-aisle aircraft are valid in this case. A flap deployment delay can be observed (see Figure 15b) in both the minimum noise and the minimum fuel solution with respect to the reference one. The engine operating points from x = −25 km to x = −10 km are lower than the reference procedure, as depicted in Figure 15b, and this behaviour justifies the reduction of the 60 dBA contour (see Figure 14b) together with the greater aircraft altitude.

5. Sound-Quality-Based Decision Making

It is known that the single-objective optimisation problem leads to the identification of a unique design space point corresponding to the minimum of the objective function, which can be considered by the designer as the optimal solution. In contrast, the multiobjective problem solution consists of a set of non-dominated optimal solutions that constitute the Pareto frontier; therefore, from the designer's point of view the choice of the optimum among the Pareto front plays a key role. It is worth noting that the Pareto front could be a mathematically consistent but technically irrelevant entity, and this occur when the non-dominated solutions are close to each other both in the domain and in the codomain: in this case, the designer can consider the whole set of optimal solutions as a unique optimum. On the other hand, when the solutions are far both in the domain and in the codomain (or even close in the domain and far away in the codomain), the need to establish a criterion for decision making has a paramount relevance.

In the view of sustainable airport scenarios, a viable strategy is including the quality of the noise within the decision making process: this can be done by comparing the noise produced by the aircraft and a weakly annoying target sound. Such an approach is based on the measure of the matching of the noise emitted by the operation under analysis with a previously defined weakly annoying sound. The noise reaching the virtual microphone during a flight operation is characterised by a specific spectrogram, which provides the amplitude of the noise event in the time-frequency domain. Thus, let us consider the vector space defined by the difference between the spectrogram S_c related to the current flight path and the spectrogram S_t related to the target sound

$$\Delta = S_c(f,t) - S_t(f,t) \tag{11}$$

The norm of Δ in the L^p space can be used as a metric to quantify the similarity of the actual spectrogram with respect to the target one. Accordingly, the sound-matching index $\mathcal{I}_{SM}$ can be formalised as follows

$$\mathcal{I}^p_{SM} = ||\Delta||_p = \left[\frac{1}{T}\frac{1}{F}\int_{t_1}^{t_2}\int_{f_{min}}^{f_{max}} |S_c(f,t) - S_t(f,t)|^p df dt\right]^{\frac{1}{p}} \tag{12}$$

where $T = t_2 - t_1$ and $F = f_{max} - f_{min}$. The target sounds used in this work were synthesised on the basis of psychometric test campaigns performed within the projects SEFA and COSMA [14,17]. It is worth noting that the choice of p can be exploited by the designer to focus the metric on tonal broadband or tonal component dissimilarities (low and high values of p, respectively); indeed, low values of p enhance the contribution of distributed differences and high values of p emphasise the local ones. Optimisation problems involving $\mathcal{I}_{SM}$ have been successfully used in both the aircraft conceptual design and single-point low-noise flight path optimisation [16,22].

Figure 16 shows a pictorial representation of the sound-matching concept, and depicts the spectral comparison of two signals (solution A and B) with respect to a target spectrum for a certain time instant.

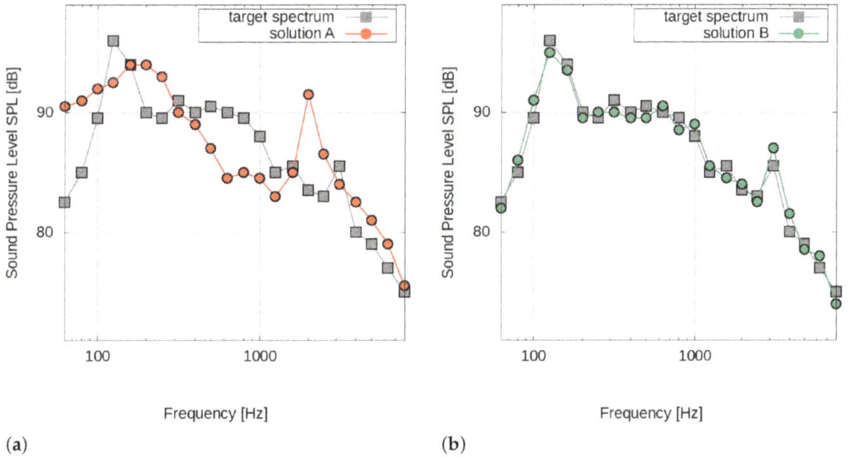

(**a**) (**b**)

Figure 16. Example of spectral matching. (**a**) solution A: poor matching. (**b**) solution B: good matching.

The spectral properties related to solution A of Figure 16a are quite different with respect to the target spectrum: in this case, Equation (12) (in this example with $p = 2$) returns a high value of $\mathcal{I}_{SM}$. In contrast, solution B (see Figure 16b) turns out to have frequency characteristics similar to the target spectrum; therefore, the $\mathcal{I}_{SM}$ value is low. In the case of an entire manoeuvre simulation, the $\mathcal{I}_{SM}$ value is simply a cumulative value over all the analysed time instants.

Pareto Solution Analysis and Designer's Choice

Let us consider the optimisation results reported in Section 4. Figures 5a and 11a (single- and twin-aisle takeoff operations) show the solutions related to the different minimum noise and minimum fuel produce trajectories in terms of geometric, kinematic and procedural variables; even in the codomain, the two solutions turn out to be distant from each other as the contours of the 60 dBA areas have different shapes (see Figures 5b and 11b). The minimum noise and the minimum fuel approach operations of the twin-aisle aircraft depicted in Figure 14a are quite different paths, but Figure 14b shows that there are no relevant discrepancies between the 60 dBA contours (both different form the reference flight path). As said, this also applies to the 60 dBA contours related to the single-aisle aircraft solutions (in Figure 8b); in this case, the similarity between the noise footprints is justified by two practically identical trajectories, depicted in Figure 8a.

It must be noted that the behaviour of the solutions for the approach manoeuvres (in Figures 8 and 14) is not surprising since the entire last part of the flight path is constrained by strict regulations that impose a -3° ramp angle. For the sound quality assessment, let us consider a virtual microphone placed under the flight path at x = −7.5 km (see Figures 8 and 14) that represents an urbanised location close to the airport boundary; it should be expected that there are no large variations in the time-frequency contributions of noise in landing operations, and this is confirmed by Figure 17, which depicts the mapped approximated Pareto optimal solutions with the sound-matching index $\mathcal{I}_{SM}$ as a parameter for the approach operations.

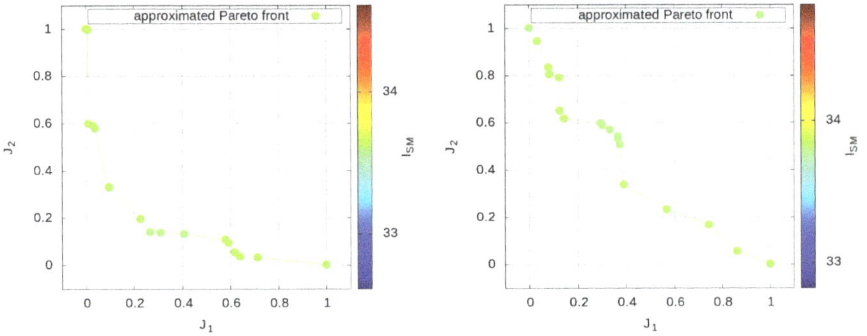

(**a**) (**b**)

Figure 17. Solutions of the multiobjective problems for the approach operations: mapped approximated Pareto optimal solutions with the sound-matching index $\mathcal{I}_{SM}$ as parameter. The solutions with the minimum $\mathcal{I}_{SM}$ cannot be found. (**a**) single–aisle, (**b**) twin–aisle.

The designer should therefore choose the trajectory related to the lowest fuel consumption, as each Pareto solution, despite being better than the reference one, is very similar to the others in terms of noise footprint and spectral components

Regarding the takeoff operations, the choice of the optimal solution among the Pareto front is addressed by means of the sound-matching index $\mathcal{I}_{SM}$ using $p = 2$ (see Equation (12)). Let us consider a virtual microphone close to the airport boundary x = 4.5 km; Figure 18 shows the mapped approximated Pareto optimal solutions with the sound-matching index $\mathcal{I}_{SM}$ as parameter related to the takeoff operation for both the single- and the twin-aisle aircraft.

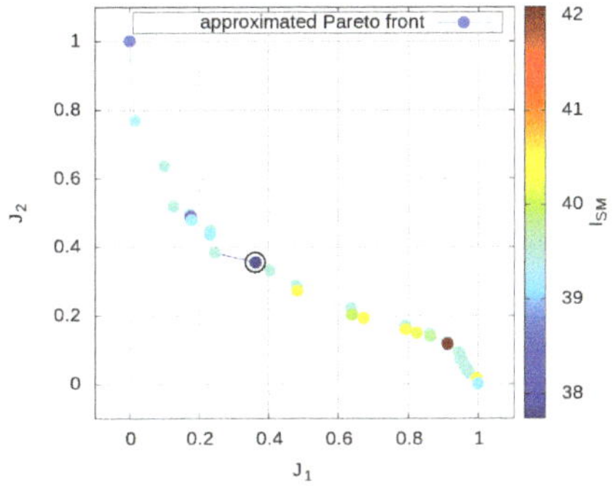

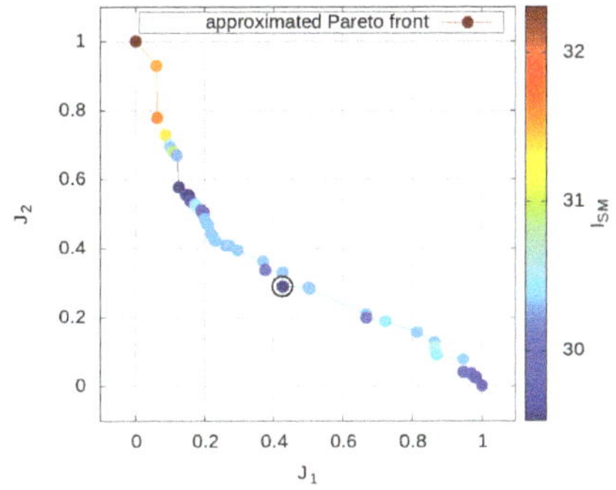

Figure 18. Solutions of the multiobjective problems for the takeoff operations: mapped approximated Pareto optimal solutions with the sound-matching index $\mathcal{I}_{SM}$ as parameter. The marked solutions are the ones with the minimum $\mathcal{I}_{SM}$. (**a**) single–aisle, (**b**) twin–aisle.

The analysis of Figure 18 highlights that there is not a specific functional dependency between the 60 dBA contour, the fuel consumption and the spectral content of the noise: indeed, the distribution of $\mathcal{I}_{SM}$ turns out to be in apparently random locations along the Pareto front. It is worth noting that the great variability of the $\mathcal{I}_{SM}$ demonstrates how this approach guarantees the designer an additional degree of freedom in the takeoff optimisation; this could be explained by considering the great sensitivity of the noise tonal components with respect to the variations in the engine operating point during the takeoff operations.

6. Conclusions

Nowadays, the introduction of alternative paradigms in the management of airport noise is becoming an imperative need. This work presents a multiobjective optimisation of takeoff and approach concerning low-noise procedures of single- and twin-aisle commercial aircraft with special attention paid to the sound-quality assessment of the MOP solutions. To simultaneously mitigate both the acoustic and chemical emissions, two merit functions are minimised through a multiobjective and multidisciplinary optimisation problem. The first objective function is the SEL 60 dBA contour area, and the amount of fuel burn is used as the second objective. The optimisation problems are carried out within the in-house MCRDO framework FRIDA. The results related to the approach operations show that, due to the strict regulations that impose a -3° ramp angle during the last phase of the manoeuvre, the 60 dBA contour can only be minimally decreased; thus, the designer's choice can fall on the solution related to minimum fuel consumption. The takeoff problems show a large margin of choice among the Pareto solutions, which is why an additional objective function to select the final configuration must be introduced. The latter is defined as the norm of the difference between the noise produced by the configuration under analysis and a weakly annoying target sound, and the designer is free to choose the solution corresponding to the best sound quality. The results demonstrate that the sound-quality assessment can successfully help the designer to select the optimal flight path by providing an additional degree of freedom within the optimisation loop.

Author Contributions: Conceptualisation, U.I. and F.C.; methodology, U.I. and F.C.; software, F.C.; validation, F.C.; formal analysis, U.I. and F.C.; investigation, F.C.; writing, original draft preparation, F.C.; writing, review and editing, U.I. and F.C.; supervision, U.I.; project administration, U.I.; funding acquisition, U.I. All authors have read and agreed to the published version of the manuscript.

Funding: This research has received funding from the European Union's Horizon 2020 research and innovation programme under project ANIMA, grant agreement No. 769627.

Institutional Review Board Statement: Not applicable.

Informed Consent Statement: Not applicable.

Conflicts of Interest: The authors declare no conflict of interest.

Abbreviations

The following abbreviations are used in this manuscript:

ANIMA	Aviation Noise Impact Management through Novel Approaches
BWB	Blended Wing Body
CDA	Continuous Descent Approach
COSMA	Community Noise Solutions to Minimise aircraft noise Annoyance
DPSO	Deterministic PSO
FAA	Federal Aviation Administration
FRIDA	Framework for Innovative Design in Aeronautics
ICAO	International Civil Aviation Organization
MOP	Multiobjective Optimisation Problem
MCRDO	Multidisciplinary Conceptual Robust Design Optimisation
MTOW	Maximum Takeoff Weight
NAP	Noise Abatement Procedure
OEW	Operating Empty Weight
PSO	Particle Swarm Optimisation
SEFA	Sound EngineeringFor Aircraft
SEL	Sound Exposure Level
TAS	True Air Speed
UAM	Urban Air Mobility

Appendix A. FRIDA

The tool FRIDA (Framework for Innovative Design in Aeronautics) is the Multidisciplinary Conceptual Robust Design Optimisation (MCRDO) framework developed by the Aerospace Structures and Design group of the Roma Tre University. The framework is characterised by a collection of simulation modules and a library of optimisation methods that can be called in the appropriate sequence to execute direct simulations, optimisation, quantification of uncertainties, design-space exploration, surrogate modelling or any combination these fundamental tasks (see Figure A1 and Knobloch et al. [1] for details).

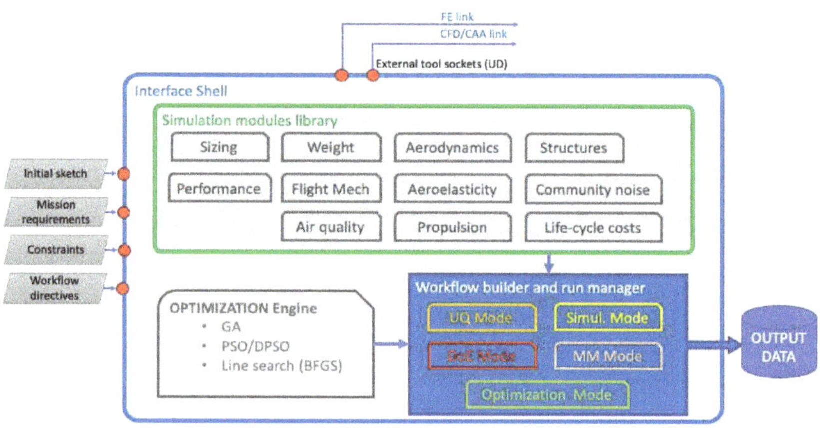

Figure A1. Conceptual layout of the FRIDA tool.

The specific workflow needed to attain a particular objective can be easily built taking advantage of the structure of the tool. Figure A2 depicts the block diagram of the workflow adopted for the optimisation of procedures presented in this work. It is worth noting how the constraints due to regulation (highlighted in yellow in Figure A2) strongly influence the variance of the non-dominated set. Indeed, the limitations imposed for the approach procedures restrict the maximum variation of the Pareto solutions to less than 10 kg of fuel and 10 km^2 of $|A_{60}|$.

The estimate of the engine noise sources and the fuel consumption requires the knowledge of the engine operating points at each trajectory sample. With this aim, a semi-empirical turbofan model was implemented. For a prescribed flight condition, the model provides the percentage of throttle, $t_{\%}$, as a function of the flight mechanics variable $\mathbf{X}_{fm}$ (altitude, drag force, actual aircraft weight, acceleration of the aircraft, etc.) and the engine characteristics $\mathbf{X}_{eng}$ (engine pitch, bypass ratio, maximum thrust at sea level, etc.).

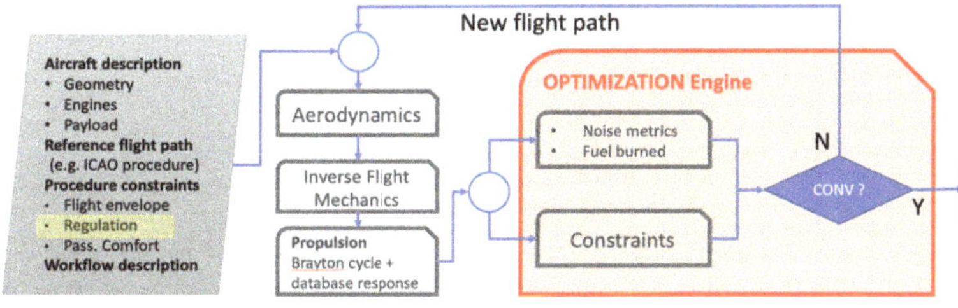

Figure A2. FRIDA workflow for the procedure optimisation.

Starting from the throttle percentage, the rotational speeds $N1$ and $N2$ of low-pressure and high-pressure spools, respectively, are evaluated. The jet velocities are then calculated through the momentum equation and their temperatures are estimated with the energy balance. The airframe and the engine noise are estimated within the aeroacoustics module. The noise of lifting surfaces, tail, high-lift devices and landing gears is based on semi-empirical functions according to Fink's model [26,27]. The engine noise estimate is based on the Morfey and Fisher model [28,29] for the buzz-saw noise and Heidmann's model [30] for the fan and the compressor noise. The jet noise is calculated by means of an interpolating metamodel built from available data. The calculation of the one-third octave band Sound Pressure Level (SPL) takes into account the Doppler effect, the atmospheric absorption [31] and the ground reflection. The Sound Exposure Level (SEL) and the Effective Perceived Noise Level (EPNL) are also estimated by means of a suitable postprocessing. More recently, surrogate models of the engine noise shielding effects for unconventional configurations (like, e.g., the Blended Wing Body) have been implemented. A financial module, which allows the estimation of financial implications from an airline company perspective [20], is also included in FRIDA. The positive cash flows (related to revenues) and the negative ones (fuel and maintenance costs, and social costs related to noise pollution) are calculated and actualised to evaluate the airliner net present value.

References

1. Knobloch, K.; Manoha, E.; Atinault, O.; Barrier, R.; Polacsek, C.; Lorteau, M.; Iemma, U.; Centracchio, F.; Rossetti, M.; Cioffi, I.; et al. Future Aircraft and the Future of Aircraft Noise. In *Aviation Noise Impact Management: Technologies, Regulations, and Societal Well-Being in Europe*; Leylekian, L., Covrig, A., Maximova, A., Eds.; Springer International Publishing: Cham, Switzerland, 2022; pp. 117–139. [CrossRef]
2. Dancila, R.; Botez, R. New flight trajectory optimisation method using genetic algorithms. *Aeronaut. J.* **2021**, *125*, 618–671. [CrossRef]
3. Gardi, A.; Sabatini, R.; Ramasamy, S. Multi-objective optimisation of aircraft flight trajectories in the ATM and avionics context. *Prog. Aerosp. Sci.* **2016**, *83*, 1–36. [CrossRef]
4. Zhang, M.; Filippone, A.; Bojdo, N. Multi-objective optimisation of aircraft departure trajectories. *Aerosp. Sci. Technol.* **2018**, *79*, 37–47. [CrossRef]
5. Prats, X.; Puig, V.; Quevedo, J. A multi-objective optimization strategy for designing aircraft noise abatement procedures. Case study at Girona airport. *Transp. Res. Part Transp. Environ.* **2011**, *16*, 31–41. [CrossRef]
6. Behere, A.; Mavris, D.N. Optimization of Takeoff Departure Procedures for Airport Noise Mitigation. *Transp. Res. Rec.* **2021**, *2675*, 81–92. [CrossRef]

7. Khardi, S.; Abdallah, L. Optimization approaches of aircraft flight path reducing noise: Comparison of modeling methods. *Appl. Acoust.* **2012**, *73*, 291–301. [CrossRef]
8. Camilleri, W.; Chircop, K.; Zammit-Mangion, D.; Sabatini, R.; Sethi, V. Design and Validation of a Detailed Aircraft Performance Model for Trajectory Optimisation. In Proceedings of the AIAA Modeling and Simulation Technologies Conference, Minneapolis, MN, USA, 13–16 August 2012. [CrossRef]
9. Dancila, B.D.; Botez, R.; Labour, D. Fuel burn prediction algorithm for cruise, constant speed and level flight segments. *Aeronaut. J.* **2013**, *117*, 491–504. [CrossRef]
10. Delfs, J.; Bertsch, L.; Zellmann, C.; Rossian, L.; Kian Far, E.; Ring, T.; Langer, S.C. Aircraft Noise Assessment—From Single Components to Large Scenarios. *Energies* **2018**, *11*, 429. [CrossRef]
11. Elango, P.; Mohan, R. Trajectory optimisation of six degree of freedom aircraft using differential flatness. *Aeronaut. J.* **2018**, *122*, 1788–1810. [CrossRef]
12. Centracchio, F.; Rossetti, M.; Iemma, U. Approach to the Weight Estimation in the Conceptual Design of Hybrid-Electric-Powered Unconventional Regional Aircraft. *J. Adv. Transp.* **2018**, *2018*. [CrossRef]
13. Burghignoli, L.; Rossetti, M.; Centracchio, F.; Palma, G.; Iemma, U. Adaptive RBF with hyperparameter optimisation for aeroacoustic applications. *Int. J. Aeroacoust.* **2022**, *21*, 22–42. [CrossRef]
14. European Commission. SEFA: Sound Engineering for Aircraft. Community Research and Development Information Service, from 1 February 2004 to 30 June 2007. Available online: https://cordis.europa.eu/project/id/502865/reporting/fr (accessed on 25 February 2022).
15. Diez, M.; Iemma, U.; Marchese, V. A sound-matching-based approach for aircraft noise annoyance alleviation via MDO. In Proceedings of the 13th AIAA/CEAS Aeroacoustics Conference, Rome, Italy, 21–23 May 2007.
16. Diez, M.; Iemma, U. Multidisciplinary conceptual design optimization of aircraft using a sound–matching-based objective function. *Eng. Optim.* **2012**, *44*, 591–612. [CrossRef]
17. European Commission. COSMA: Community Oriented Solutions to Minimise Aircraft Noise Annoyance. Community Research and Development Information Service, from 1 June 2009 to 31 March 2013. Available online: https://trimis.ec.europa.eu/project/community-oriented-solutions-minimise-aircraft-noise-annoyance (accessed on 25 February 2022).
18. Iemma, U. Multi-disciplinary, community-oriented design of low-noise aircraft: The COSMA project. *Noise Mapp.* **2016**, *3*, 59–70. [CrossRef]
19. European Commission. ANIMA: Aviation Noise Impact Management through Novel Approaches. Community Research and Development Information Service, from 1 October 2017 to 31 December 2021. Available online: https://cordis.europa.eu/project/id/769627 (accessed on 25 February 2022).
20. Iemma, U.; Pisi Vitagliano, F.; Centracchio, F. A multi-objective design optimisation of eco-friendly aircraft: The impact of noise fees on airplanes sustainable development. *Int. J. Sustain. Eng.* **2017**, *11*, 122–134. [CrossRef]
21. Bernardini, G.; Centracchio, F.; Gennaretti, M.; Iemma, U.; Pasquali, C.; Poggi, C.; Rossetti, M.; Serafini, J. Numerical Characterisation of the Aeroacoustic Signature of Propeller Arrays for Distributed Electric Propulsion. *Appl. Sci.* **2020**, *10*, 2643. [CrossRef]
22. Centracchio, F.; Burghignoli, L.; Iemma, U. Multiobjective optimisation of flight paths for noise level mitigation and sound quality improvement. *Noise Mapp.* **2021**, *8*, 268–280. [CrossRef]
23. Kennedy, J. Particle swarm optimization. In Proceedings of the 1995 IEEE International Conference on Neural Networks, Perth, Australia, 27 November–1 December 1995; Volume 4, pp. 1942–1948. [CrossRef]
24. Campana, E.; Diez, M.; Fasano, G.; Peri, D. Initial particles position for PSO, in Bound Constrained Optimization. In *Advances in Swarm Intelligence*; Lecture Notes in Computer Science; Tan, Y., Shi, Y., Mo, H., Eds.; Springer: Berlin/Heidelberg, Germany, 2013; Volume 7928, pp. 112–119.
25. Pellegrini, R.; Serani, A.; Leotardi, C.; Iemma, U.; Campana, E.F.; Diez, M. Formulation and parameter selection of multi-objective deterministic particle swarm for simulation-based optimization. *Appl. Soft Comput.* **2017**, *58*, 714–731. [CrossRef]
26. Fink, M.R. Approximate prediction of airframe noise. *J. Aircr.* **1976**, *13*, 833–834. [CrossRef]
27. Fink, M.R. Airframe noise prediction. In *Federal Aviation Administration Report FAA–RD–77–29*; Federal Aviation Administration: Washington, DC, USA, 1977.
28. Morfey, C.; Fisher, M. Shock–Wave radiation from a supersonic ducted rotor. *Aeronaut. J.* **1970**, *74*, 579–585.
29. McAlpine, A.; Fisher, M. On the prediction of buzz–saw noise generated by an aero–engine. In Proceedings of the 6th Aeroacoustics Conference and Exhibit, Lahaina, HI, USA, 12–14 June 2000.
30. Heidmann, M. *Interim Prediction Method for Fan and Compressor Source Noise*; NASA TM-X-71763; NASA: Washington, DC, USA, 1979.
31. Sutherland, L.; Piercy, J. Method for calculating the absorption of sound by the atmosphere. In Proceedings of the 88th Meeting of the Acoustical Society of America, Washington, DC, USA, 22–24 April 1974.

aerospace MDPI

Article

Low-Noise Design of Medium-Range Aircraft for Energy Efficient Aviation

Vincent Domogalla [1,2,*,†], Lothar Bertsch [1,2,†], Martin Plohr [3], Eike Stumpf [4] and Zoltán S. Spakovszky [5]

1 German Aerospace Center (DLR), Institute of Aerodynamics and Flow Technology, Bunsenstraße 10, 37073 Göttingen, Germany; lothar.bertsch@dlr.de
2 Exzellenzcluster SE^2A—Sustainable and Energy-Efficient Aviation, Hermann-Blenk-Str. 42, 38108 Braunschweig, Germany
3 German Aerospace Center (DLR), Institute of Propulsion Technology, Engine, 51147 Köln, Germany; martin.plohr@dlr.de
4 Institute of Aerospace Systems (ILR), RWTH Aachen University, Wuellnerstr. 7, 52062 Aachen, Germany; stumpf@ilr.rwth-aachen.de
5 Gas Turbine Laboratory (GTL), Massachusetts Institute of Technology, 70 Vassar Street, Cambridge, MA 02139, USA; zolti@mit.edu
* Correspondence: vincent.domogalla@dlr.de
† These authors contributed equally to this work.

Abstract: Promising low-noise aircraft architectures have been identified over the last few years at DLR. A set of DLR aircraft concepts was selected for further assessment in the context of sustainable and energy-efficient aviation and was established at the TU Braunschweig in 2019, the Cluster of Excellence for *Sustainable and Energy-Efficient Aviation* (SE^2A). Specific Top-Level aircraft requirements were defined by the cluster and the selected DLR aircraft designs were improved with focus on aircraft noise, emissions, and contrail generation. The presented paper specifically addresses the reduction of aviation noise with focus on noise shielding and modifications to the flight performance. This article presents the state of the art of the simulation process at DLR and demonstrates that the novel aircraft concepts can reduce the noise impact by up to 50% in terms of sound exposure level isocontour area while reducing the fuel burn by 6%, respective to a conventional aircraft for the same mission. The study shows that a tube-wing architecture with a top-mounted, forward-swept wing and low fan pressure ratio propulsors installed above the fuselage at the wing junction can yield significant noise shielding at improved low-speed performance and reduce critical fuel burn and emissions.

Keywords: low-noise aircraft design; center of excellence; *Sustainable and Energy-Efficient Aviation* (SE^2A); forward swept wing; PANAM; PrADO; RCE

Citation: Domogalla, V.; Bertsch, L.; Plohr, M.; Stumpf, E.; Spakovszky, Z.S. Low-Noise Design of Medium-Range Aircraft for Energy Efficient Aviation. *Aerospace* **2022**, *9*, 3. https://doi.org/10.3390/aerospace9010003

Academic Editor: Wing Chiu

Received: 26 October 2021
Accepted: 10 December 2021
Published: 22 December 2021

1. Introduction

In order to fulfill the goals of the Flight Path 2050 initiative by the European Union, it is necessary to reduce both the emissions and noise of future aircraft [1]. In this context, the Cluster of Excellence for *Sustainable and Energy-Efficient Aviation* (SE^2A) was established at the TU Braunschweig in 2019. The overall goals of the SE^2A focus on "the challenge to the structure of modern society, as transportation consumes high amounts of energy, produces high amounts of pollutants and noise, and thereby threatens our resource base, environment, and climate" [2]. To address these challenges, different aircraft architectures and technologies are under consideration with the intention of meeting the requirements for short, mid, and long-range applications. New technologies will be developed with a focus on electrification or alternative fuel concepts to assess their overall impact on the sustainability and energy efficiency of the entire air transport system.

Any new technology as introduced into the market will be assessed for its impact on overall aircraft noise. Currently, only the impact on noise certification levels according to ICAO's Annex 16 [3] is evaluated. In the near future, the reduction of community noise will remain a key issue. At this point, DLR Göttingen is contributing to SE^{2}A by tackling the noise challenge based on its long-time expertise and available simulation capabilities [4]. Especially since the growth of air transport, noise reduction technologies and their environmental impact are of great importance. Various authors have already shown that worldwide air traffic plays an important role in environmental pollution [5–7]. The aviation sector's CO_2 contribution is estimated at 2.5% of the anthropogenic CO_2 emission worldwide [8]. Furthermore, the contributions of nitrogen oxide (NO_x) emissions and contrails to global warming have become the focus of investigations in recent years. The climate impact of contrails could exceed the impact of CO_2 by a factor of three or more in terms of radiative forcing (RF) [9].

As part of the activities in the cluster, this study scrutinizes the impact of low atmospheric emission technologies on the noise impact on the ground of mid-range aircraft. As low emission technologies, a geared turbofan (GTF) with an ultra-high bypass ratio (BPR = 12) in combination with a forward-swept wing (FSW) were chosen. For this study, carbon dioxide, nitrogen oxide, and nvPM (non-volatile Particulate Matter) emissions are evaluated along with the sound exposure level (SEL) noise contours and 1000 m sideline levels. Additional low noise technologies such as landing gear fairings are not assessed in this study.

It can be demonstrated that the FSW in combination with a GTF and an over the wing shielding architecture reduces the fuel burn of a tube-wing mid-range aircraft by 6% while also reducing the noise impact in terms of SEL area by up to 50%.

2. Environmental Considerations

2.1. Aircraft Noise

During departure, when engines operate near or at maximum thrust, engine noise dominates. For aircraft with turbofan engines, one can further divide the engine noise sources into dominating contributors, i.e., tonal and broadband fan noise and broadband jet noise. The noise contribution of other engine components, e.g., the combustion noise contribution, can be neglected within this study since their contribution to the standard noise descriptors is not significant for the conventional turbofan engines under consideration [10]. The tonal fan noise depends on the actual blade loading, the rotational speed, and the geometry of the rotor (including stator design). At takeoff, the fan tip can be exposed to supersonic velocities, resulting in significant additional noise, i.e., a shock system at the fan blade tips can yield so-called *buzz-saw* noise or multiple pure tone noise (MPT). Broadband contribution is mainly dependent on the specific fan blade geometry. Jet noise is caused by the mixing of the two exhaust streams (core and bypass) with one another and the free stream. In these mixing regions, the temperature and velocity gradients are large and strong shear layers develop with strong turbulent fluctuations. Broadband jet noise is emitted and depends mainly on the jet velocities but also on the mixing characteristics of the different flow regimes [11]. The relevance of fan and jet noise can vary along the flight path with segments dominated by fan or jet noise, respectively.

During approach, the most significant noise contribution can be attributed to the airframe, because the engines operate at reduced thrust settings or in high idle [12]. In particular, the high lift devices and the landing gear are the major noise contributors. With retracted landing gear and high lift devices, the wing and tailplane trailing edge noise dominates the aircraft noise. Deploying the landing gear causes a strong turbulent wake system, impinging on deployed high-lift elements downstream of the gear. In certain flight conditions, this can cause significant additional noise that is referred to as installation noise [13]. In general, installation noise describes source mechanisms caused by the effect of other aircraft parts upstream of the flow, altering the flow condition at the noise source. Moreover, the overall aircraft noise can be significantly increased if so-called parasitic noise sources are present. For example, wing fuel overpressure vents can act as cavities and emit tone noise [14].

2.2. Atmospheric Emissions

The aircraft's environmental impact strongly depends on its emissions. During the combustion process in jet engines, those emissions are generated and emitted into the atmosphere as exhaust gas. They can either be the natural outcome of the combustion process (CO_2, H_2O) or the result of an incomplete combustion process (CO, UHC, SO_x, NO_x, nvPM). CO_2 and H_2O are well known as greenhouse gases (GHG). They influence the natural heat exchange of the earth by preventing low-frequency infrared radiation, emitted by the earth's surface, from escaping the atmosphere while letting higher frequent solar radiation from the sun passing. Thus, the equilibrium temperature of the atmosphere is pushed to higher temperatures, proportional to the amount of GHG in the atmosphere.

NO_x obtains its global warming potential from its effect on the ozone cycle. The emission in the lower atmosphere leads to an increased formation of ozone, where it affects the air quality due to its highly oxidizing nature and therefore the hazardous effect on the human body. When emitting ozone into the upper atmosphere, an opposing effect takes place. Because of the additional presence of high-energy radiation ozone is split irreversibly into oxygen. This leads to a depletion of the ozone layer, which protects the earth against high energetic radiation [15].

2.3. Contrails

Contrails can contribute similarly to global warming as GHG. They block long-wave radiation from the earth's surface from being emitted back to space and trapping it into the atmosphere while being relatively transparent for short-wave solar radiation coming from the sun [9]. The formation takes place within the exhaust plume of the engine. Small particles act as condensation nuclei forcing the water vapor emitted by the engine to condensate. Therefore the amount of emitted nvPM and H_2O is an important indicator for the tendency to form contrails [16].

3. SE^{2}A Mid-Range Aircraft Applications

Future long-range aircraft might use the Blended Wing Body (BWB) architecture, leaving the tube and wing architecture of special interest for the short- and mid-range designs. Furthermore, market forecasts for the upcoming decades foresee a significant increase in demand for short- and mid-range aircraft [17–19]. Therefore, this work will focus on the mid-range aircraft with a conventional tube and wing architecture. The TLAR's are based on the performance characteristics of an Airbus A320 aircraft, as one of two market leading options in this segment.

In addition to the mission requirements provided in Table 1, various technologies are proposed for mid-range vehicles, i.e., hybrid-electric propulsion (HEP) and boundary layer ingestion (BLI) concepts. Electric propulsion is among the considered technologies because of the latest improvements in battery technology. They enable the application of electric drives at some point to at least partially replace conventional gas turbine engines in flight operation. The significant increase in design space for such aircraft can also reduce local emissions. The fact that energy storage, generation, and propulsion units can be spatially separated opens up an extensive solution space for novel ideas. The separation and distribution of subsystems allow for a tailored arrangement of propulsion units maximizing aerodynamic performance. Furthermore, propulsors could be installed in a way that potential noise shielding effects can be exploited. The weight of electric motors scales differently than conventional gas turbines. One way to exploit this is to use multiple small propulsors operating at high rotational speeds instead of having a few large engines that operate at low rotational speeds [20]. A potential weight reduction with HEP can have beneficial effects on both cruise performance and noise generation throughout departure and landing. The boundary layer ingestion promises a significant reduction in overall aircraft drag and hence can yield significant fuel savings [21]. As recent publications showed, however, a mid-range application of battery-based technologies in near future is unlikely [22,23].

Table 1. Top-Level Aircraft Requirements (TLAR) for SE^2A mid-range aircraft.

Parameter	Requirement
No. of PAX	150
Range	4500 km
Cruise Mach	0.78
Initial cruise alt.	11,000 m

Simulation capabilities to assess the noise reductions with these technologies are far from mature and need more fundamental research. For example, BLI might lead to significant excess noise that would counteract any other noise reduction measure and increase ground noise levels [24]. The noise generation mechanics for these technologies are still not well understood and addressed through efforts led by SE^2A.

Additional concepts and ideas toward mitigation of the emissions and the contrail generation are addressed in the following section. These mitigation strategies are not applied within this study but could be considered for future research activities to ultimately address all three SE^2A challenges. Some proposed measures will have advantageous effects on all three challenges, i.e., simultaneously reducing noise, emissions, and contrail generation.

3.1. Noise and Emission Mitigation Strategies

To reduce environmental pollutants, the overall fuel consumption must be decreased. A reduction in fuel burn can be achieved either by reducing the thrust needed, which is equivalent to reducing the overall drag, and/or by improving the efficiency of the propulsion system. Furthermore, advanced thermodynamic engine cycle choices can also reduce emissions. Promising technologies are the lean premixed pre-vaporized (LPP) combustor, the staged combustor, and the rich burn quick lean (RQL) combustor which could play a role in future concepts [25]. Furthermore, previous DLR studies pointed out further reductions in the environmental impact of emissions from changes in aircraft operations [25]. In particular, the effects of ozone (O_3) and water vapor of contrails on the RF are sensitive to flight operation, i.e., region and altitude of the actual flight path. Avoiding certain flight altitudes and regions is estimated to reduce the climate impact by 30–60% while resulting in about 8–30% increased direct operating costs [26–28]. In this context, Koch et al. [25] and Grewe et al. [29] proposed a set of design rules. In order to avoid excessive pollution in higher altitudes, it is necessary to design the aircraft for a cruise flight at lower altitudes. This also leads to a lower optimal cruise Mach number due to the higher air density, which possibly results in a reduced wing sweep, a higher aspect ratio, and an increased wing thickness [25]. According to [9] sustainable aviation fuels (SAF) can yield short-term mitigation of CO_2 and contain smaller amounts of sulfur and aromatic species, which can reduce ice and nvPM formation. However, the use of SAF is currently constrained by safety regulations which will not allow the usage of SAF blends larger than 50% [30].

More design rules and mitigation concepts are described by Farokhi [15], who proposes to focus on aerodynamic efficiency, the reduction of fuel consumption, and the reduction of harmful emissions. To improve the aerodynamic efficiency, several technologies including laminar flow control, exploitation of lifting surfaces (lifting fuselage), high aspect ratio wings (folding wings), tailless configurations, application of boundary layer ingestion, and sophisticated propulsion system integration are proposed. To reduce the fuel burn, suggested technologies include ultra-high bypass ratio engines, hybrid-electric propulsion, extensive use of lightweight materials, and fully electric operation at the gate and during taxi. With respect to harmful emissions, alternative biofuels with lower life-cycle emissions, staged combustion designs, and (fully) electric propulsion concepts are proposed. These mitigation strategies support the goals of the SE^2A efforts and should be introduced as early as within the conceptual design of new SE^2A vehicles.

4. DLR Simulation Process

More than a decade ago, DLR and TU Braunschweig launched a joint research effort to reduce exterior aircraft noise by assessing promising reduction measures [4]. An automated simulation process for the conceptual design of such low-noise aircraft was established, i.e., see Ref. [31], and is ready to be used by SE^{2}A. The latest version of the simulation process, as described in Ref. [32], is comprised of the aircraft design synthesis tool PrADO [33,34] (TU Braunschweig), the noise shielding simulation tool SHADOW [35] (DLR), the flight simulation tool FLIPNA [11], and the aircraft system noise prediction tool PANAM [31,32]. Dedicated interfaces allow direct processing of external data from measurements or high-fidelity simulation. For example, measured weights, aerodynamic information, and delta noise levels of novel high-lift concepts can be processed and influence the final results [32].

Certain early design choices within the conceptual design phase can have an extensive impact on the acoustic performance of the vehicle. These are depicted in Figure 1. The most relevant effects must be captured adequately by any reliable simulation process. As shown in Figure 1, the sound intensity I_x from most of the source mechanisms scale with the local flow velocity v_x, i.e., airframe noise scales approximately with the power of six with the flow velocity. The only exception, in this case is, the effect of altitude, which scales by distance squared. At the subsequent design stages, these parameter choices are fixed and cannot be modified without major implications that require a revisit to the conceptual design stage. Based on findings from previous studies, simple guidelines for low-noise aircraft design can be deduced and are summarized in Refs. [4,32]. According to these rules, interaction noise sources should be avoided, e.g., jet-flap interaction noise can be avoided by selecting an over-wing engine position. Gear wake impingement on the extracted flaps is avoided by fuselage-mounted landing gears. Noise shielding shall be exploited, e.g., over-the-wing engine positioning can result in advantageous noise shielding of the forward fan noise emission. If available, low-noise technology shall be applied to selected components, e.g., the application of acoustic lining to the engine inlet and exhaust ducts [36]. The overall aircraft design shall be adapted as early as within the conceptual phase. Only then can advantageous propulsion integration concepts or low-noise flight performance be achieved, e.g., the latter can be influenced by modifications to the wing planform and thereby reduction in the minimum approach speed of the aircraft. Finally, the approach and departure trajectories should be adapted and tailored to the individual aircraft design under consideration [11].

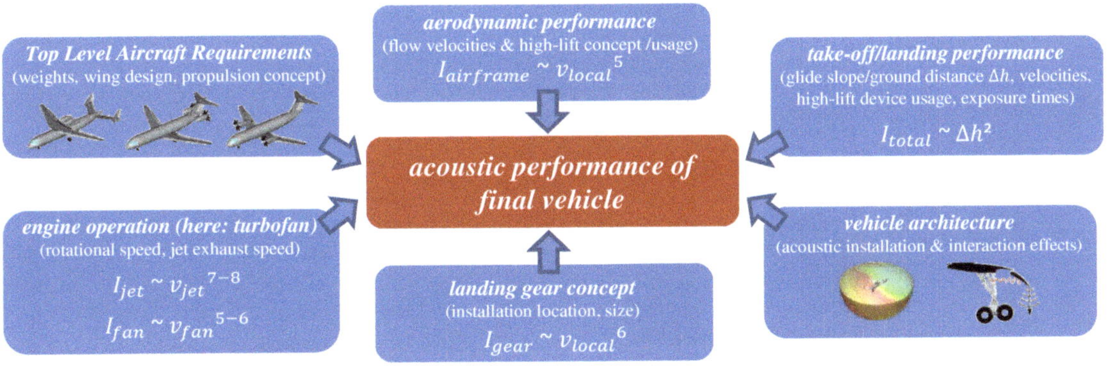

Figure 1. Influence of design options on acoustic performance of resulting vehicle (changed, originally from Ref. [4]).

4.1. Aircraft and Engine Design Tools

The initial workflow of the SE^2A noise assessment simulation process is depicted in Figure 2. The toolchain is integrated into DLR's RCE [37] framework to make these tools available in a more convenient way with the ability of distributed simulation.

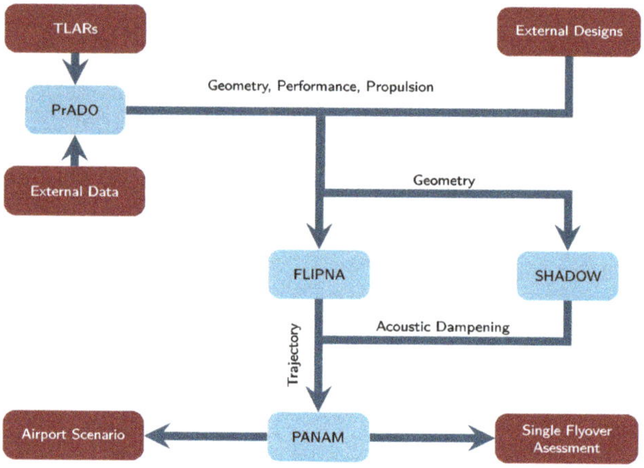

Figure 2. RCE toolchain.

All vehicles in this study were designed with TU Braunschweig's PrADO [33] tool. PrADO is comprised of individual modules each dedicated to individual tasks in the conceptual design process of an aircraft; additional information about PrADO funcionality can also be found in Ref. [34]. For example, specific modules evaluate the weights of the fuselage, wing, and propulsion system. These weights are then subsequently processed by another module to determine the flight performance of the aircraft. Hereby, modules of different fidelity are available for certain tasks, i.e., ranging from fully empirical to analytical models. Furthermore, dedicated interfaces allow processing external data from experiments or numerical investigations by replacing the corresponding simulation module with external data [32].

PrADO is run iteratively until certain predefined parameters converge. Each design modification or alternative engine option impacts the components and thus the overall system yielding snow-ball effects. The result of the overall PrADO simulation process is a physics-based aircraft design [33]. The final vehicle resulting from the PrADO design process is then simulated along departure and approach trajectories to provide all required input data for the subsequent simulation of the ground noise impact. PrADO has the option to run internal flight simulation modules [31] or as selected within this study to process external data, e.g., from DLR's FLIPNA tool [11].

4.2. Engine Model

For this study, the PrADO internal engine modules are replaced by external high-fidelity simulation results of the DLR Institute of Propulsion Technology's GTlab tool [38]. It is a component-based engine design platform, which allows the virtual design and optimization of aircraft engines and stationary gas turbines. The performance calculations include the thermodynamic cycle design and off design simulations, both for ate and transient conditions.

The emission indices for CO_2 and H_2O were calculated under the assumption of complete combustion of the jet fuel, due to the high burn-out rates of modern aircraft engine combustors. For an average Kerosene sum formula of $C_{12}H_{23}$, this gives EI CO_2 = 3157.3 g/kg and EI H_2O = 1237.2 g/kg for all operating conditions [39].

Emissions of NO_x and Particulates are more dependent on the engine operating condition. Sea level static (SLS) emission indices for all commercial engines currently in service with a rated thrust greater than 26.7 kN are provided in the ICAO Engine Exhaust Emissions Databank [40]. To calculate altitude emissions based on these data, the Boeing Fuel Flow Method 2 [41] was recommended by ICAO. A similar fuel flow correlation method for NO_x emissions has also been published by DLR [42]. For calculating cruise nvPM emission indices, DLR has developed the so-called DLR Soot Correlation [43] which was applied here. For the advanced engine model, no emissions data were available in the ICAO Databank. Therefore, SLS emission indices of a similar, modern engine (the PW1100G) were used as the basis for the emissions model. These data have been corrected for the different thermodynamic engine conditions of the advanced engine, compared to a generic model of the PW1100G model. This correction was performed with the P3T3 method for NO_x [41,44] and the DLR Soot Correlation (s.o., [43]) for nvPM. The P3T3 method usually assumes a fuel air ratio (FAR) exponent of 0, because in a rich–quench–lean (RQL) combustor, which is currently the most frequent combustion technology in aircraft engines, there is always a region with stoichiometric conditions which dominates the NO_x production. However, in this case the combustor overall design FAR was different for the reference and the advanced engine model. Therefore, an FAR exponent of 0.55 was used to account for this difference (taken from an earlier investigation in [45]). The DLR soot correlation already includes an FAR correction term. Both the standard NO_x and nvPM correlation methods have then been applied with these corrected SLS reference emission indices.

4.3. Noise Shielding

Depending on engine airframe integration, acoustic interaction or installation effects can influence the resulting ground noise impact of the aircraft [31]. A dedicated DLR tool to assess these acoustic interaction effects can directly be applied to the PrADO aircraft model. The ray-tracing tool SHADOW [35] provides level differences that can directly be processed when the overall ground noise impact is assessed. In Figure 3, a representation of the data provided by SHADOW is shown. The tool delivers delta levels (AttnM-dB) on a sphere around the aircraft for the third octave bands. The effect of noise shielding clearly becomes visible below the aircraft in the green–blue area of the sphere. The pink dot represents the source position, in this case the position of the right engine, which is modeled as a monopol.

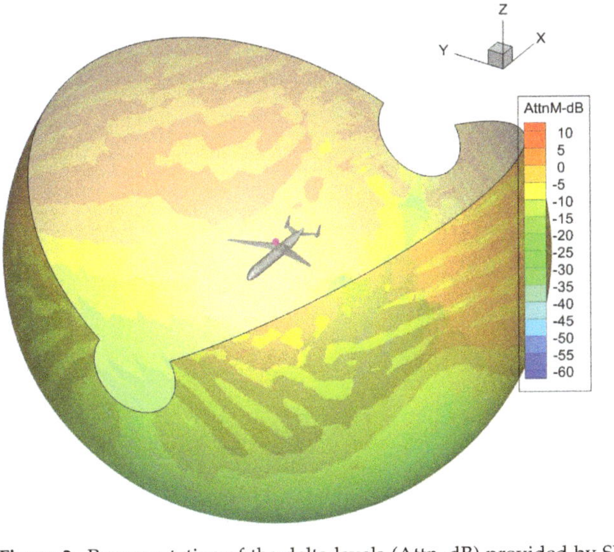

Figure 3. Representation of the delta levels (Attn–dB) provided by SHADOW, on sphere with a radius of 50 m for a frequency of 1000 Hz. The pink dot represents the simulated source position.

4.4. Flight Simulation

FLIPNA [11] is applied to simulate the individual flight tracks of each design variant. PrADO's internal simulation modules for flight simulation are bypassed and the data from FLIPNA is processed. Essential for any meaningful flight simulation is a detailed definition of the high-lift system and the corresponding aerodynamic performance. Throughout an approach trajectory, the overall ground noise impact is directly dependent on the configuration setting, i.e., the deployment of the high-lift system and the landing gear. During the departure trajectory, the flight performance and thus the noise impact can be influenced by the high-lift system. PrADO provides these data for the simulation with FLIPNA, which implements the ECAC.CEAC DOC 29 Volume 2, App. B regulations, but uses the engine and flight performance data provided by PrADO instead of performance coefficients. It allows the calculation of NADP-1/2 departure trajectories and LDLP/CDA approach trajectories. The mission analysis is carried out by PrADO internal flight simulation.

4.5. Overall Aircraft Noise Simulation

The overall aircraft noise is assessed with DLR's tool PANAM [31]. A brief description of PANAM can also be found in Ref. [46]. The final aircraft and engine design and performance parameters (including acoustic installation effects from SHADOW) are used as inputs. Specific noise source models from DLR and open literature are applied to simulated individual noise sources of the vehicle [31], references to the specific models are listed in Table 2.

Table 2. Applied simulation methods for noise assessment.

Source/Mechanism	References
trailing edge noise	DLR model [31,47–51]
leading edge noise	DLR model [31,47,48,51]
landing gear noise	DLR model [31,47,48,51]
fan noise	modified Heidmann model [52]
jet noise	modified Stone model [53]
noise shielding	DLR tool SHADOW [35,54]
atmospheric propagation	SAE standard 886A [55]
ground attenuation	SAE standard AIR1751 [56]

The noise sources can vary during the flight path (including acoustic installation effects) and are accounted for and summed up to yield the overall aircraft noise emission. The emission from each aircraft position is then propagated through the atmosphere to finally yield the ground noise impact. Standard noise metrics, e.g., SEL, $L_{A,max}$, and effective perceived noise level (EPNL), are evaluated and different observer locations are considered. Other metrics to assess sound quality are part of ongoing research activities within SE^{2}A and can be applied via an external tool by TU Braunschweig [57], which is based on psychoacoustic metrics (loudness, sharpness and tonality) and delivers a combined metric $PA_{mod,5\%}$, based on the work of More [58].

4.6. Verification of Simulated Noise Levels

Feasibility and validity of the simulated noise levels are confirmed by previous and ongoing investigations not in the scope of the presented SE2A activities. Predicted noise levels are compared to measured noise levels for certain existing aircraft in dedicated flyover campaigns orchestrated by DLR, e.g., B737-700 [59] or A319 [31]. The flight campaign with the A319 was comprised of multiple approach and departure flights. Thereby, the aircraft was flown and later simulated under various typical operating conditions along both approach and departure flights. Comparison of simulation and measurement shows good agreement for all considered operating conditions and observer locations [31]. A study with a VFW614 aircraft confirmed PANAM's simulation capabilities even along unconventional and novel approach trajectories [60]. More recently, the implemented noise models and the overall simulation process are carefully assessed for inherent simulation uncertainties, i.e., input, modeling, and propagation uncertainty [61]. The level of uncertainty that can be associated with the simulated noise levels are suitable for the comparative assessment presented in this work [61]. Last but not least, the noise simulation tool PANAM has been validated against simulation tools from ONERA and NASA [46]. The comparison is based on almost identical input data in order to clearly work out the differences due to the inherent models. This comparison indicates that the PANAM simulation results for tube-and-wing aircraft with turbofan engines are plausible and feasible if compared to other simulation tools [46]. In conclusion, previous and ongoing investigations are understood as a confirmation of the overall simulation result, i.e., the outcome of aircraft and engine design, flight simulation, and noise prediction.

4.7. Advanced Aircraft Concepts

Based on a previous DLR report [62] and dedicated low-noise design studies by DLR [4,31], a promising aircraft architecture with an alternative engine option is selected for the mid-range vehicle concepts. A tube-and-wing architecture with the engines mounted above the fuselage–wing junction is most promising for shielding engine noise. This configuration yields approx. 11 dB lower max. A-weighted sound pressure level ($L_{A,max}$) during approach and, about 3 dB lower $L_{A,max}$ during departure [31,32]. Two variants of this architecture are selected for further assessment, i.e., one concept with backward-swept wings (concept V-2 adapted from Ref. [31]) and one vehicle with forward-swept wings adapted [62]. A previous study demonstrated significant advantages for a new engine option, i.e., a geared turbofan engine with a bypass ratio of 12, even for aircraft with shielded fan noise [32]. Furthermore, a significant reduction in fuel burn was demonstrated for this engine [32].

The investigated concept vehicles are derived from available PrADO aircraft designs [31,32] and are modified to meet the requirements for a mid-range mission, in terms of number of passengers, fuselage length, design range cruise altitude, and Mach number. The evolution of the design concept is shown in Figure 4. It started with a simple engine replacement as a first modification for reduced noise. Next, more advanced design changes such as a high-mounted, forward-swept wing, based on [62], were incorporated. To further improve system performance, additional technologies will be included.

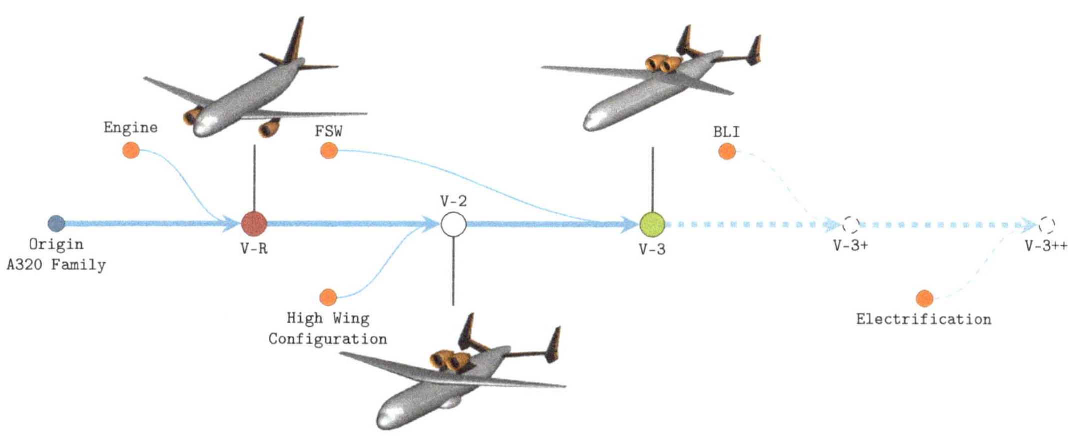

Figure 4. Evolution of low noise aircraft design at DLR.

As a reference vehicle, a standard configuration based on the Airbus A319 was chosen, representing the current state of the art. The low noise configuration V-2 has a shoulder-mounted wing to avoid possible installation effects of the gear and the high-lift system. In addition, the aspect ratio of the wing was reduced to increase the shielding area in front of the engine by increasing the chord length at the wing root [31]. The FSW wing design concept, V-3, adopts the shoulder-mounted wing concept as well as the shoulder-mounted engines but uses a forward-swept high aspect ratio wing design to increase aerodynamic efficiency and to reduce emissions [62].

As Redeker et al. [63] have shown, a laminar wing is a promising solution to reduce the airfoil drag in the transonic regime by delaying the laminar–turbulent transition of the boundary layer flow. While for lower Mach numbers laminar wings are already common, e.g., sailplanes, the boundary layer becomes increasingly unstable for higher Mach numbers. This is particularly true for swept wings which suffer from attachment–line transition (ALT) and cross-flow instability (CFI) besides the Tollmien–Schlichting instabilities (TSI). These also occur on a non swept wing [64]. In contrast to a backward swept wing, where the leading edge sweep angle increases compared to the effective sweep angle, the opposite applies to forward-swept wings. Thus, ALT and CFI, which both are sensitive to the leading edge sweep angle [63], are suppressed. Effective measures to delay transition are also explained in an earlier work by Seitz et al. [65]. The specific design of the forward-swept wing is directly derived from previous results obtained in dedicated DLR studies, i.e., projects LamAiR [66] and TuLam [67]. Seitz et al. [66,67] demonstrated promising benefits in terms of fuel-saving but at the cost of additional weight and adding complexity in the structural design of the wing and the design of the high lift devices. A lower fuel consumption also means lower emissions but might come at the cost of higher aircraft weight and reduced takeoff and landing performance increasing noise.

For a fair comparison, all aircraft have a conventional aluminum fuselage, a carbon fiber composite wing, and empennage. The wing-span is limited by size to a maximum width of 36 m, meeting the ICAO Aerodrome Reference Code element 2 'C' or ADG Group III requirements. Considering the available two engine options, a total of six PrADO aircraft were selected for this study, see Table 3. The aircraft equipped with the conventional BPR-6 turbofan are referenced as *ceo* (current engine option), whereas the aircraft equipped with the BPR 12 GTF are referenced as *neo* (new engine option). The design details are provided in Table 4. All six vehicles were simulated throughout the design mission considered in SE^2A.

Table 3. Aircraft designs considered in study.

a/c Design	Architecture	Engine Option	BPR
V-R	reference	ceo	6
V-R-g	reference	neo	12
V-2	noise-shielding	ceo	6
V-2-g	noise-shielding	neo	12
V-3	noise-shielding, foward-swept wing	ceo	6
V-3-g	noise-shielding, foward-swept wing	neo	12

The emissions from these flights can be summed up and are used for an initial environmental assessment. Furthermore, individual flight trajectories for approach and departure are computed for each vehicle so that a full noise assessment can be conducted. The aircraft noise impacts along the entirety of the approach and departure flights are then predicted and compared. The reference vehicle is referred to as V-R (Figure 5a), the modified low-noise design from [31,32] as V-2 (Figure 5b), and the new PrADO design of a forward-swept vehicle as V-3 (Figure 5c); the vehicles with the new engine option are labeled with the addition '-g'.

It is important to note that no additional low-noise technologies were applied to the vehicles at this point. Only the effect of noise shielding and the flight performance on the aircraft noise are assessed in detail here.

Table 4. Aircraft design and performance parameters.

Parameter	V-R	V-R-g	V-2	V-2-g	V-3	V-3-g
		Propulsion				
Engine BPR	6	12	6	12	6	12
Static thrust in kN	116.56	121.39	116.56	121.39	116.56	121.39
TET in K	1803.66	1883.14	1803.66	1883.140	1803.66	1883.14
OPR	29.0	34.6	29.0	34.6	29.0	34.6
FPR	1.80	1.51	1.80	1.51	1.80	1.51
TSFC in g/kN/s	11.51	7.68	11.51	7.68	11.51	7.68
Fan diameter in m	1.37	1.65	1.37	1.65	1.37	1.65
Rel. pos. X in %	28.75	28.75	47.26	47.26	51.75	51.75
Rel. pos. Y in %	33.88	33.88	7.39	7.39	7.74	7.74
		Total weights				
OEW in kg	38,721	39,907	38,317	39,494	41,256	42,512
Max. takeoff in kg	68,341	66,416	67,782	65,627	68,901	67,140
Max. landing in kg	63,779	63,192	62,820	62,098	57,512	58,143
		Component weights				
Fuselage in kg	9221	9214	10,218	10,228	9426	9425
Wing in kg	7231	7301	5689	5744	8863	8968
HTP in kg	504	504	576	576	691	691
VTP in kg	420	420	396	396	378	378
Propulsion sys. in kg	6873	8053	6873	8053	6985	8165
Landing gear in kg	2183	2167	1937	1912	2008	2022
		Fuselage				
Length in m	37.70	37.70	39.90	39.90	40.25	40.25
		Wing				
Span width in m	33.96	33.96	31.64	31.64	35.86	35.86
Reference area in m^2	122.60	122.60	125.00	125.00	126.00	126.00
Aspect ratio	9.40	9.40	8.00	8.00	10.20	10.20
c/4 sweep angle in deg	23.98	23.98	23.29	23.29	−19.32	−19.32
l. e. sweep in deg	27.48	27.48	26.91	26.91	−16.47	−16.47
		Horizontal tailplane				
Span width in m	11.83	11.83	12.25	12.25	13.23	13.23
Reference area in m^2	28.00	28.00	30.00	30.00	35.00	35.00
		Vertical tailplane				
Span width in m	5.87	5.87	4.42	4.42	4.33	4.33
Reference area in m^2	21.50	21.50	13.00	13.00	12.50	12.50
		Landing gear length				
Front gear in m	2.27	2.27	1.34	1.34	1.90	1.90
Main gear in m	2.75	2.75	1.78	1.78	2.09	2.09
		Performance along design mission				
Max. C_L (land.)	3.09	3.08	3.24	3.26	3.26	3.28
Cruise L/D	17.61	17.48	16.62	16.55	19.13	19.04
Cruise C_L	0.55	0.54	0.53	0.52	0.54	0.53
Cruise Mach	0.78	0.78	0.78	0.78	0.78	0.78
Block time in h	5.68	5.69	5.64	5.65	5.68	5.68
Fuel in kg/100 km/seat	1.80	1.43	1.91	1.52	1.69	1.33
Max. field length in m	1932	1917	1545	1449	1456	1399
DOCs in Ct/SKM	2.69	2.60	2.71	2.59	2.73	2.62

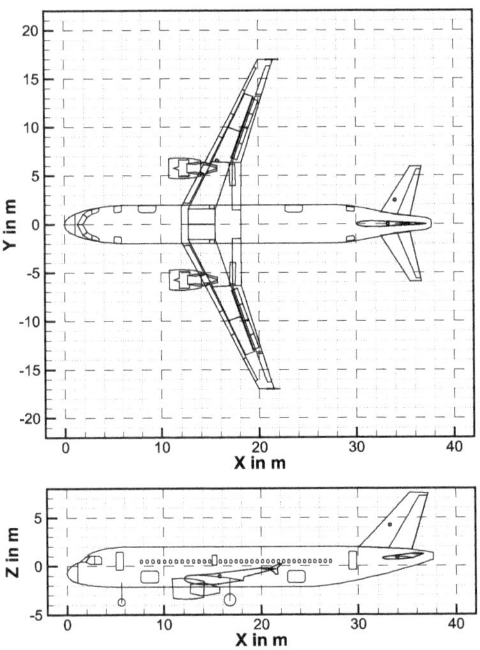

(a) V-R

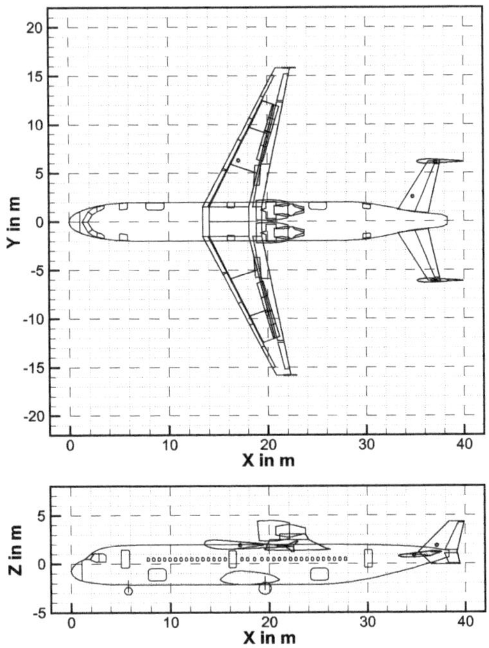

(b) V-2

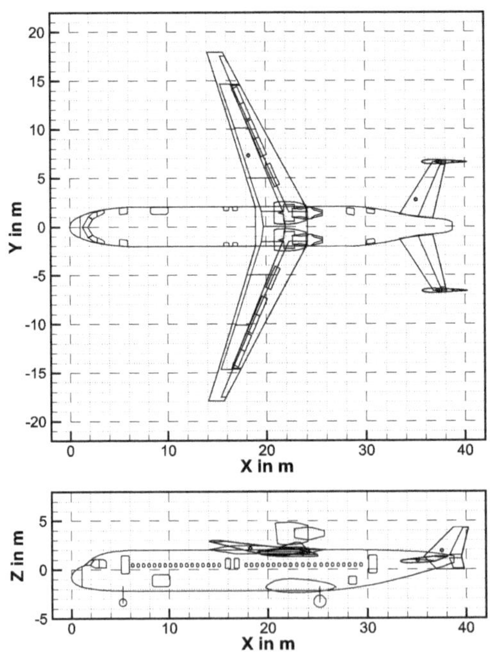

(c) V-3

Figure 5. Aircraft designs from PrADO.

5. Results

In Section 5.2 the fuel consumption and emission of CO_2, H_2O, NO_x, and nvPM (Tables 5 and 6) emitted over the design mission are presented, both separately for departure, cruise, and approach. The mission profile is depicted in Figure 6, including a departure, cruise and approach phase. Additional diversion and loiter time are also considered, as required by FAA regulations.

Table 5. Fuel consumption and emissions along the design mission (*ceo*). Advantageous change: green, disadvantageous change: red.

Vehicle	Component	Departure	Cruise	Approach	Total	%
V-R	Fuel in kg	1450	10,096	490	12,036	(-)
	CO2 in kg	4577	31,876	1549	38,001	(-)
	H2O in kg	1793.5	12,490.5	606.8	14,890.9	(-)
	NOx in kg	35.1	109.6	3.2	147.9	(-)
	nvPM in g	67.3	61.0	0.5	128.8	(-)
V-2	Fuel in kg	1468	10861	527	12,855	+6.8
	CO2 in kg	4634.2	34,290.1	1662.8	40,587.0	+6.8
	H2O in kg	1815.9	13,436.7	651.6	15,904.2	+6.8
	NOx in kg	34.6	121.5	3.9	160.1	+8.2
	nvPM in g	65.5	73.8	0.5	139.8	+8.5
V-3	Fuel in kg	1423	9386	468	11,277	−6.3
	CO2 in kg	4492.9	29,634.1	1477.6	35,605.0	−6.3
	H2O in kg	1760.6	11,612.2	579.0	13,951.8	−6.3
	NOx in kg	34.9	97.6	3.0	135.6	−8.3
	nvPM in g	67.2	42.3	0.5	110.0	−14.6

Table 6. Fuel consumption and emissions along the design mission (*neo*) and difference to the reference. Advantageous change: green, disadvantageous change: red.

Vehicle	Component	Departure	Cruise	Approach	Total	%	% V-R
V-R-g	Fuel in kg	1282	7997	373	9652	(−)	−11.6
	CO2 in kg	4046.2	25,249.3	1178.3	30,474.0	(−)	−11.6
	H2O in kg	1585.5	9894.0	461.7	11,941.3	(−)	−11.6
	NOx in kg	28.5	100.5	3.2	132.2	(−)	−18.8
	nvPM in g	99.7	146.1	3.4	249.2	(−)	48.1
V-2-g	Fuel in kg	1363	8482	410	10255	6.3	−7.2
	CO2 in kg	4304.3	26,778.7	1295.9	32,379.0	6.3	−7.2
	H2O in kg	1686.7	10,493.3	507.8	12,687.8	6.3	−7.2
	NOx in kg	29.3	109.4	3.7	142.4	7.7	−15.3
	nvPM in g	99.7	166.6	3.6	269.9	8.3	52.2
V-3-g	Fuel in kg	1219	7450	357	9027	−6.5	−14.3
	CO2 in kg	3849.7	23,521.2	1128.5	28,499.0	−6.5	−14.3
	H2O in kg	1508.5	9216.9	442.2	11,167.6	−6.5	−14.3
	NOx in kg	27.6	87.2	3.0	117.9	−10.8	−20.9
	nvPM in g	98.2	82.0	3.3	183.5	−26.4	46.1

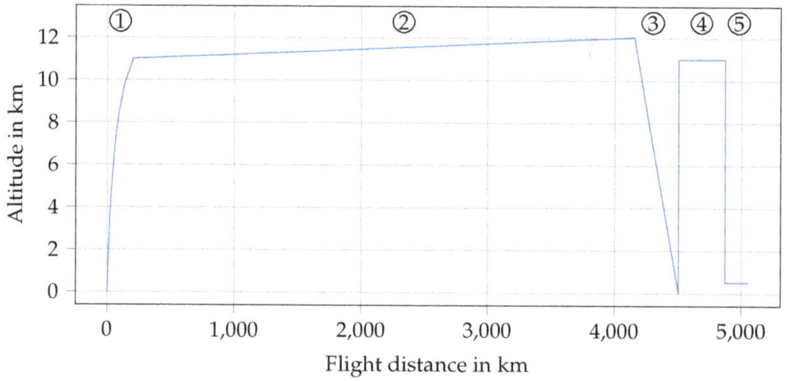

Figure 6. Mission profile for the design mission, including: departure (1), cruise (2), approach (3), diversion (4), loiter (5).

Then the aircraft noise is assessed by evaluating the contour areas of predefined SEL levels (Tables 7 and 8), which was chosen because of its easy translation into $L_{A,eq}$ levels, which are the foundation of noise prediction zones in Germany. For a more detailed evaluation, the contour plots (Figures A1 and A2) are also presented together with the corresponding flight trajectories (Figures 7–10) and the A-weighted sound pressure level of a sideline observer track, 1000 m lateral to the groundtrack (Figures 11–14). The lateral displacement of 1000 m was chosen because the advanced shielding of the V-2 and V-3 designs is most effective directly beneath the flight path but loses its efficiency quickly in the lateral direction. Typically, air traffic routing is adapted to an airport's surroundings to avoid direct flyover events of highly populated areas. So, the selected observer locations are more representative compared to observers aligned directly under the flight path. Moreover, the unshielded levels are plotted for departure, where engine noise dominates, in order to give a better noise understanding of the effectiveness of the shielding architecture.

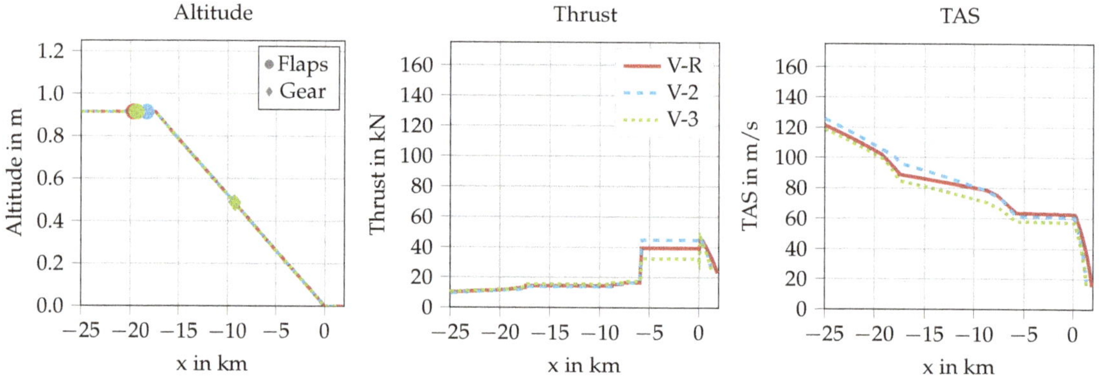

Figure 7. Approach trajectories: current engine option (BPR = 6).

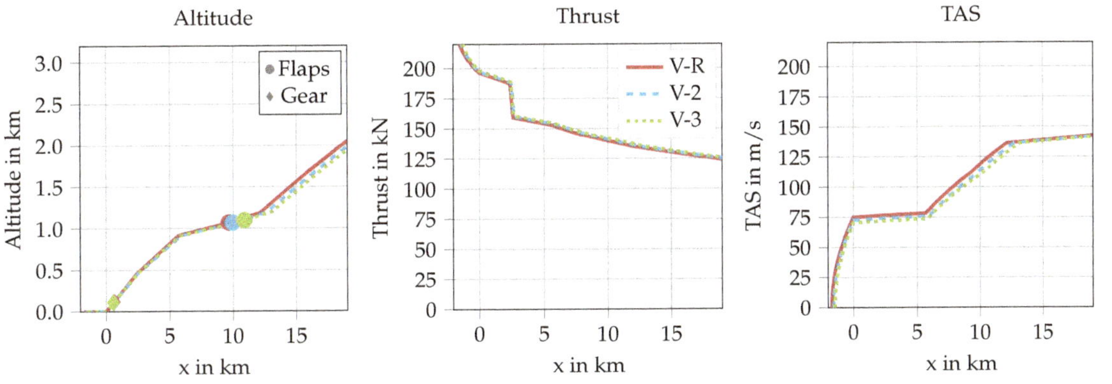

Figure 8. Departure trajectories: current engine option (BPR = 6).

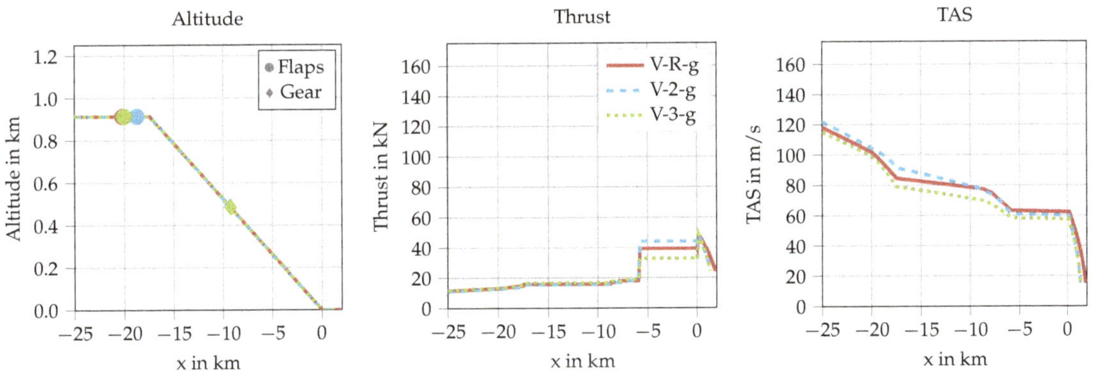

Figure 9. Approach trajectories: new engine option (BPR = 12).

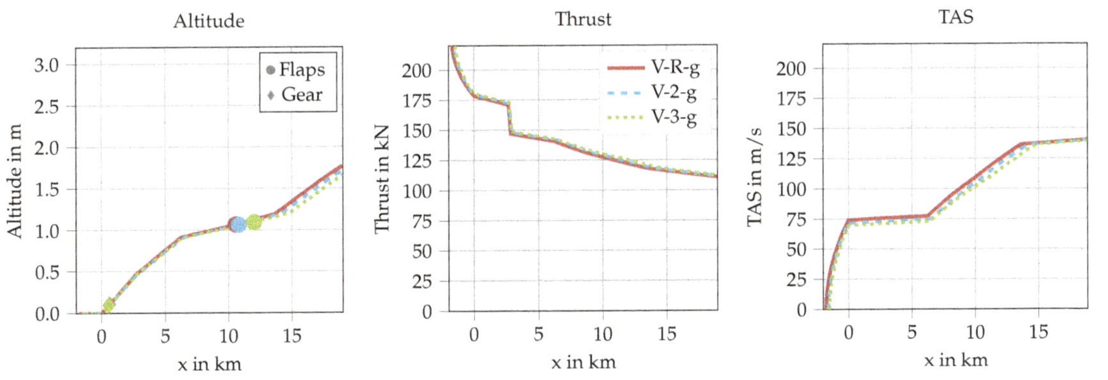

Figure 10. Departure trajectories: new engine option (BPR = 12).

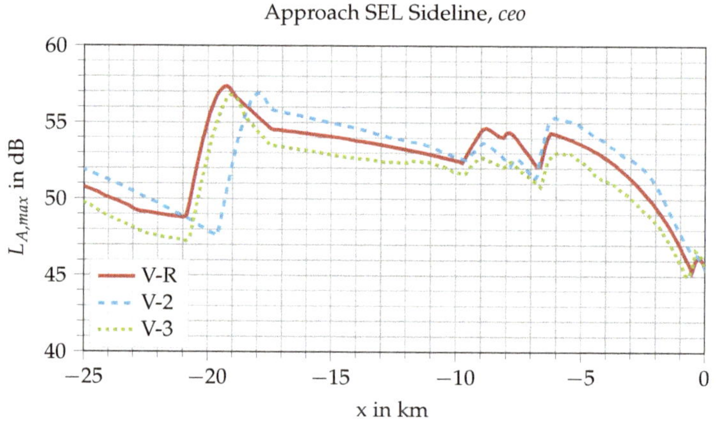

Figure 11. $L_{A,max}$ level 1000 m sideline during approach (*ceo*).

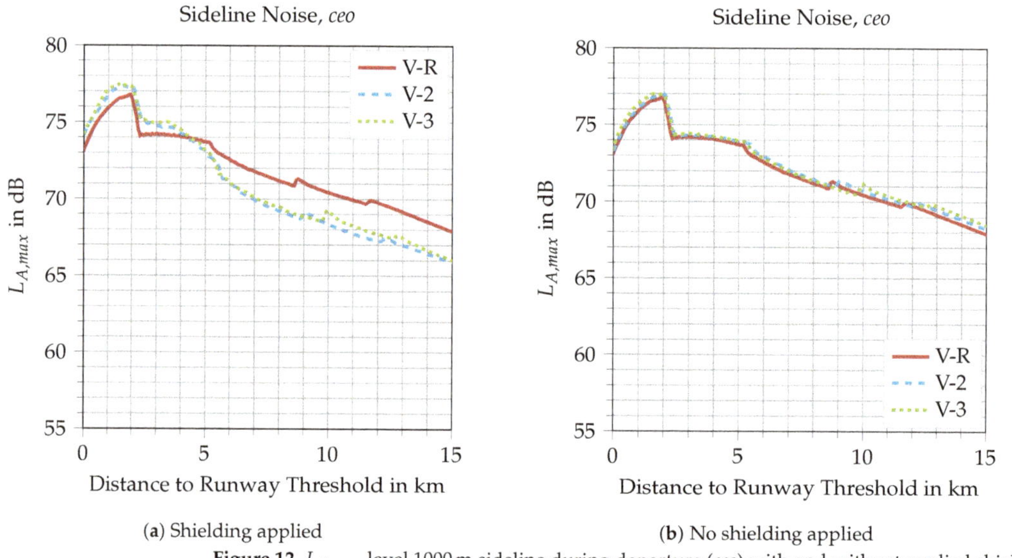

Figure 12. $L_{A,max}$ level 1000 m sideline during departure (*ceo*) with and without applied shielding.

Table 7. Sizes of iso-contour areas in km^2, *ceo*.

SEL Level in dB	V-R	V-2	V-3
	approach		
75	12.18	10.97	7.45
80	4.57	3.75	2.57
85	1.87	1.17	1.00
90	0.18	0.11	0.12
	departure		
75	–	–	113.52
80	66.60	53.15	51.17
85	28.67	22.90	21.93
90	11.34	8.32	8.27

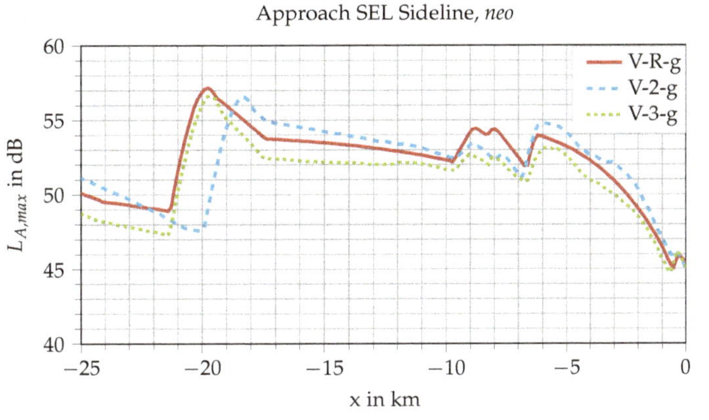

Figure 13. $L_{A,max}$ level 1000 m sideline during approach (*neo*).

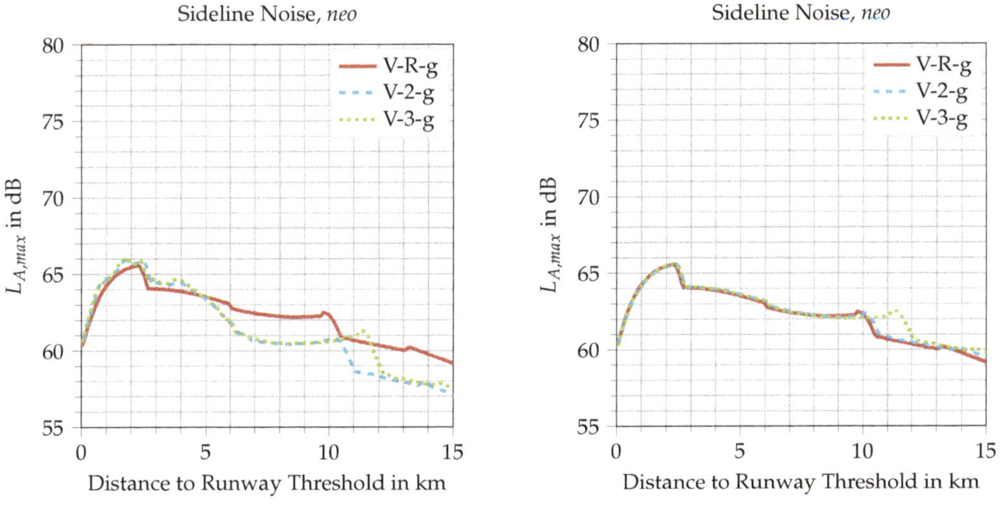

(**a**) Shielding applied (**b**) No shielding applied

Figure 14. $L_{A,max}$ level 1000 m sideline during departure (*neo*) with and without applied shielding.

Table 8. Sizes of isocontour areas in km^2, *neo*. Relative difference to *ceo* counterparts in brackets (advantageous change: green, disadvantageous change: red).

SEL Level in dB	V-R-g	V-2-g	V-3-g
	approach		
75	10.21 (−16.2)	8.57 (−21.9)	6.81 (−8.6)
80	3.26 (−28.7)	2.38 (−36.5)	2.14 (−16.7)
85	1.57 (−16.0)	0.84 (−28.2)	0.72 (−28.0)
90	0.27 (50.0)	0.14 (27.3)	0.19 (58.3)
	departure		
75	30.34 (–)	21.43 (–)	21.88 (−80.7)
80	11.76 (−82.3)	7.01 (−86.8)	6.69 (−86.9)
85	4.76 (−83.4)	3.18 (−86.1)	3.02 (−86.2)
90	2.11 (−81.4)	1.46 (−82.5)	1.33 (−83.9)

5.1. Aircraft Design

The most important parameters of the six aircraft designs are listed in Table 3. The geared turbofan was designed with a slightly higher static thrust at sea level (121 vs. 117 kN) to avoid implications caused by the higher thrust lapse of the GTF. Because of the higher bypass ratio, these engines are more sensitive to the decreasing air density with increasing altitude, resulting in a higher thrust loss. The modern GTF design can sustain higher temperatures at turbine entry (TET) and achieves a higher overall pressure ratio (OPR), leading to a higher efficiency, e.g., less thrust specific fuel consumption (TSFC). It also has a larger fan diameter and therefore a smaller fan pressure ratio (FPR).

The integration of the GTF also results in an overall heavier aircraft, caused by the heavier propulsion system. It is noticeable that the V-2 design has a lighter wing than the V-R and V-3, which is the result of the smaller wing span. The heavier wing of the V-3 is the result of the forward sweep and larger wing span, requiring the wing to sustain higher bending moments [66]. The fuselage of the V-2 and V-3 are longer than the V-R's (39.90 m/40.25 m vs. 37.70 m) in order to reduce the empennage size and reduce trim drag, again leading to a heavier fuselage compared to the V-R. Due to the slightly different engine position, the vertical tail size of V-3 is smaller and lighter than the V-2's tail. The V-3 also uses a different engine pylon, resulting in a slightly heavier weight of the propulsion system for the V-3 than for the V-2. Although the GTF designs are heavier than their counterparts, their landing gear weight is lower. This is caused by less necessary landing weight. Because the GTF burns less fuel, the required reserve fuel is lower, which outweighs the increase in OEW.

In terms of aircraft performance, the V-3 design shows the advantage of the FSW. While cruises C_L are on a comparable level, the L/D of the V-3 is higher compared to those of V-R and V-2, which is the result of the lower drag of the extended laminar flow. The maximum C_L in landing configuration shows opposing trends for the high and low engine position. Because the shown value is calculated in the trimmed condition, the overall weight and center of gravity position influences the maximum lift due to different deflection angles of the elevator. In the case of the top mounted engine, the *neo* designs require a higher deflection angle compared to their *ceo* counterparts. For the low mounted engines of the V-R, the effect is the opposite and therefore also the influences the maximum lift coefficient.

5.2. Emissions and Fuel Burn Along Design Mission

Each vehicle was simulated with the defined design mission. Emission indices were multiplied by the actual fuel consumption for each flight segment and integrated along the entire trajectory in order to yield the overall emission. For the six vehicles, the estimated fuel consumption and emission values are listed in Tables 5 and 6. Additionally, the difference to the reference design V-R is also shown in Table 6.

The fuel consumption shows significant advantages of the FSW design over the conventional wing designs, for both the current engine option and the new engine option, as a result of the lower cruise L/D. The fuel consumption of the V-2(-g) configuration increases compared to the V-R(-g) by 6.8% for the current engine option and 6.2% for the new engine option. It decreases by 6.3% and 6.5%, respectively, for the V-3(-g) compared to the V-R(-g). This trend is mostly due to the change in cruise L/D, which increases by 5.6% (*ceo*) and 5.3% (*neo*) for the V-2 and decreases for the V-3 by 8.6% and 8.9%, respectively. Using the BPR 12 GTF engine reduces fuel consumption by approximately 20% for all designs, compared to their *ceo* counterparts.

The improved fuel consumption directly translates to a lower CO_2 and H_2O emissions, which are calculated for stoichiometric combustion [39]. The emissions of NO_x and nvPM are calculated from the engine map based on thrust, airspeed, and altitude based on the trajectory of the design mission. The total NO_x and nvPM emissions for the conventional wing design V-2 exceed those of the reference design (V-R) by 8.2% (NO_x) and by 8.5% (nvPM). For the new engine option V-2-g the emissions in comparison to the reference

V-R-g decrease as well by 8.3% and 14.6%, respectively. The FSW design(V-3) with the current engine option V-3 achieves savings of 8.3% in terms of NO_x emissions and 14.6% in terms of nvPM emissions. The new engine option achieves even greater savings in comparison to the reference V-R-g, 10.8% and 26.4%, respectively. When comparing the nvPM emissions of the *ceo* and *neo* designs, it is noticeable that all *neo* designs have a significantly higher nvPM emission. This is a result of the combustor design choice used for this version of the GTF engine, which is not optimized for low nvPM emissions. In future revisions of the engine design, the combustor design will be updated and nvPM emissions are expected to decrease below *ceo* levels.

5.3. *Aircraft Noise Assessment along Approach and Departure*

The noise assessment methods used for this study are listed in Table 2. Each vehicle is simulated for its individual approach and departure trajectory, i.e., the resulting operating conditions along the simulated flight are defined by the individual flight performance of each aircraft. At this point, no modification or low-noise optimization to the flight procedures is investigated. The vehicles are simulated according to fixed procedure definitions, i.e., a low-drag–low-power approach and a standardized NADP-1 departure. Noise footprints or isocontour areas are evaluated and compared for the different vehicles under consideration along their individual flight trajectories. The flight trajectories show visible differences mainly in velocity and thrust profile due to changes in flight performance of each individual aircraft. The resulting approach and departure trajectory of the three vehicles with the current engine option are depicted in Figures 7 and 8, respectively. Results for the new engine option are depicted in Figures 9 and 10. Depicted are the three most relevant parameters for the noise contribution on ground, altitude, thrust, and true air speed (TAS).

5.3.1. Current Engine Option (BPR = 6)

Approach: The approach trajectories are depicted in Figure 7. The V-3 vehicle is specifically designed for reduced flight velocities along with the final approach segments. As a consequence, a longer distance is required for decelerating the aircraft to its final approach speed. The reduced flight velocity is clearly visible in the velocity profiles for the V-3 aircraft compared to the other two vehicles.

These effects have direct implications on the predicted noise contours shown in Figure A1. Noise contours are shown for the final segment of the approach procedure. For x = 25 km all aircraft are already in the horizontal flight segment but prior to high-lift deployment. The corresponding changes in noise contour areas are shown in Table 7. Significantly reduced areas are predicted for the V-2 and V-3 vehicle compared to the other vehicles due to the noise shielding (V-2 and V-3) and reduced flight velocity along the trajectory (V-3).

Data are extracted from the noise contour plots to evaluate the effects in more detail. Noise levels along the 1000 m sideline are depicted in Figure 11 and highlight the influence of the flight speed, comparing V-2 and V-3.

The effects of the initial high-lift deployment at around 20 km prior to the runway threshold are clearly visible and lead to an increase of up to 10 dB along the sideline, depending on the design. It is also visible that the V-2 designs allow a comparably late high-lift deployment, resulting in a local difference of up to 9 dB. At approximately 9.5 km before touchdown, a double peak of 2–3 dB in the sideline levels is visible, caused by the final high-lift and landing gear deployment. Finally at around 6–7 km before touchdown the increase in thrust to maintain the approach velocity results in the third rise in sideline levels of 2–3 dB, depending on the design.

In conclusion, the noise shielding architecture of the V-2 and V-3 show the strongest effect whenever the engine noise is dominant, i.e., visible in the 80 dB and 85 dB contours along the very final flight segment of the approach. This was also expected because at this point in the approach the engine operates at the highest thrust setting in the final approach segment—even if still significantly lower than during takeoff. This is also the reason why shielding is not very effective here. The V-2 vehicle profits the most from its reduced landing gear noise, which becomes noticeable from 10 km before touchdown, but is mostly overshadowed by the increased engine noise because of the higher thrust setting. Significantly more advantageous is the effect of the lower approach speed of the V-3 vehicle. Noise levels are reduced throughout the entire approach trajectory, compared to the V-R and the V-2 vehicles, resulting in a reduction in the SEL isocontour areas of up to 46% for V-3. In comparison, the V-2 vehicle only achieves a reduction of up to 39%. In particular, the areas further away from the airport benefit more. This confirms the hypothesis that modifications of the flight performance can enable quieter flight operation.

Departure: The departure trajectories of the *ceo* designs are shown in Figure 8. The NADP-1 procedure provides a thrust cutback at 1500 ft altitude (approx. 2.5 km after take-off), which was chosen to be 6% below takeoff thrust. At 3000 ft the aircraft enters an acceleration profile, where the climb rate is reduced and the available thrust is mostly converted into speed. After reaching 250 kts (129 m/s) indicated airspeed (IAS)–TAS, as depicted in Figure 8 is higher—the aircraft continues the climb with constant N1 as steep as possible. Significant noise reduction for the low-noise vehicles is achieved along the entire observer sideline, as depicted in Figure 12. Comparing Figure 12a,b, the influence of the shielding architecture becomes obvious. During departure, both low-noise designs reach reductions of 2–3 dB of the sideline levels, which would not be possible without the dedicated shielding design of the vehicles. The differences along the sideline between the V-2 and V-3 can mostly be attributed to the different points in time, where the flaps are retracted, and to the 3.5 m/s lower flight speed of the V-3 design, compared to the V-2. The 3 dB drop of the sideline levels 2 km after takeoff, observable for all designs, is caused by the thrust cutback which is defined in the NADP-1 departure procedure, while the steep rise before is the result of the diminishing influence of the ground effect due to the increasing altitude.

The SEL isocontour area in Figure A1 shows the differences for the three vehicles for their individual flights. Compared to the V-R, both low-noise vehicles show significantly lower noise levels, i.e., especially close to the flight ground track, due to maximum fan noise shielding into this area because of the fuselage–wing engine installation. This engine arrangement results in excessive noise shielding toward observers directly below the aircraft, where a reduction of 10–12 dB in the $L_{A,max}$ can be achieved. This effect becomes especially noticeable from the dovetail shape in the contour area directly under the flight path (y = 0 km), as seen in Figure A1b,c. This is mainly due to extensive directivity characteristics of the noise emission because of shielding. This advantage is reduced with increasing distance to the flight ground track which is clearly visible in the lobe-shaped contours.

In conclusion, the effect of the selected noise shielding architecture results in significant noise reduction during departure. The SEL isocontour areas can be reduced by around 20–25% for the V-2 and around 23–25% for the V-3. Both vehicles, i.e., V-2 and V-3, show excessive reduction close to the flight ground track. The advantage is diminished with increasing distance to the flight ground track. After all, the advantageous flight performance of the V-3, i.e, the higher L/D and higher $C_{L,max}$, does not result in significant advantages in terms of noise emission because of its higher takeoff mass.

5.3.2. New Engine Option (BPR = 12 GTF)

Approach: During approach, the new engine option is not expected to make a relevant difference. The engine operates at low power settings and the dominant noise source is the airframe, most sensitive to approach speed.

Comparing the flight trajectories of the designs with BPR 12 GTF engines (Figure 9) and the trajectories of the BPR 6 designs (Figure 7), the BPR 12 designs can achieve a roughly 5 m/s lower approach velocity than their BPR 6 counterparts at the same distance to the runway (compared 19 km before touchdown, after the first flaps are set). This is an effect of the higher minimum thrust setting of the *neo* engines, which leads to a longer and therefore earlier deceleration segment (not visible in the depicted altitude profile).

The lower approach speed results in lower airframe noise levels. The SEL contour areas can be reduced by about 30–50% each (see Tables 7 and 8). Evaluating the levels of the sideline observers support these findings, the new engine option results in an approximately 2 dB lower $L_{A,max}$ level during most of the approach phase.

Comparing the BPR 12 designs among themselves in terms of contour size yields results similar to those of the BPR 6 designs. The reference design leads to the largest contour areas, while the V-2-g design as well as the V-3-g design are able to reduce the contour areas significantly by 20–40% compared to the reference V-R-g. Compared to their lower bypass ratio counterparts, which reduce the contribution of the engine to the overall noise impact on the ground, the difference is smaller and can even lead to larger contour areas for SEL levels above 90 dB (Tables 7 and 8).

Departure: During departure, the geared turbofan engines with ultra-high bypass ratios operate at high thrust and the airspeed is lower than during approach, hence the engine is the most dominant noise source.

In comparison to the *ceo* variants, the evaluation of the sideline observers demonstrates a strong reduction in noise levels of more than 10 dB at peak levels for the reference design as well as the low noise variants. Both of the low noise designs V-2-g and V-3-g yield significantly reduced contour sizes compared to the reference V-R-g. The heavier FSW design leads to slightly larger 75 dB SEL contours than the V-2-g design due to its earlier acceleration segment. The contour also shows the influence of the engine noise shielding close to the flight ground track, which is still relevant even for the quieter *neo* design. In Figure A2c a 'hole' with reduced noise levels is visible in the SEL 75 contour just before the flaps are retracted 4–9 km after takeoff. In contrast to the V-2-g the 75 dB contour of the V-3-g closes again at around 10 km after takeoff. The effect occurs directly below the flight path due to the extensive shielding of the engine noise and the broader directivity pattern of the V-2 design compared to the V-3.

While the *neo* designs yield overall reduced engine noise, the shielding effectiveness starts to decrease, noticeably at the end of the acceleration segment. On the one hand, as depicted in Figure 14, the shielded $L_{A,max}$ levels experienced by the sideline observers (Figure 14a) exceed the reference aircraft for the V-3. Due to the comparable late flap retraction, airframe noise becomes partially relevant in this area. On the other hand, the difference to the unshielded values (Figure 14b) during the acceleration segment is within 2 dB of the *ceo* levels.

6. Discussion

For an initial assessment of the designs, certain parameters were chosen and evaluated, describing aircraft performance, noise impact, and atmospheric emissions. In Figure 15 those parameters are shown for the *ceo* engine and in Figure 16 for the *neo* designs, each compared to the reference design V-R or V-R-g, respectively. Evidently, the fuel consumption of the V-2 designs increases compared with the respective reference designs, which is consistent with the findings in [31,32] and the overall fuel-savings for the *new engine option* of 20 % does as well. Nevertheless, an 6.8 % increase in fuel burn for the V-2 (Figure 15) appears comparatively high but can be attributed to the lower installed engine thrust and the higher cruise speed of the current designs. This also applies for the cruise L/D, which decreases for both, the V-2 and V-2-g, by 5–6%. Ultimately the reason for this is the increased design Mach number, compared to the original design [31]. Since the foremost purpose of this design was a significant system noise reduction, this shortcoming was accepted. The maximum C_L in landing configuration increases along with the drag, mainly because of the reduced wing sweep. The operating empty weight remains almost the same for the V-2 but slightly decreases for the *neo* design, resulting from the lower wing weight due to design and lower fuel weight and therefore lower structural load.

For the V-3 and V-3-g, the fuel consumption was reduced significantly by the higher cruise lift-to-drag ratio [66,67], which in turn is mostly the result of significantly reduced drag. For the V-2 designs, the maximum C_L also increases due to the reduced wing sweep. In this case, the increase in the total drag is avoided by the extended laminar flow over the wing. The operating empty weight of both FSW designs increases due to additional wing weight. However, the results of Seitz et al. [67] indicate that even a reduction in the OEW could be achieved in further design optimization loops that have not been in the scope of this study.

Both novel designs reduce the noise impact significantly during approach and departure compared to the reference, e.g., as demonstrated for the 80 dB SEL contour area. Looking at all contour areas, it is noticeable that the 80 dB SEL area is particularly sensitive to noise shielding. This area has been specifically selected as a design objective because it is the largest closed contour area available for a comparison between *ceo* and *neo* designs. Evaluating the effect of the shielding architecture, the findings correspond well with the results in [12,31], where a reduction of up to 11 dB for departure and up to 3 dB for approach were found, similar to the results of the present study. The 80 dB SEL isocontour area shows, accordingly, significant reductions, compared to the respective reference aircraft. They show a reduction of 26 % for the V-2 and 49 % for the V-3 design during the approach. For departure, the area was reduced by 25 % and 28 %, respectively. The *neo* designs, equipped with GTF, achieve even greater reductions of 26 % (V-2-g) and 26 % (V-3-g) for approach, where especially the V-2(-g) design profits from the reduced engine noise. In case of the V-3(-g) design the engine noise contribution is already on such a low level, that further reductions in the source emissions are not noticeable. For departure both, designs benefit the most from the reduced source emissions of the GTF, hence the isocontour areas can be more than halved by 52 % (V-2-g) and 54 % (V-3-g).

Figures 15 and 16 only show CO_2, H_2O, and NO_x emissions. Since CO_2 and H_2O are calculated for a stoichiometric combustion, both directly scale with fuel consumption. The NO_x and nvPM emissions also change proportional with fuel consumption, except for the V-3-g design, which achieves lower NO_x and nvPM emissions, compared to the V-R-g and V-2-g, because of the different operating condition of the engines.

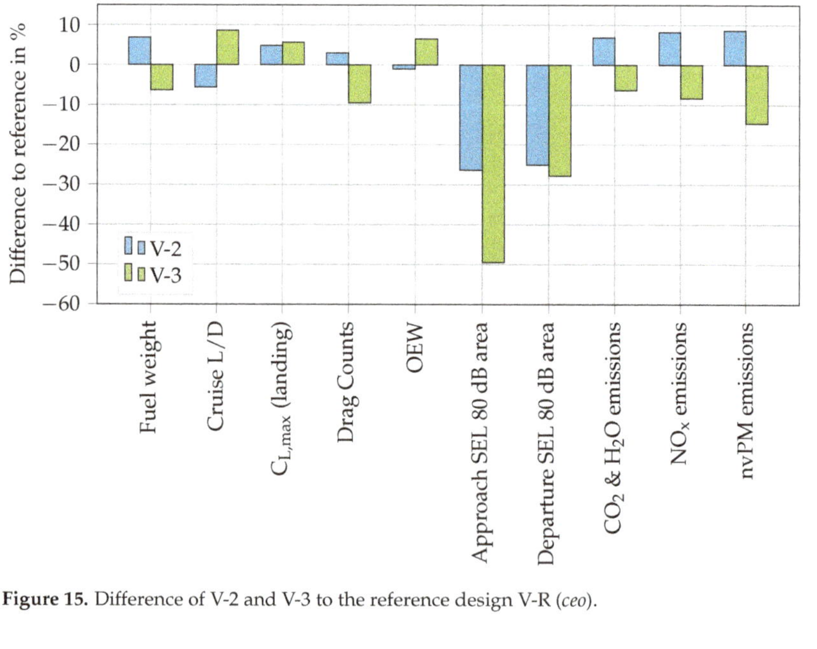

Figure 15. Difference of V-2 and V-3 to the reference design V-R (*ceo*).

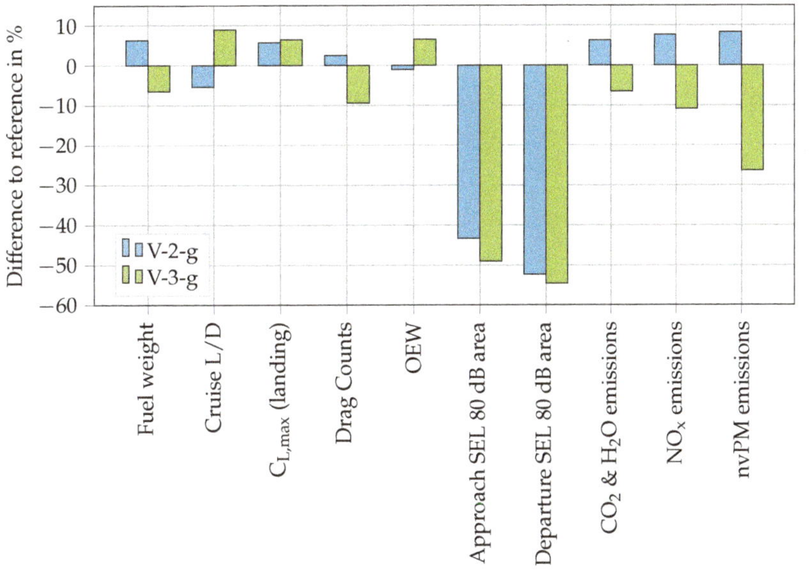

Figure 16. Difference of V-2-g and V-3-g to the reference design V-R-g (*neo*).

7. Summary

The overall goal behind the presented activities is a novel low-noise aircraft-design with reduced environmental impact. The idea behind the V-3 concept, i.e., a tube-and-wing design with a high-mounted FSW and above the wing-mounted engines, is to tailor the overall flight performance to improve its resulting low-noise characteristics. It is designed for a low approach speed and achieves a significant reduction in fuel consumption for improved sustainability and energy efficiency. Furthermore, two different engine options (*ceo* and *neo*), of which *neo* is a geared turbofan with BPR of 12, were investigated. To evaluate the impact of these technologies, the V-3(-g) design is compared to a conventional reference aircraft design V-R(-g) and a dedicated low-noise design V-2(-g), i.e., both equipped with the same engine.

The evaluation shows significant improvements due to shielding, provided by V-2(-g) and V-3(-g) designs, yielding a reduction in the SEL isocontour area of 25–50% relative to the reference aircraft. While the V-2(-g) designs can achieve this only at some cost in terms of their environmental impact, the results of this study show that a FSW design can yield similar or better noise reduction while simultaneously reducing the environmental impact of the aircraft by 5–25%. Furthermore, this study highlights the importance of engine technology on both noise and environmental impact.

In conclusion, this work supports the hypothesis that a low-noise architecture can be combined with fuel-saving technologies for a mid-range tube and wing aircraft, without compromising the direct impact of the selected technologies. An assessment of the general applicability to short and long-range designs is yet to be conducted. It was also found that a truly multidisciplinary and holistic approach is required using methodologies and design systems which simultaneously addresses all of the meanings and modeling requirements in order to achieve a step change in noise and environmental impact.

Future research activities will focus on the modeling and assessment of additional technologies to improve the environmental performance, as described in Section 3.1. Furthermore, the application of novel technologies to reduce airframe noise sources will be investigated in more detail. At the same time, novel SE^2A technologies and aircraft concepts will be developed. Consequently, the presented DLR simulation framework will be upgraded with new models and interfaces to process external data and high-fidelity simulation data as they become available within the SE^2A. Ultimately, the DLR simulation framework will support a decision-making process within the SE^2A in order to identify the most promising technologies and aircraft concepts toward a sustainable and energy-efficient aviation.

Author Contributions: Conceptualization, V.D. and L.B.; methodology, V.D. and L.B.; software, V.D., L.B. and M.P.; validation, V.D., L.B., M.P., E.S. and Z.S.S.; formal analysis, V.D.; investigation, V.D. and L.B.; resources, V.D. and L.B.; data curation, V.D. and L.B.; writing—original draft preparation, V.D. and L.B.; writing—review and editing, V.D. and M.P. and Z.S.S.; visualization, V.D. and L.B.; supervision, L.B., E.S. and Z.S.S.; project administration, L.B.; funding acquisition, L.B. All authors have read and agreed to the published version of the manuscript.

Funding: We would like to acknowledge the funding by the Deutsche Forschungsgemeinschaft (DFG, German Research Foundation) under Germany´s Excellence Strategy—EXC 2163/1- Sustainable and Energy Efficient Aviation—Project-ID 390881007.

Acknowledgments: The authors acknowledge the support of Jason Blinstrub with respect to the flight simulation of the novel vehicles. Furthermore, the authors thank Wolfgang Heinze for his expertise and advice with respect to aircraft design and the support with respect to PrADO.

Conflicts of Interest: The authors declare no conflict of interest.

Abbreviations

The following abbreviations are used in this manuscript:

ALT	Attachment-line transition
BLI	Boundary layer ingestion
CFI	Cross-flow instability
CO	Carbon monoxide
CO_2	Carbon dioxide
FAR	Fuel to air ratio
FLIPNA	Flight simulation code, DLR
FPR	Fan pressure ratio
FSW	Forward swept wing
GHG	Green house gases
GTF	Geared turbofan engine
H_2O	Water
HEP	Hybrid electric propulsion
IAS	Indicated air speed
ICAO	International Civil Aviation Organization
LPP	Lean premixed pre-vaporized
MPT	Multiple pure tone noise
NO_x	Nitrogen oxides
nvPM	non-volatile Particulate Matter
O_3	Ozone
OPR	Overall pressure ratio
PANAM	Overall aircraft noise prediction code, DLR
PrADO	Aircraft design synthesis code, TU BS
RF	Radiative forcing
RQL	Rich quench lean
SAF	Sustainable aviation fuel
SE^2A	Sustainable and energy-efficient aviation
SHADOW	Noise shielding prediction code, DLR
SLS	Sea level static
SO_x	Sulfur oxides
TAS	True air speed
TET	Turbine entry temperature
TLAR	Top-Level Aircraft Requirements
TSI	Tollmien–Schlichting instability

UHC Unburned hydrocarbons
SKM Seat kilometer
Noise metrics
EPNL Effective perceived noise level, [EPNdB]
I Sound intensity, [W/m^2]
L_A A-weighted sound pressure level, [dB]
$L_{A,eq}$ equivalent continuous sound pressure level, [dB]
SEL Sound exposure level, [dB]
Nomenclature
L/D Lift to drag ratio
C_L Lift coefficient

Appendix A. Contour Plots

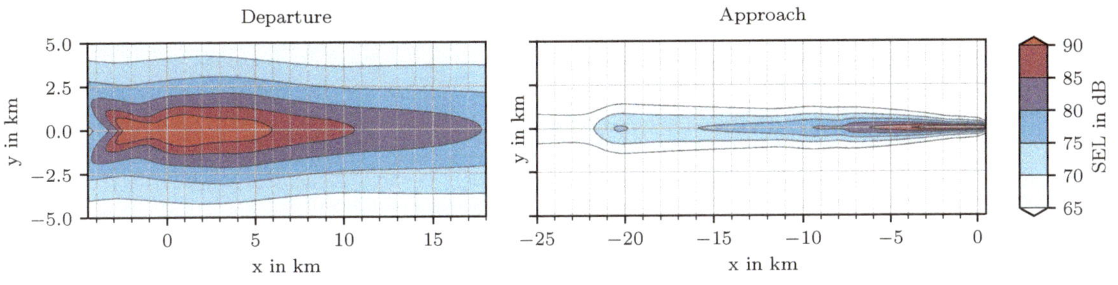

(**a**) V-R

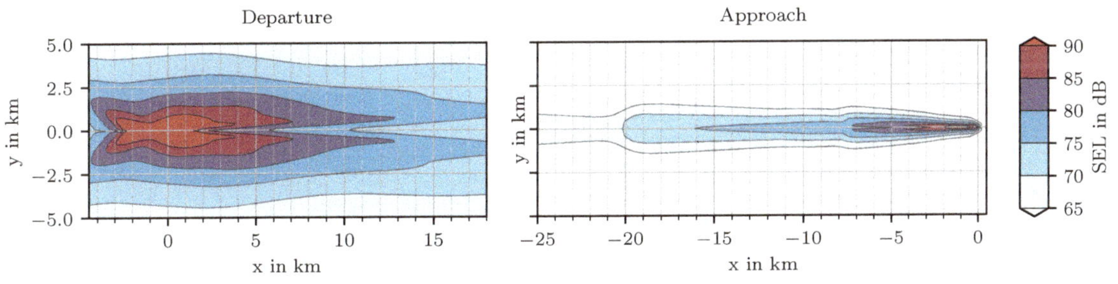

(**b**) V-2

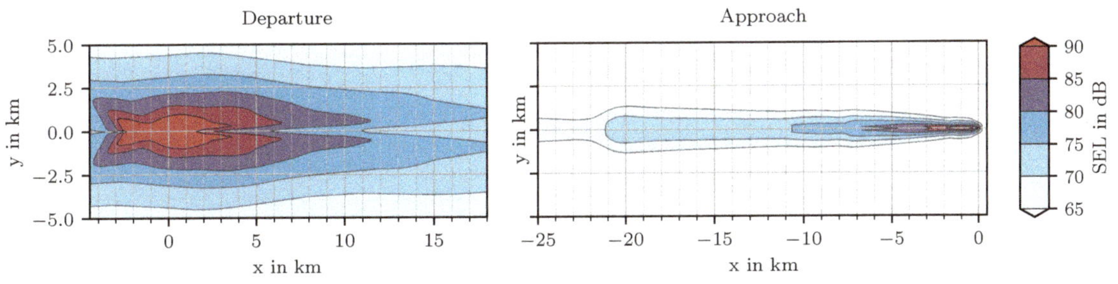

(**c**) V-3

Figure A1. SEL contour of the *ceo* designs.

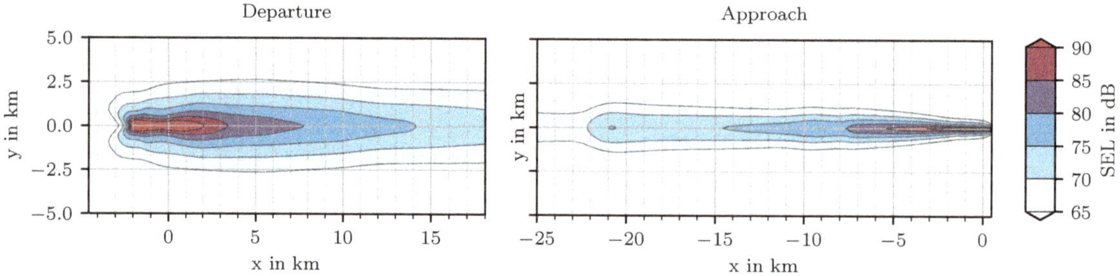

(**a**) V-R-g

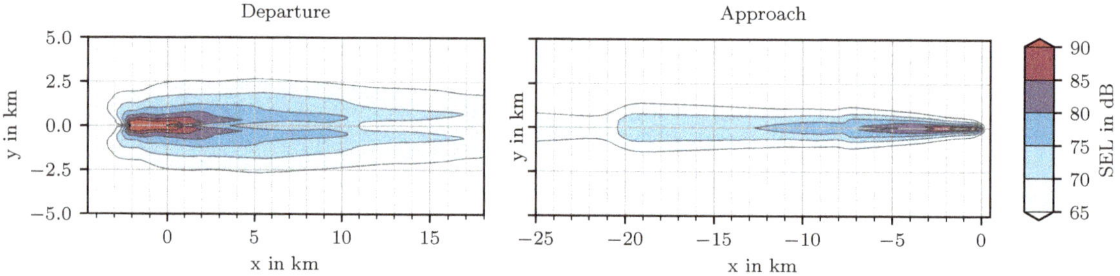

(**b**) V-2-g

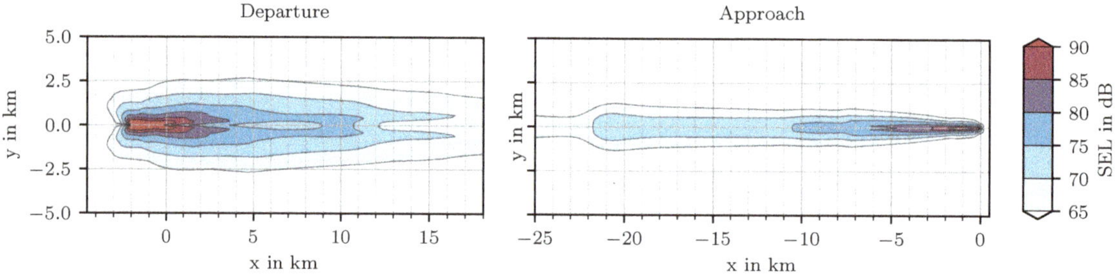

(**c**) V-3-g

Figure A2. SEL contour of the *neo* designs.

References

1. High Level Group on Aviation Research. *Flightpath 2050—Europe's Vision for Aviation*; Technical Report; European Commission: Brussels, Belgium, 2011. [CrossRef]
2. Friedrichs, J. *SE2A—Sustainable and Energy-Efficient Aviation, Cluster of Excellence, Full Proposal*; Technical Report; TU Braunschweig: Braunschweig, Germany, 2018.
3. International Civil Aviation Organization. *Annex 16 to the Convention on International Civil Aviation: Environmental Protection: Volume I—Aircraft Noise*, 5th ed.; AN 16-1, International Civil Aviation Organization (ICAO): Montreal, Canada, 2008.
4. Bertsch, L.; Heinze, W.; Guerin, S.; Lummer, M.; Delfs, J. 10 years of joint research at DLR and TU Braunschweig toward low-noise aircraft design-what did we achieve? *Aeronaut. Aerosp. Open Access J.* **2019**, *3*, 89–105. [CrossRef]
5. Lee, D.S.; Fahey, D.W.; Forster, P.M.; Newton, P.J.; Wit, R.C.N.; Lim, L.L.; Owen, B.; Sausen, R. Aviation and global climate change in the 21st century. *Atmos. Environ.* **2009**, *43*, 3520–3537. [CrossRef] [PubMed]
6. International Civil Aviation Organization (ICAO). *ICAO Global Environmental Trends–Present and Future Aircraft Noise and Emissions*; Technical Report; International Civil Aviation Organization (ICAO): Montreal, Canada, 2019.
7. Graver, B.; Rutherford, D.; Zheng, Z. *CO2 Emissions from Commercial Aviation: 2013, 2018, and 2019*; Technical Report; International Council on Clean Transportation (ICCT): Washington, DC, USA, 2020.
8. Graver, B.; Zhang, K.; Rutherford, D. *CO2 Emissions from Commercial Aviation, 2018*; Technical Report 16; International Council on Clean Transportation (ICCT): Washington, DC, USA, 2019.
9. Kärcher, B. Formation and radiative forcing of contrail cirrus. *Nat. Commun.* **2018**, *9*, 1824. [CrossRef] [PubMed]

10. Isermann, U.; Bertsch, L. Aircraft Noise Immission Modeling. *CEAS Aeronaut. J.* **2019**, *10*, 287–311. [CrossRef]
11. Blinstrub, J. *Immission-Based Noise Reduction within Conceptual Aircraft Design*; Technical Report DLR-FB-2019-12; DLR: Göttingen, Germany, 2019.
12. Bertsch, L.; Clark, I.; Thomas, R.; Sanders, L.; LeGriffon, I. The Aircraft Noise Simulation Working Group (ANSWr)–Tool Benchmark and Reference Aircraft Results. In *25th AIAA/CEAS Aeroacoustics Conference*; AIAA: Delft, The Netherlands, 2019. [CrossRef]
13. Dobrzynski, W. Almost 40 Years of Airframe Noise Research: What Did We Achieve? *J. Aircr.* **2010**, *47*, 353–367. [CrossRef]
14. Pieren, R.; Bertsch, L.; Lauper, D.; Schäffer, B. Improving future low-noise aircraft technologies using experimental perception-based evaluation of synthetic flyovers. *Sci. Total. Environ.* **2019**, *692*, 68–81. [CrossRef] [PubMed]
15. Farokhi, S. *Future Propulsion Systems and Energy Sources in Sustainable Aviation*; John Wiley & Sons, Inc.: Hoboken, NJ, USA, 2020.
16. Burkhardt, U.; Kärcher, B.; Schumann, U. Global Modeling of the Contrail and Contrail Cirrus Climate Impact. *Bull. Am. Meteorol. Soc.* **2010**, *91*, 479–484. [CrossRef]
17. Airbus, Global Market Forecast 2021–2040. Available online: https://www.airbus.com/en/products-services/commercial-aircraft/market/global-market-forecast (accessed on 17 November 2021).
18. Boeing, Commercial Market Outlook 2021–2040. Available online: https://www.boeing.com/commercial/market/commercial-market-outlook/ (accessed on 17 November 2021).
19. Federal Aviation Administration. *FAA Aerospace Forecast Fiscal Years 2021–2041*; Technical Report; Federal Aviation Administration: Washington, DC, USA, 2021.
20. Raymer, D.P. *Aircraft Design: A Conceptual Approach*, 6th ed.; AIAA Education Series; American Institute of Aeronautics and Astronautics Inc.: Reston, VA, USA, 2018.
21. Habermann, A.L.; Bijewitz, J.; Seitz, A.; Hornung, M. Performance bookkeeping for aircraft configurations with fuselage wake-filling propulsion integration. *CEAS Aeronaut. J.* **2020**, *11*, 529–551. [CrossRef]
22. Aigner, B.; Nollmann, M.; Stumpf, E. Design of a Hybrid Electric Propulsion System within a Preliminary Aircraft Design Software Environment. In Proceedings of the Luft- und Raumfahrt—Digitalisierung und Vernetzung: Deutscher Luft- und Raumfahrtkongress 2018, Bonn, Germany, 4–6 September 2018. [CrossRef]
23. Karpuk, S.; Elham, A. Influence of Novel Airframe Technologies on the Feasibility of Fully-Electric Regional Aviation. *Aerospace* **2021**, *8*, 163. [CrossRef]
24. Staggat, M.; Moreau, A.; Guérin, S. Analytical prediction of boundary layer ingestion noise for an integrated turbofan. In Proceedings of the 26th International Congress on Sound and Vibration (ICSV), International Institute of Acoustics and Vibration, Montreal, QC, Canada, 7–11 July 2019; Volume 26, pp. 1304–1311.
25. Koch, A.; Nagel, B.; Grewe, V.; Dahlmann, K.; Schumann, U.; Gollnick, V.; Kärcher, B. Integrated Analysis and Design Environment for a Climate Compatible Air Transport System. In Proceedings of the 9th AIAA Aviation Technology, Integration, and Operations Conference (ATIO), American Institute of Aeronautics and Astronautics, Reston, VA, USA, 21–23 September 2009. [CrossRef]
26. Dahlmann, K. Eine Methode zur Effizienten Bewertung von Maßnahmen zur Klimaoptimierung des Luftverkehrs. Ph.D. Thesis, Ludwig–Maximilians–Universität München, München, Germany, 2011. [CrossRef]
27. Koch, A. Climate Impact Mitigation Potential Given by Flight Profile and Aircraft Optimization. Ph.D. Thesis, Technischen Universität Hamburg-Harburg, Hamburg, Germany, 2013. [CrossRef]
28. Husemann, M.; Schäfer, K.; Schültke, F.; Aigner, B.; Stumpf, E. *A Summary of the Climate Change Mitigation Potential through Climate-Optimized Aircraft Design*; Technical Report; Institut für Luft- und Raumfahrtsysteme (ILR), RWTH Aachen University: Aachen, Germany, 2017. [CrossRef]
29. Grewe, V.; Linke, F. Eco-efficiency in aviation. *Meteorol. Z.* **2017**, *26*, 689–696. [CrossRef]
30. D02 Committee. ASTM D7566-16: Specification for Aviation Turbine Fuel Containing Synthesized Hydrocarbons. Available online: https://www.astm.org/d7566-09.html (accessed on 17 November 2021).
31. Bertsch, L. *Noise Prediction Within Conceptual Aircraft Design*; Technical Report DLR-FB-2013-20; DLR, Deutsches Zentrum für Luft- und Raumfahrt (DLR): Göttingen, Germany, 2013. [CrossRef]
32. Bertsch, L.; Wolters, F.; Heinze, W.; Pott-Pollenske, M.; Blinstrub, J. System Noise Assessment of a Tube-and-Wing Aircraft with Geared Turbofan Engines. *J. Aircr.* **2019**, *56*, 1577–1596. [CrossRef]
33. Heinze, W. Ein Beitrag zur quantitativen Analyse der technischen und wirtschaftlichen Auslegungsgrenzen verschiedener Flugzeugkonzepte für den Transport grosser Nutzlasten. Technische Universität Braunschweig. In *Zentrum für Luft- und Raumfahrttechnik*; Inst. f. Flugzeugbau u. Leichtbau d. TU: Braunschweig, Germany, 1994.
34. Werner-Spatz, C.; Heinze, W.; Horst, P. Improved Representation of High-Lift Devices for a Multidisciplinary Conceptual Aircraft Design Process. *J. Aircr.* **2009**, *46*, 1984–1994. [CrossRef]
35. Lummer, M. Maggi-Rubinowicz Diffraction Correction for Ray-Tracing Calculations of Engine Noise Shielding. In *14th AIAA/CEAS Aeroacoustics Conference*; AIAA: Vancouver, BC, Canada, 2008. [CrossRef]
36. Koch, M.; Bertsch, L. *Engine Noise Source Placement for Shielding Calculation*; Inter Noise; Spanish Acoustical Society—SEA: Madrid, Spain, 2019.
37. Seider, D.; Litz, M.; Schreiber, A.; Fischer, P.M.; Gerndt, A. Open source software framework for applications in aeronautics and space. In Proceedings of the 2012 IEEE Aerospace Conference, Big Sky, MT, USA, 3–10 March 2012; IEEE: Piscataway, NJ, USA, 2012; pp. 1–11. [CrossRef]

38. Reitenbach, S.; Vieweg, M.; Becker, R.; Hollmann, C.; Wolters, F.; Schmeink, J.; Otten, T.; Siggel, M. Collaborative Aircraft Engine Preliminary Design using a Virtual Engine Platform, Part A: Architecture and Methodology. In *AIAA Scitech 2020 Forum*; American Institute of Aeronautics and Astronautics: Reston, VA, USA, 2020. [CrossRef]
39. Rachner, M. Die Stoffeigenschaften von Kerosin Jet A-1. Technical Report. 1998. Available online: https://elib.dlr.de/3185/ (accessed on 17 November 2021).
40. EASA. ICAO Engine Exhaust Emissions Databank. Available online: http://easa.europa.eu/document-library/icao-aircraft-engine-emissions-databank (accessed on 17 November 2021).
41. DuBois, D.; Paynter, G.C. *"Fuel Flow Method2" for Estimating Aircraft Emissions*; SAE Technical Paper Series; SAE International: Warrendale, PA, USA, 2006. [CrossRef]
42. Döpelheuer, A.; Lecht, M. Influence of Engine Performance on Emission Characteristics. In *Gas Turbine Engine Combustion, Emissions and Alternative Fuels*; NATO, Ed.; AC/323(AVT)TP; NATO Research and Technology Organization: Neuilly-sur-Seine, France, 1999.
43. Döpelheuer, A. *Anwendungsorientierte Verfahren zur Bestimmung von CO, HC und Ruß aus Luftfahrttriebwerken*; Technical Report FB-2002-10; Institut für Antriebstechnik: Köln, Germany, 2002.
44. Madden, P.; Park, K. *Methodology for Predicting NOx Emissions at Altitude Conditions from Ground Level Engine Emissions and Performance Test Information*; Technical Report DNS 90713; Rolls-Royce: Westhampnett, UK, 2003.
45. Plohr, M.; von der Bank, R.; Schilling, T. Vergleich des Emissionsverhaltens effizienter Hochbypasstriebwerke mittlerer Schubgröße für den ICAO LTO-Zyklus und Flugmissionen. In Proceedings of the Deutscher Luft- und Raumfahrtkongress, München, Germany, 17–20 November 2003.
46. Bertsch, L.; Sanders, L.; Thomas, R.H.; LeGriffon, I.; June, J.C.; Clark, I.A.; Lorteau, M. Comparative Assessment of Aircraft System Noise Simulation Tools. *J. Aircr.* **2021**, *58*, 867–884. [CrossRef]
47. Dobrzynski, W.; Pott-Pollenske, M. Slat Noise Source Studies for Farfield Noise Prediction. In *7th AIAA/CEAS Aeroacoustics Conference*; AIAA: Maastricht, The Netherlands, 2001. [CrossRef]
48. Pott-Pollenske, M.; Dobrzynski, W.; Buchholz, H.; Gehlhar, B.; Walle, F. Validation of Semiempirical Airframe Noise prediction Method through Dedicated A319 Flyover Noise Measurements. In *8th AIAA/CEAS Aeroacoustics Conference*; AIAA: Breckenridge, CO, USA, 2002. [CrossRef]
49. Rossignol, K.S. Development of an empirical prediction model for flap side-edge noise. In *16th AIAA/CEAS Aeroacoustics Conference*; AIAA: Stockholm, Sweden, 2010. [CrossRef]
50. Rossignol, K.S. Empirical Prediction of Airfoil Tip Noise. In *17th AIAA/CEAS Aeroacoustics Conference*; AIAA: Portland, ON, USA, 2011. [CrossRef]
51. Dobrzynski, W.; Chow, L.; Guion, P.; Shiells, D. A European Study on Landing Gear Airframe Noise Sources. In *6th AIAA/CEAS Aeroacoustics Conference*; AIAA: Lahaina, HI, USA, 2000. [CrossRef]
52. Heidmann, M.F. *Interim Prediction Method for Fan and Compressor Source Noise*; Technical Report NASA TMX-71763; NASA Lewis Research Center: Cleveland, OH, USA, 1979.
53. Stone, J.; Groesbeck, D.; Zola, C. Conventional profile coaxial jet noise prediction. *AIAA J.* **1983**, *21*, 336–342. [CrossRef]
54. Lummer, M.; Hepperle, M.; Delfs, J. Towards a Tool for the Noise Assessment of Aircraft Configurations. In Proceedings of the 8th ASC-CEAS Workshop, Aeroacoustics of New Aircraft and Engine Configurations, Budapest, Hungry, 11–12 November 2004.
55. A-21 Aircraft Noise Measurement Aviation Emission Modeling. *Standard Values of Atmospheric Absorption as a Function of Temperature and Humidity*; Technical Report; SAE International: Warrendale, PA, USA, 1975. [CrossRef]
56. A-21 Aircraft Noise Measurement Aviation Emission Modeling. *Prediction Method for Lateral Attenuation of Airplane Noise during Takeoff and Landing*; Technical Report Aerospace Information Report, AIR 1751; SAE International: Warrendale, PA, USA, 1991. [CrossRef]
57. Felix Greco, G.; Bertsch, L.; Ring, T.P.; Langer, S.C. Sound quality assessment of a medium-range aircraft with enhanced fan-noise shielding design. *CEAS Aeronaut. J.* **2021**, *12*, 481–493. [CrossRef]
58. More, S.R. Aircraft Noise Characteristics and Metrics. Ph.D. Thesis, Purdue University, West Lafayette, IN, USA, 2011.
59. Bertsch, L.; Guerin, S.; Looye, G.; Pott-Pollenske, M. The Parametric Aircraft Noise Analysis Module-status overview and recent applications. In *17th AIAA/CEAS Aeroacoustics Conference (32nd AIAA Aeroacoustics Conference)*; American Institute of Aeronautics and Astronautics: Reston, VA, USA, 2011. [CrossRef]
60. Bertsch, L.; Looye, G.; Anton, E.; Schwanke, S. Flyover Noise Measurements of a Spiraling Noise Abatement Approach Procedure. *J. Aircr.* **2011**, *48*, 436–448. [CrossRef]
61. Bertsch, L.; Schäffer, B.; Guérin, S. Uncertainty Analysis for Parametric Aircraft System Noise Prediction. *J. Aircr.* **2019**, *56*, 529–544. [CrossRef]
62. Hepperle, M. Environmental Friendly Transport Aircraft. *Notes Numer. Fluid Mech. Multidiscip. Des.* **2004**, *87*, 26–33. [CrossRef]
63. Redeker, G.; Wichmann, G. Forward sweep—A favorable concept for a laminar flow wing. *J. Aircr.* **1991**, *28*, 97–103. [CrossRef]
64. Risse, K. Preliminary Overall Aircraft Design with Hybrid Laminar Flow Control. Ph.D. Thesis, RWTH Aachen, Aachen, Germany, 2016.
65. Seitz, A.; Horstmann, K.H. Design Studies on NLF and HLFC Applications at DLR. In *ICAS 2010*; Grant, I., Ed.; International Council of the Aeronautical Sciences: Nizza, France, 2010; Volume 27.

66. Seitz, A.; Kruse, M.; Wunderlich, T.; Bold, J.; Heinrich, L. The DLR Project LamAiR: Design of a NLF Forward Swept Wing for Short and Medium Range Transport Application. In *29th AIAA Applied Aerodynamics Conference*; American Institute of Aeronautics and Astronautics: Reston, VA, USA, 2011. [CrossRef]
67. Seitz, A.; Hübner, A.; Risse, K. The DLR TuLam project: Design of a short and medium range transport aircraft with forward swept NLF wing. *CEAS Aeronaut. J.* **2019**, *11*, 449–459. [CrossRef]

aerospace

MDPI

Article

Simulation of Landing and Take-Off Noise for Supersonic Transport Aircraft at a Conceptual Design Fidelity Level

Michel Nöding [1,*], Martin Schuermann [2], Lothar Bertsch [1,*], Marc Koch [1], Martin Plohr [3], Robert Jaron [4] and Jeffrey J. Berton [5]

[1] Institute of Aerodynamics and Flow Technology, German Aerospace Center (DLR), 37073 Göttingen, Germany; marc.koch@dlr.de
[2] Aerospace Engineer and Scientist, 22589 Hamburg, Germany; schuermann-martin@gmx.de
[3] Institute of Propulsion Technology, German Aerospace Center (DLR), 51147 Köln, Germany; martin.plohr@dlr.de
[4] Institute of Propulsion Technology, German Aerospace Center (DLR), 10623 Berlin, Germany; robert.jaron@dlr.de
[5] Glenn Research Center, National Aeronautics and Space Administration (NASA), Cleveland, OH 44135, USA; jeff.berton@nasa.gov
* Correspondence: michel.noeding@dlr.de (M.N.); lothar.bertsch@dlr.de (L.B.)

Abstract: The German Aerospace Center has launched an internal project to assess the noise impact associated with supersonic transport aircraft during approach and departure. A dedicated simulation process is established to cover all relevant disciplines, i.e., aircraft and engine design, engine installation effects, flight simulation, and system noise prediction. The core of the simulation process is comprised of methods at the complexity and fidelity level of conceptual aircraft design, i.e., typical overall aircraft design methods and a semi-empirical approach for the noise modeling. Dedicated interfaces allow to process data from high fidelity simulation that will support or even replace initial low fidelity results in the long run. All of the results shown and discussed in this study are limited to the fidelity level of conceptual design. The application of the simulation process to the NASA 55t Supersonic Technology Concept Aeroplane, i.e., based on non-proprietary data for this vehicle, yields similar noise level predictions when compared to the published NASA results. This is used as an initial feasibility check of the new process and confirms the underlying methods and models. Such an initial verification of the process is understood as an essential step due to the lack of available noise data for supersonic transport aircraft in general. The advantageous effect of engine noise shielding on the resulting system noise is demonstrated based on predicted level time histories and certification noise levels. After this initial verification, the process is applied to evaluate a conceptual supersonic transport design based on a PhD thesis with two engines mounted under the wing, which is referred to as aircraft *TWO*. Full access to this vehicle's design and performance data allows to investigate the influence of flight procedures on the resulting noise impact along approach and departure. These noise results are then assembled according to proposed Federal Aviation Agency regulations in their Notice of Proposed Rulemaking, e.g., speed limitations, for Supersonic transport noise certification and the regulations from Noise Chapters of the Annex 16 from the International Civil Aviation Organization in order to evaluate the resulting levels as a function of the flight procedure.

Keywords: aircraft noise prediction; conceptual aircraft design; noise certification; supersonic transport aircraft; NASA STCA; ICAO Annex 16; PANAM; FAA NPRM

Citation: Nöding, M.; Schuermann, M.; Bertsch, L.; Koch, M.; Plohr, M.; Jaron, R.; Berton, J.J. Simulation of Landing and Take-Off Noise for Supersonic Transport Aircraft at a Conceptual Design Fidelity Level. *Aerospace* **2022**, *9*, 9. https://doi.org/10.3390/aerospace9010009

Academic Editor: Wing Chiu

Received: 11 November 2021
Accepted: 20 December 2021
Published: 23 December 2021

1. Resurrection of Civil Supersonic Air Transport

There have been two commercial supersonic aircraft in operation the past century. The Tupolev TU-144 in the Soviet Union and Aérospatiale/BAC Concorde in the western world. For the certification of these supersonic vehicles, the noise certification regulations mainly addressed en route sonic boom noise. Part 36 [1] of the Federal Aviation Agency's

(FAA) rules as well as ICAO Annex 16 Vol. 1 does not provide special standards for landing and take-off (LTO) noise limits for supersonic aircraft. Additionally, most countries ban supersonic flight over land for commercial aviation or at least do not accept a perceptible sonic boom on the ground. Consequently, the sonic boom will remain a major challenge, but LTO noise has to be addressed as well. Up to this day no officially approved noise certification standard for the future generation of supersonic business jets has been defined, i.e., sonic boom and LTO noise. Amid ongoing discussions with respect to novel regulations, several commercial projects aim at bringing supersonic transport aircraft back into service within the next decade, e.g., Boom Technology. Obviously, a major goal of all novel concepts is to tackle the LTO and sonic boom noise challenge. Boom Technology, Inc., known as Boom Supersonic, aims at bringing a commercial airliner into service called "Overture". It is designed to transport up to 88 passengers over a range of 4250 nautical miles (transatlantic capability) and offers the possibility to operate at Mach 1.7 powered by three engines. The aircraft is to be operated with 100% sustainable aviation fuel. No final specifications will be presented here since they are subject to change during the evolution of the final design. Yet, Boom aims at full compliance with strict noise regulations of Chapter 14 of the ICAO Annex 16 by advanced engine technology, high performance aerodynamics, and application of variable noise reduction systems (VNRS). Japan Airlines entered into a still active strategic partnership with Boom Technology Inc. The company put down a deposit for 20 aircraft [2]. Furthermore, United Airlines signed an order for 15 Overture aircraft and 35 options [3]. The companies looking to build supersonic transport aircraft are putting pressure on regulators as their designs advance and they are pushing towards an entry into service. Consequently, the International Civil Aviation Organization (ICAO) has committed itself to a new international standard for noise certification of supersonic aircraft, including flight over land and LTO noise. FAA has proposed a first set of regulations, i.e., Notice of Proposed Rulemaking (NPRM), for the noise certification of next generation supersonic aircraft.

Scope of This Article

The focus of this study lies on the LTO noise and the issue of sonic boom during cruise is not addressed here. Since all supersonic transport (SST) aircraft have to operate at typical airports, understanding and assessing noise generation and ground impact during subsonic flight phases is of utmost importance. Furthermore, other important aspects such as economical implications and climate effects are not in the scope of this study. Two specific SST aircraft concepts are under investigation, i.e., aircraft and engine design, flight performance and operation, and noise prediction. Both concepts are assessed under the proposed NPRM by FAA, which are described below and the Noise Chapters of ICAO Annex 16. The findings of this assessment support rule makers and SST designers in the context of minimum LTO noise for aircraft certification.

2. LTO Noise Challenge

When it comes to LTO noise of supersonic aircraft, some general difficulties and challenges are inherent to SST aircraft design. When compared to conventional aircraft, supersonic aircraft are subject to increased drag, including wave drag, due to their high flight velocities. Therefore, the thrust requirement is derived from cruise flight rather than take-off conditions, resulting in higher specific thrust compared to conventional aircraft. Consequently, due to higher specific thrust and poor propulsive efficiency, supersonic aircraft consume more fuel than conventional aircraft and are more expensive to operate. Significant fuel requirements of such vehicles can furthermore result in increasing vehicle weight. Furthermore, the high specific thrust leads to a high jet velocity and thus high jet noise. Due to the supersonic cruise flight, wing area and airfoil shapes are significantly different compared to subsonic designs. One common solution is the delta wing shape. It will deteriorate the aircraft's capabilities for slow speed approaches. This problem is furthermore intensified by slim supersonic airfoils that complicate any installation of a high-lift system and limit available options. All these aspects result in adverse effects on the LTO noise. The required high specific thrust directly contributes to excessive engine noise during take-off under existing certification regulations, i.e., specified full engine thrust setting along take-off. The increased aircraft weight will negatively influence the flight performance of the vehicle especially during take-off due to reduced climb-out capabilities.

2.1. Towards Novel Certification Regulations

The FAA NPRM on Noise Certification of Supersonic Airplanes proposes to add an appendix C to Part 36 of the noise certification requirements of the United States Code of Federal Regulations (CFR) Title 14 [4] by applying existing NASA ideas [5] to these new SST vehicles. The NPRM contains a detailed description of proposed noise limits for the three individual noise reference measurement points. Limits are defined depending on number of engines and maximum take-off mass of the supersonic aircraft to be certified. Figure 1 summarizes the limits for the individual reference measurement points as proposed in §C36.5(a), C36.5(b) and C36.5(c) of the NPRM, which is identical to the values from Chapter 3 of ICAO Annex 16 Vol. 1. Modifications to Chapter 3 are prescribed in §C36.5(d), i.e., no airplane may exceed the noise limits defined in §C36.5(a–c) at any measurement point. Furthermore, the cumulative margin with respect to the cumulative Chapter 3 level must be at least 13.5 EPNdB. The proposed regulation changes by FAA need to be carefully assessed for such SST vehicles. Certain existing regulations for subsonic aircraft pose significant challenges and leave unsolvable hurdles for any new SST concept, e.g., the requirement of maximum thrust setting during departure prior to the pilot-initiated cutback could result in excessive SST noise at the lateral point. Furthermore, the before mentioned differences in flight performance of SST could require different speed limitations during certification in order to meet any reasonable noise limit. As a consequence, FAA's proposed modifications to existing regulations for subsonic aircraft seem necessary and are under consideration by many research groups with the main focus on departure procedures. Several operational rule changes are under discussion within ICAO, so that SST vehicles stand a chance in any noise certification process. Potential modifications are mainly assessed by NASA for their impact based on existing noise regulations for subsonic aircraft. These NPRM definitions are applied within the presented study to evaluate the calculated levels. In addition, levels are compared to rules of the ICAO Noise Chapters, which are the current standard for noise certification of subsonic aircraft.

M = Maximum take-off mass in 1000 kg	0 – 28.615	28.615 – 35.0	35.0 – 48.125	48.125 – 68.039	> 68.039
Lateral full-power noise level (EPNdB) All aeroplanes	94 EPNdB		80.87 + 8.51 log(M)		96.5 EPNdB
Approach noise level (EPNdB) All aeroplanes	98 EPNdB		86.03 + 7.75 log(M)		100.2 EPNdB
Flyover noise level (EPNdB) 2 engines	89 EPNdB			66.65 + 13.29 log(M)	94.0 EPNdB
Flyover noise level (EPNdB) 3 engines	89 EPNdB	69.65 + 13.29 log(M)			91.0 EPNdB

Figure 1. Noise limits on individual reference noise measurement points as proposed by FAA as a function of number of engines and maximum take-off mass (M). Table from the descriptions in Ref. [4].

2.2. Activities in the Research Community

In the past, the sonic boom has been identified as the major bottleneck for any potential SST product, hence any low-boom technologies would improve the business case. Ever since, the sonic boom reduction is a major topic and challenge. NASA carries out its Low-Boom Flight Demonstration program and tasked Lockheed Martin to build the X-59 experimental plane which is called Quiet SuperSonic Technology (QueSST). The aircraft shall be delivered by 2023 and the produced data will be made available to regulators. The Carpet Determination In Entirety Measurements flight series, or CarpetDIEM [6], is paving the way for the acoustics measurements of the X-59 [7]. In recent years, it became most obvious, that not only the sonic boom but also the noise generated during landing and take-off could possibly be a potential show-stopper for the introduction of novel SST concepts.

Consequently, NASA has launched dedicated research activities with the focus on LTO noise based on a novel SST aircraft concept, i.e., NASA 55t Supersonic Technology Concept Aeroplanes (STCA), as early as in 2017 [8–10].

In Europe, the large H2020 project SENECA has been recently initiated and is funded by the European Union. The focus lies on simulation of SST LTO noise including assessment of gaseous emissions in the vicinity of airports and the impact of supersonic travel on the global climate.

Similar research activities have been launched recently at major research establishments around the world, e.g., Japan Aerospace Exploration Agency (JAXA) [11] and the Central Aerohydrodynamic Institute of Russia (TsAGI) [12,13].

2.3. DLR Research Activities

To specifically address the issue of LTO noise for SST vehicles and to understand proposed modifications to the regulations, DLR has initiated an internal research project. The project Estimation of Landing and Take-off Noise for Supersonic Transport Aircraft (ELTON_SST) joins different DLR institutes and brings together experts from different fields in order to capture all relevant aspects and interdependencies. The work presented in this article is part of the ELTON_SST project.

In ELTON_SST a dedicated simulation process has been established that is described in the next section. In a first step, the core simulation process comprises simulation tools at a low to mid fidelity level but interfaces for the integration of high fidelity simulation results are under development. The core process will be applied to a concept of NASA for verification of the methods, i.e., the STCA [9]. Furthermore, the process is applied to a SST design as described in Ref. [14] for a more detailed assessment. The findings from these initial applications shall provide insights and contribute toward novel regulations for SST noise certification.

3. Simulation Process

An existing process for automated simulation of noise certification of subsonic aircraft described in Reference [15] has been extended to assess future civil supersonic aircraft. Dedicated interfaces to external simulation tools are implemented into the existing process to enable future research activities based on higher fidelity simulation results. Furthermore, the existing process is upgraded to calculate SST certification levels at the three reference points according to the suggested rule changes by FAA [4].

A schematic illustration of the novel process is depicted in Figure 2. The simulation core is already described in detail in Reference [15]. Dashed lines indicate the interfaces to external high fidelity simulation tools. Design aspects as well as operational conditions that have to be provided to these external tools as inputs are generated within the simulation core. Based on this input from the low to mid fidelity tools, complementary simulation on a higher fidelity level can then be initiated to confirm earlier assumptions and results. For example, integration of aerodynamic data from CFD simulations with the DLR TAU code supports results from flight simulation at lower fidelity levels. Ultimately, CFD calculations can provide aerodynamic maps in the form of look-up tables for a more physics-based flight simulation of approach and departure.

The process in its current status enables an automated calculation of certification levels at the level of conceptual design. The results shown and discussed in this study are generated by using the simulation core with additional input from complementary simulation results as described in the following section.

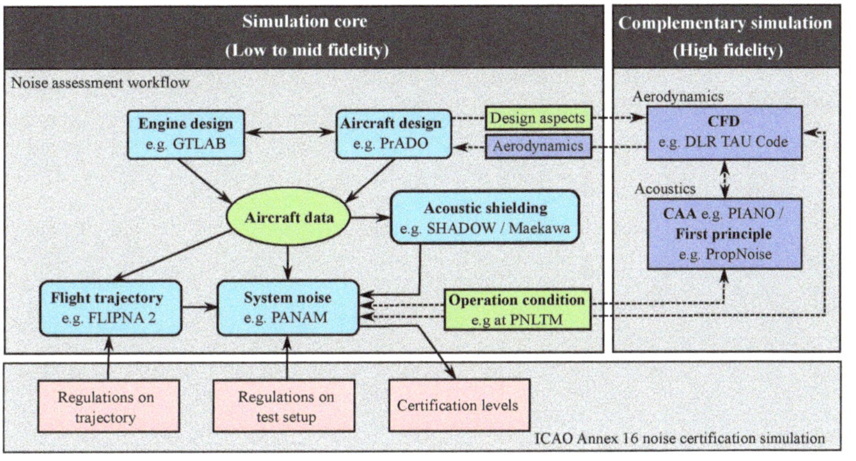

Figure 2. Process chain with simulation core for assessment of noise certification of supersonic aircraft (low to mid fidelity) and interfaces to complementary simulations (high fidelity). Extension of the illustration from Ref. [15].

3.1. Aircraft and Engine Design

All the required aircraft data, e.g., geometries, weights, and engine specifics, are available as direct input or based on engineering judgement for the NASA 55t STCA. For the aircraft *TWO*, the aircraft design is elaborated by the tool PrADO [16], a development of the Technical University of Braunschweig. PrADO consists of individual modules, each of which represents one discipline of the overall design process (e.g., aerodynamics). Through an iterative process, the interactions between different disciplines are calculated, taking into account positive and negative effects of design aspects on other sub-disciplines, i.e., considering so called snowball effects. Modules with different fidelity levels are available, but the individual module can also be replaced by external results of higher fidelity simulations. For example, the presented application example incorporates high-lift and cruise aerodynamics from the DLR CFD code TAU [17]. Furthermore, the PrADO engine design module is replaced by external engine data for this SST study. DLR experts have delivered performance and geometrical data for the SST engines under consideration for direct implementation into the process. The DLR tool GTlab (Gas Turbine Laboratory) was applied to design the SST engine model and to calculate required data [18]. The tool is an interactive, cross-platform simulation and preliminary design environment for aero engines and gas turbines. The object-oriented software concept in C++ and the use of standardized libraries and procedures ensure a high degree of usability, extensibility and flexibility. The GTlab Software Suite consists of the three main modules Performance, Sketchpad and Designer, which are adapted to the application spectrum of Engine Performance and Preliminary Design as needed.

3.2. Flightpath Simulation

For the detailed simulation of approach and departure flight paths a dedicated tool has been developed, i.e., FLIPNA [19]. FLIPNA can directly operate based on detailed engine maps and aerodynamics provided by PrADO and/or GTlab. The underlying flight performance calculation is based on the method described in ECAC Doc. 29. Procedural profile steps for approach and departure can be generated. The resulting trajectory simulated by FLIPNA consists of a series of flight points with parameters describing engine and aircraft operation conditions, e.g., setting of the engine thrust and the high-lift system. The resulting trajectory is physics-based and yields all the required input parameters for a subsequent system noise prediction.

3.3. Noise Shielding

Different software tools are available at DLR to compute the effect of noise shielding from the airframe. Different fidelity levels can be assessed but within this study for a quick estimate on potential noise shielding effects a simple method is applied. The method of Maekawa [20] is applied here with some additional modifications as described in Ref. [21] to reproduce the published NASA results [9,10]. Available and more advanced noise shielding simulation tools could be applied in the future, if access is granted to 3D-geometry data of the NASA 55t STCA. On the other hand, Maekawa can directly be applied to the available and very simplified description of the aircraft. Based on wing planform and engine locations, the Maekawa method can predicts noise shielding effects for all vehicles within this study.

3.4. Aircraft System Noise Prediction

At this point, all the required input data is available to initiate a system noise prediction, i.e., aircraft and engine parameters, prevailing operating conditions (flight trajectories), and potential noise shielding factors. For the system noise prediction the DLR code Parametric Aircraft Noise Analysis Module (PANAM) has been selected. PANAM is a componential and parametrical simulation tool and has been validated against measurements, e.g., Reference [22], and other simulation tools like ANOPP2 (NASA) or CARMEN (ONERA), e.g., Reference [23]. PANAM has been upgraded with additional noise source models in order to better capture the selected SST engines, in particular a modified jet [24] and a combustion noise model [25]. Both were verified against experiment in the original literature. The adjustment of the noise source models promise a more accurate representation of the low bypass engines of the supersonic aircraft during landing and take-off. Overall, appropriate simulation methods are selected and incorporated into the process, i.e., noise source modelling, noise shielding, and propagation attenuation. The applied methods are summarized in Table 1. All required input data for a system noise simulation are automatically generated and provided to PANAM within the process as depicted in Figure 2. PANAM outputs standard single event noise metrics, e.g., EPNL, SEL, and LAMAX, at arbitrary observer locations along approach and departure.

Table 1. PANAM system noise assessment: source models selected for this study.

Noise Source/Element	Model
airframe noise models (airf)	
trailing edge	DLR [22,26–30]
leading edge	DLR [22,26–28]
main landing gear	DLR [22,26–28]
nose landing gear	DLR [22,26–28]
engine noise models (eng)	
fan broadband & tonal	modified Heidmann [31]
jet	Stone 2 [24]
combustion	Emmerling [25]
noise shielding effects (PAA)	
-	SHADOW [32]
sound propagation effects	
-	ISO 9613 [33]

Table 1. *Cont.*

Noise Source/Element	Model
ground attenuation effects	
-	SAE AIR 1751 [34]

3.5. Interfaces to Complementary Simulation

The low to mid fidelity simulation core is supplemented by interfaces to high fidelity tools of DLR, which are indicated by dashed lines in Figure 2. As described before, aerodynamic and engine performance data is directly processed and incorporated into the overall simulation process.

The high fidelity tools are also used to validate the results of the simulation core. This is part of ongoing research activities and not within the scope of the presented study. Mainly, simulated aerodynamic data from low to mid fidelity will be subject to a high fidelity verification for selected operating conditions. Based on initial results from the simulation core, input parameters describing a selected operating condition can directly be processed by the complementary simulation tools. For example, a CFD RANS analysis of the flight conditions attributed with the PNLTM (maximum PNLT along 10-dB downtime) can be initiated. The operating conditions at this point are used as an input and resulting high fidelity predictions can be used for validation purposes. Furthermore, fan noise predictions will be compared to results from the DLR tool PropNoise [35] at characteristic operating points. Furthermore, simplified shielding assessment can be compared to corresponding simulation results from available high fidelity tools.

4. Verification

Due to the lack of available and non-proprietary data for SST vehicle, i.e., including noise data, the available NASA 55t STCA data from Refs. [9,10] was selected for an initial verification benchmark. This activity is understood as a feasibility assessment of the simulation process in its status quo. A more detailed comparison was not in the scope of the presented study but is scheduled for the near future. More work is required to assimilate the simulation methods and input parameters at DLR in order to match the NASA simulations and enable a real benchmark test, e.g., as described in Reference [23] for subsonic transport aircraft.

4.1. Aircraft Design and Noise Shielding Effects

The aircraft geometry is adapted from the published NASA geometries. Rough estimates of wing shape, high-lift concept, landing gear and engine installation have been derived from available information as provided in the available NASA publications [9,10]. The Maekawa noise shielding method [20] was implemented in the DLR simulation process. Based on the rough geometry estimates, a simplified representation of the vehicle can be derived as input for Maekawa, as depicted in Figure 3.

The gray colored area indicates the 55t STCA geometry based on available NASA publications and the colored lines show the selected representation of the 55t STCA geometry within this study. The colored symbols indicate the selected locations of inlet fan noise source and rear engine sources. The sound transmission is predicted for the center engine as well as for the two over-wing mounted engines. Fan noise contribution emitted from the inlet is shielded by the wing. Furthermore, the fuselage is approximated and considered as an additional shielding surface. All areas considered for their potential shielding effect are shown in Figure 3. Note: As the planform of the shielding surface has to be trapezoid for the Maekawa method, small deviations between the NASA 55t STCA representation and the modeled shielding planform are present but have no relevant influence on the predicted shielding. Based on the Maekawa tool, shielding factors are predicted for all emission angles and for the relevant frequency bands. These factors are then directly accounted for in the subsequent system noise prediction and applied to the predicted fan noise emission.

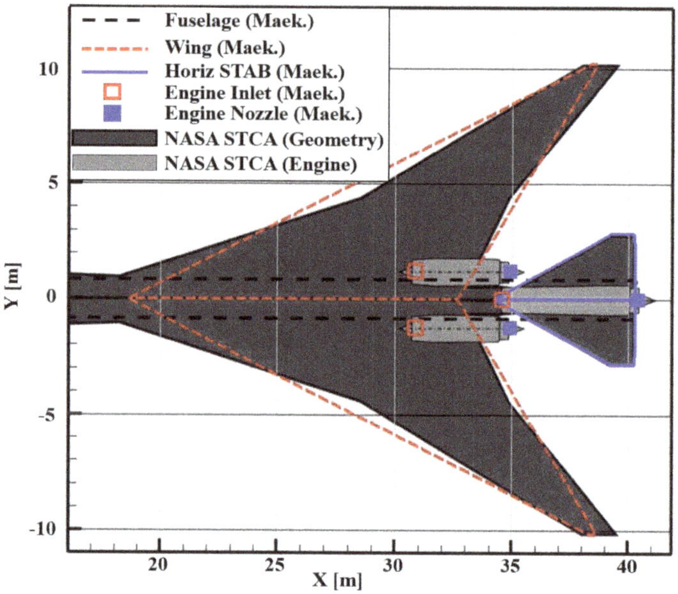

Figure 3. Representation of NASA 55t STCA as input for the noise shielding prediction.

4.2. Engine Redesign

The engine model used in this study was redesigned with GTlab to match publicly available data of the 55t STCA engine design [9]. Similar to the NASA engine model, the high pressure system (high pressure compressor, combustor and high pressure turbine) was based on a pre-existing, generic model of the CFM56-7B engine. The fan and the low-pressure turbine were newly designed as fast rotating components. Due to the lower overall pressure ratio, compared to the underlying CFM 56 engine, no booster is required on the low pressure spool. For the divergent part of the nozzle a variable geometry concept was used to allow for adjusting the engine configuration to different operating conditions.

The DLR engine model is designed to match the thrust of the NASA 55t STCA engines at the three published design points, e.g., cruise design point, end of field, and static take-off. Table 2 shows a comparison of engine performance parameters for the three design points between the NASA 55t STCA engine and the DLR redesign. The data for the cruise design point show very good agreement, with slightly higher values for the pressure ratio of compressor and nozzle and a slightly lower value of the extraction ratio. Some deviations are found for the two sea level operating conditions. These deviations can probably be attributed to the use of generic component maps with presumably different efficiency characteristics, causing deviations in the engine component off-design behavior.

4.3. Flight Procedures

Flight trajectories with all required operational parameters as required for a noise simulation are provided by NASA through one of the authors of this paper. They are used as direct input (see Ref. [10], Figure 3). No dedicated flight simulation with DLR tool FLIPNA is performed at this point due to the lack in aerodynamic data to model the flight operation of the aircraft. Only the provided fixed flight procedures can be processed and alternative flight tracks can not be assessed at this point. From NASA, one conventional approach track and two departure tracks are provided for an assessment and initial comparison. For noise certification a straight flight path is prescribed for both landing and take-off procedures.

Table 2. Engine performance summary compared to data published by NASA [9].

Parameter	Unit	Cruise Design Point NASA	Cruise Design Point DLR	End of Field NASA	End of Field DLR	Static Take-Off NASA	Static Take-Off DLR
Altitude	m	15,240	15,240	0.0	0.0	0.0	0.0
Mach	-	1.4	1.4	0.25	0.25	0.0	0.0
ΔT_{ISA}	K	0	0	15	15	15	15
Net thrust	N	14,812.6	14,812.6	62,897.9	62,897.9	73,929.4	73,929.4
SFC	g/kN/s	26.71	26.74	16.66	17.42	13.57	14.37
BPR	-	2.90	2.90	2.90	2.98	3.00	2.96
T4	K	1833.3	1833.3	1750.0	1789.7	1738.9	1781.0
TET	K	1766.7	1766.7	1688.9	1728.2	1677.8	1719.8
T3	K	805.6	805.5	800.0	813.0	794.4	807.4
Overall Pressure Ratio	-	22.00	22.00	21.00	22.20	21.00	22.50
Fan Pressure Ratio	-	2.00	2.00	1.90	1.97	1.90	1.99
Compressor Pressure Ratio	-	11.20	11.36	11.10	11.38	11.20	11.44
Extraction Ratio	-	1.10	0.94	1.10	0.94	1.10	0.94
Nozzle Pressure Ratio	-	5.90	6.09	1.90	2.01	1.80	1.94

4.3.1. Departure Procedures

The selected departure tracks to investigate the noise certification comprise a standard and an advanced flight procedure. The advanced departure procedure features a VNRS concept as described in Reference [9]. The VNRS concept under consideration is the so called Programmed Lapse Rate (PLR) [36] during take-off. The PLR refers to an additional thrust reduction before reaching the cutback altitude and in addition to the typical pilot-initiated cutback. The main reason for the application of the PLR is to achieve some additional level reduction at the lateral measurement point which is exposed to increasing noise from SST vehicles according to initial noise predictions by NASA [9]. The PLR is automatically initiated after the initial maximum thrust setting during the acceleration phase on the runway and until reaching the obstacle height. An Automatic Take-off Thrust Control System (ATTCS) would be used to initiate the PLR's proposed 10% thrust reduction so that the pilot's workload is not affected and to ensure that the PLR is always flown and certification levels are thus complied with. So the PLR could lead to an additional responsibility of the ATTCS. Another change to the standard flight procedure is a proposed increase in flight speed during take-off in combination with a delayed rotation take-off, which is referred to as high speed climb-out. This modification of flying fast at low altitudes aims to avoid flying in the region of reversed command or in the so-called "back part of the power curve" as already explained in Reference [9]. This consequently reduces thrust requirement, which is associated with a decrease in engine noise.

4.3.2. Approach Procedure

The proposed certification rules for the approach of supersonic aircraft do not differ from the regulations for subsonic aircraft, i.e., the approach will be flown with the typical 3 degree glide path and a selected speed of Vref + 10 kn or ≈84.9 m/s (165 kt) with the maximum landing weight. The selected approach speed depends on the flapped aerodynamics and stall speed, among others. As prescribed, the aircraft is above the approach measurement point at an altitude of 120 m. A flap deflection study was performed by NASA to maximize the approach speed and thus minimize engine thrust. In the relevant approach segment, the aircraft is in stabilized flight condition with landing gear extended and a constant total thrust of 71.1 kN.

4.4. Noise Assessment

Based on the simplified aircraft design, the detailed engine redesign, the Maekawa shielding approximation, and the flight trajectories, the DLR system noise prediction is initiated.

Figure 4 shows PNLT over time for DLR simulations calculated with PANAM as well as NASA data (red) extracted from Reference [10] (Figures 4–7).

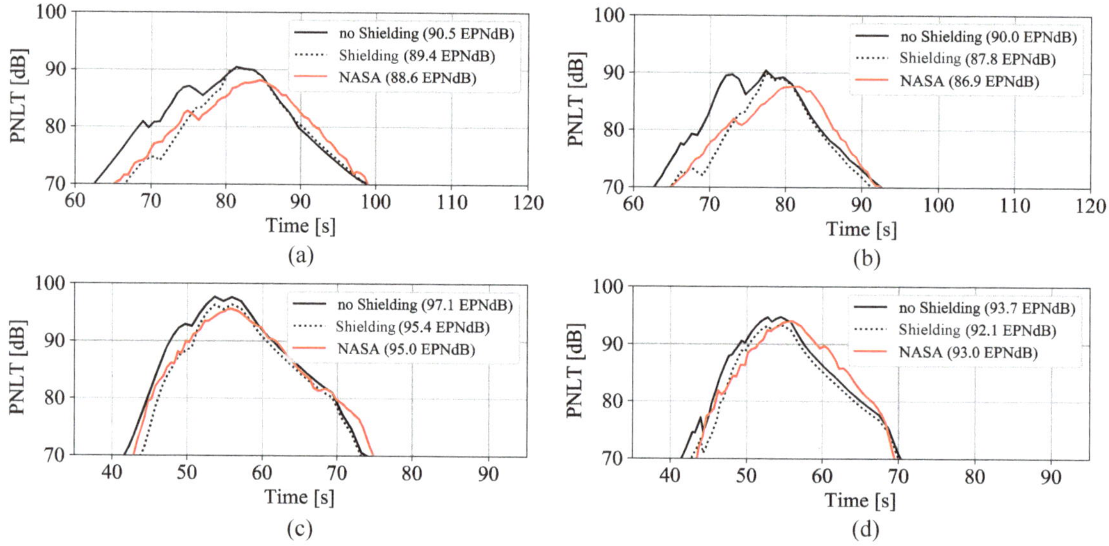

Figure 4. PNLT time history: NASA 55t STCA. Flyover measurement point for standard (**a**), advanced procedure (**b**) and lateral measurement point for standard (**c**) and advanced procedures (**d**). PANAM simulation results with engine noise shielding by using Maekawa (dotted black line), and by assuming no engine noise shielding (solid black line) as well as NASA data (solid red line) are shown.

The flyover and the lateral measurement points are assessed for two different departure procedures. For the lateral assessment, the distances of measurement point to Brake-Release (3756 m for standard procedure and 4130 m for advanced procedure) are fixed and adopted from the NASA study. This location was chosen because at this point the aircraft has climbed to about 1000 ft, where the effects of lateral attenuation have abated, but before the airplane climbs higher and recedes from observers on the ground.

In addition to the level time histories, the certification levels (EPNL) are listed in Table 3 together with limits of noise certification for the NASA 55t STCA study. A reduction in certification levels by applying the advanced procedure for both the flyover and lateral point can be confirmed. It can be seen that the benefits for the lateral point are larger with reductions around 3.3 EPNdB than for the flyover point with around 1.5 EPNdB which are both similar to the reductions in the NASA investigations (see Ref. [9]: 2.0 EPNdB for lateral point and −1.6 EPNdB for flyover). For the approach, where no procedure adjustment is sought, there is a deviation of 2.4 EPNdB from the NASA comparison data. Thereby, the jet noise is no longer the dominant sound source, so that the differences is due to the dominant fan noise. While in the NASA studies a margin on Chapter 4 of 1.6 EPNdB was observed by applying the advanced procedure, here the cumulative result is an exceedance of the required limit by 2.6 EPNdB. Consequently, noise certification to FAA NPRM or even Chapter 14 requires further level reduction through noise mitigation measures.

In order to investigate the shielding effect of the engines, a comparative calculation was also performed without taking noise shielding effects into consideration. An effective reduction in the certification level for the NASA 55t STCA due to over-the-wing engine integration is evident when the certification levels of the two setups are compared. This is particularly visible for the forward fan noise, where a system noise reduction of more than 8.5 dB can be identified under certain operating conditions. An under-the-wing configuration would result in higher levels under the same operating conditions.

Table 3. Simulated certification levels at individual measurement points (flyover, lateral and approach) and cumulative level for NASA 55t STCA. Limits of individual measurement points and cumulative from ICAO Annex 16 and FAA NPRM for Noise Certification of Supersonic Airplanes.

Procedure	Facility	Shielding	Flyover	Lateral	Approach	Cumulative
Standard	DLR	Maekawa	89.4	95.4	98.8	285.5
Advanced	DLR	Maekawa	87.8	92.1	98.8	280.6
Standard	DLR	no Shielding	90.5	97.1	102.4	290.0
Advanced	DLR	no Shielding	90.0	93.7	102.4	286.1
Standard	NASA	Maekawa	88.6	95.0	96.4	280.0
Advanced	NASA	Maekawa	87.0	93.0	96.4	276.4
Noise regulations *			**Flyover**	**Lateral**	**Approach**	**Cumulative**
Limit (Chapter 3)			92.8	95.7	99.5	288.0
Limit (Chapter 4)			92.8	95.7	99.5	278.0
Limit (Chapter 14)			91.8	94.7	98.5	271.0
Limit (FAA NPRM)			92.8	95.7	99.5	274.5

* Caution: additional rules are applicable according to the ICAO Noise Chapters/NPRM, e.g., no trading Refs. [1,37].

In addition, it is shown that without engine noise shielding, there is even an increase in PNLT of 2.5 dB in the maximum of forward directed fan noise (e.g., see Figure 4, flyover advanced at t = 73 s) when the advanced procedure is applied and the engine is operated at lower thrust setting. This is attributed to the effect of relative blade tip mach number on predicted noise levels according to the applied buzz-saw noise model Ref. [38] as already shown in Ref. [19].

A comparison of the PNLTM for the lateral position shows a good agreement with the NASA data when using the Maekawa model. Remaining discrepancies, i.e., flyover in Figure 4 can be attributed to the fan forward noise prediction in combination with the Maekawa shielding. PNLTM/EPNL prediction differences between NASA and DLR are below 2.3 dB/0.9 EPNdB for flyover and approx. 0.8 dB/0.9 EPNdB for lateral location. Unfortunately, the differences for approach are higher. Overall, the agreement is considered satisfying and can be traced back to remaining differences in simulation methods, implementation of the models, and input parameters between NASA and DLR.

The results confirm the benefits of an adapted take-off procedures for the STCA concept. Feasibility of the findings is demonstrated when comparing these findings with available NASA results. A final and detailed tool comparison is still pending in order to answer remaining open questions and to rule out a close agreement on certain result parameters by simple coincidence. Overall it can be concluded, that the novel process chain can reasonably be applied to other preliminary concepts of similar supersonic aircraft designs of which one is presented in the next section.

5. Application on a Twin-Engine Aircraft

Current concepts from manufacturers usually feature under-the-wing engine installations, where advantageous noise shielding through the wing and fuselage cannot be exploited. Therefore, a further configuration is investigated that addresses this aspect in order to estimate the magnitude of noise certification levels of such configurations. Based on an available SST aircraft design [14], the influence of the flight procedure under the proposed FAA certification rules are assessed, i.e., via dedicated parameter sensitivities. The DLR simulation process can generate arbitrary noise metrics, e.g., SPL(t), SEL, LAMAX, EPNL, and even Sound Quality metrics [39,40]. With the focus on certification, the assessment within this study is limited to EPNL. Levels at selected observer locations and the actual noise distribution along the entire flight trajectory are investigated.

The aircraft and engine details are described in the following section. Thereafter, the flight simulation and noise prediction results are presented.

5.1. Aircraft and Engine Design

The basic aircraft that is used for the study presented in this publication has been derived within the frame of conceptual/preliminary design. The selected aircraft is well described in Ref. [14] and its geometrical data is provided in Table A1. The results of the aircraft design process are given in Table A3. The aircraft data presented here originates from this source except of an updated engine design. The aircraft *TWO* is equipped with a scaled version of the DLR engine described in Table 2. The aircraft is designed to carry eight passengers, equal to 726 kg of payload, along a range of 7500 km (4050 nautical miles), thus providing transatlantic capability. Its maximum fuel volume is ($V_{fuel,max}$) is 44.7 m^3.

However one significant change has been made for this study by reducing cruise speed at the design point of the aircraft from Ma = 1.6 to Ma = 1.4. The change to a lower Mach number in this publication is a result of current trends in the design of supersonic business jets as outlined in the introduction. Consequently the resulting aircraft does have a lower Maximum Take-off Mass (m_{MTOM}) than the basic design from Ref. [14] and it thus requires engines with less static thrust. Keeping in mind that this publication focuses on acoustics, the Ma = 1.4 design promises a more favorable starting point with respect to engine noise compared to an aircraft with a design speed of Ma = 1.6.

An extract to the requirements is given in Table A2.

5.2. Flight Simulation and Noise Prediction

To determine the trajectories for standard and advanced take-off, an optimization was performed taking into account the ICAO noise certification regulations, which includes a variation of airspeed, thrust and cutback height. The standard definition of NASA was chosen as a reference Reference [9]. The aim of this optimization was to minimize the certification level at the individual noise measurement points. The results are shown in the form of EPNL at the flyover measurement point above relative engine speed in Figure 5 for the standard (bottom) and the advanced departure procedure (top). In addition, a figure of all trajectories evaluated for the variation is included in the Appendix A in Figure A2. The noise certification regulations limit the ranges of the individual flight path parameters, e.g., minimum climb rate (4%), minimum cutback altitude for two engines (300 m), and maximum cutback altitude, respectivly the requirement that the thrust reduction of the cutback must be completed before the measurement of the EPNL at the flyover point. The application of the highspeed climbout allows the thrust for the advanced procedure to be reduced further after the pilot-initiated cutback, due to the decreased thrust requirement as described above. Obviously, the lowest certification level at the flyover measurement point results for the highest cutback altitude still allowed, whereas this altitude is larger for the standard procedure, since there is no additional thrust reduction through PLR and thus a faster increase in flight altitude. For both procedures, the lowest EPNL occurs in the range of low relative engine speeds. As can be seen, it is also confirmed for the aircraft *TWO* that an advanced procedure or a highspeed climbout results in a reduction at the flyover measurement point. In the following, the two procedures with the lowest EPNL are analyzed in more detail.

Figure 6 provide the resulting trajectories as simulated for this vehicle, i.e., take-off (left) and approach (right), respectively. The proposed SST rule changes refer to the take-off situation only, so that it is sufficient to assess one approach procedure here.

Since the engines are mounted under the wing in the configuration investigated, no effective shielding of the engine noise can be achieved here, as was observed with the NASA 55t STCA. For this reason, the modeling of the shielding effect was neglected in these investigations. The lateral measurement position was calculated separately for each simulated trajectory and the position with the highest EPNL was selected for noise assessment.

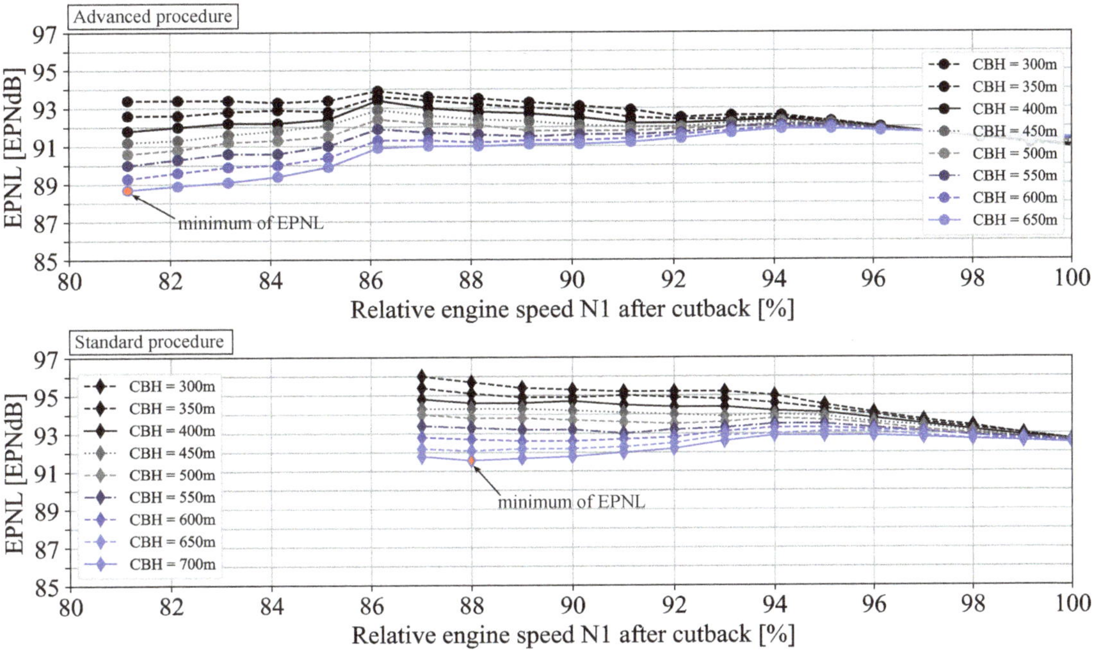

Figure 5. EPNL at flyover measurement point over relative engine speed N1 after pilot-initiated cutback of flightpath variation for advanced (**top**) and standard departure procedure (**bottom**) in the allowed range of certification regulations in terms of cutback height (CBH) (≥300 m) and climb rate (≥4%) for the aircraft *TWO*.

Although the maximum take off mass is very similar, the aircraft *TWO* has a higher thrust excess compared to NASA 55t STCA due to the smaller number of engines. The reason for this is that, airworthiness regulation requires an aircraft to be able to take-off with one engine inoperative. This means that the aircraft *TWO* requires a shorter runway length and can gain altitude more quickly. The characteristic take-off speed V2, which determines the speed limitation in the departure procedure of the noise certification, is lower for the aircraft *TWO* than for the NASA 55t STCA. Nevertheless, the speed differences between standard and advanced procedure are the same.

In order to achieve a higher flightspeed (V2 + 35 kn) for the highspeed climbout, a delayed rotation take-off is used in the advanced take-off, which results in a significantly longer runway length of approx. 1690 m compared to the standard take-off (approx. 1410 m) with lower flightspeed (V2 + 20 kn). As a result, the aircraft has a higher altitude along the standard take-off, which can be advantageous on the ground due to atmospheric attenuation and distance. However, due to the variable position of the lateral measurement point, there is a shift of the highest EPNL and thus the selected lateral measurement point position relative to the brake release point changes from 2300 m to 2700 m. It is observed that flight altitude above the lateral measurement point as well as the relevant range of the trajectory related to altitude is very similar for the two procedures.

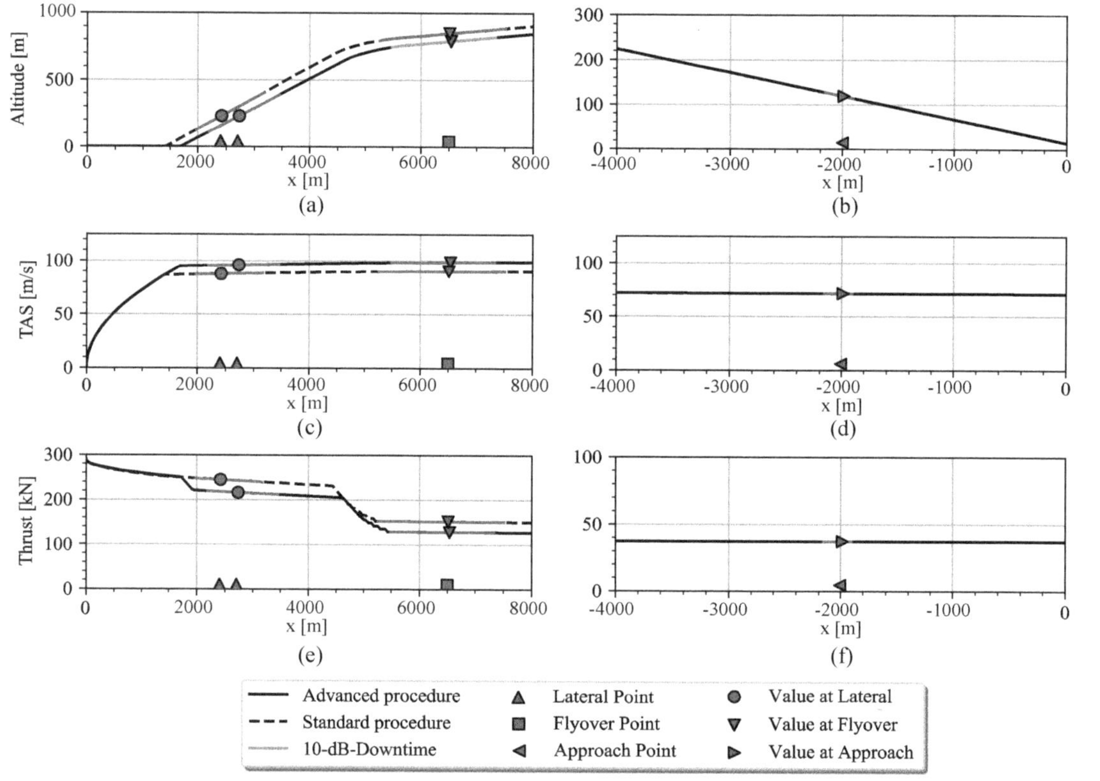

Figure 6. Take-off (**left**) and approach (**right**) flightpaths (altitude (**a**,**b**), calibrated airspeed (**c**,**d**) and total thrust (**e**,**f**)) as result of optimization for standard and advanced procedure. Flyover, lateral and approach measurement points as well as operation condition above measurement points and relevant part of flightparameters for EPNL calculation (10-dB-Downtime) are marked. The x position is specified relative to the brake-release point for take-off and relative to the runway threshold for approach.

For the approach simulation a 3° glide path with a height of 120 m above the measurement point is assessed, i.e., with the aircraft in a stable flight condition. For the aircraft *TWO*, this results in an airspeed of 71.48 m/s above the measurement point with a constant total thrust of 37,160 N over the relevant area of the flight path according to the specifications with Vref+10 knts.

5.3. Noise Assessment

In order to assess the advanced and standard departure and the approach procedure, the tone corrected perceived noise data (PNLT) for the total aircraft as well as for individual noise components (jet, fan, airframe), are depicted in Figures 7 and 8. In addition, the area of the 10 dB-down time relevant for the calculation of the EPNL and the maximum of the tone corrected perceived noise level (PNLTM) are highlighted in the figures. Predicted EPNL are summarized in Table 4.

In both PNLT plots (standard and advanced), there are two distinct local peaks, which can be attributed to forward and backward radiated engine noise contribution. The first peak is associated with the fan contribution, cf. see red dotted lines. The second peak is at a significantly higher level and can be associated with the rearward radiated jet contribution. The following differences in noise source ranking along standard and advanced procedure can be identified for the aircraft *TWO*. Jet noise is dominating the fan noise at the lateral measurement point by +5 dB along the standard and by +3.5 dB along the advanced flight procedure, respectively. At the flyover measurement point the contribution of both noise sources, i.e., jet and fan, is of similar magnitude due to the reduced thrust setting (thrust cutback) and the resulting reduction in jet exhaust velocity. In general, the influence of the fan on total aircraft noise is higher along the advanced take-off compared to the standard procedure due to the PLR and the highspeed climbout. Fan noise influence is increased due to a lower jet noise contribution which is a direct consequence of the PLR thrust reduction and the increased flight speed along the highspeed climbout as already mentioned in Ref. [10]. Consequently, the effectiveness of fan noise shielding is increased along the advanced procedure as can be demonstrated for NASA 55t STCA. The highspeed climbout has a larger effect on the flyover measurement point, which also confirms the observations of NASA.

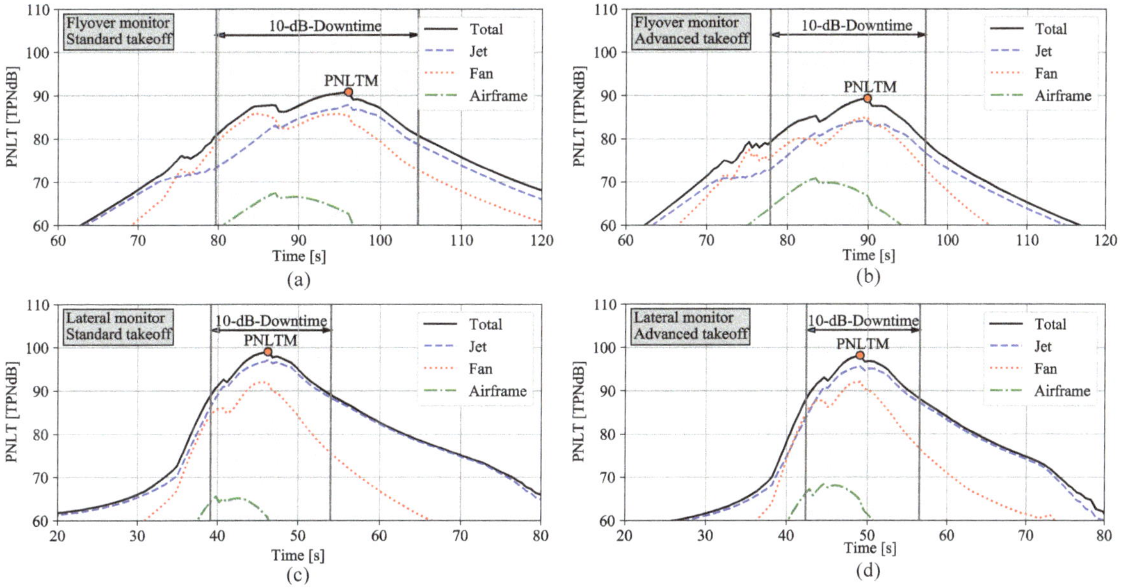

Figure 7. Calculated PNLT over time at flyover measurement point for standard (**a**) and advanced take-off (**b**) and at lateral measurement point standard (**c**) and advanced take-off (**d**). 10 dB-down time is filled grey and PNLTM is marked. 10 dB-down time represents the relevant time of the PNLT course for the calculation of the EPNL.

Only one approach flight is simulated for the aircraft *TWO* as described above. The predicted PNLT is plotted in Figure 8. Jet noise contribution is significantly decreased, revealing a dominant fan noise. Airframe noise levels are still insignificant even for this approach situation. The predicted EPNL are listed in Table 4 and no additional approach contour plots are shown in this paper. These plots are omitted since only one approach flight procedure is considered in the study.

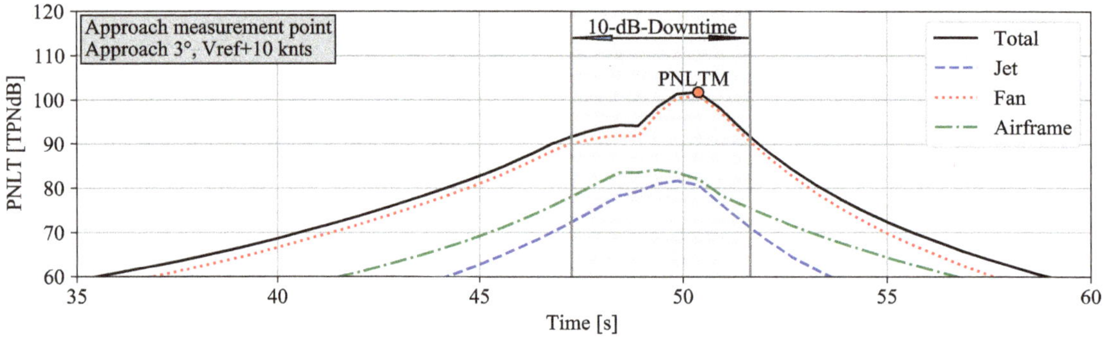

Figure 8. Calculated PNLT of the aircraft *TWO* over time for approach reference noise measurement point. 10 dB-down time is filled grey and PNLTM is marked. 10 dB-down time represents the relevant time of the PNLT course for the calculation of the EPNL.

Table 4. Simulated certification levels at individual measurement points (flyover, lateral and approach) and cumulative level for aircraft TWO. Limits of individual measurement points and cumulative from ICAO Annex 16 and FAA NPRM for Noise Certification of Supersonic Airplanes.

Procedure	Facility	Shielding	Flyover	Lateral	Approach	Cumulative
Standard	DLR	no Shielding	91.6	97.4	94.3	283.3
Advanced	DLR	no Shielding	88.7	96.4	94.3	279.4
Noise Regulations *			**Flyover**	**Lateral**	**Approach**	**Cumulative**
Limit (Chapter 3)			89.8	95.7	99.5	285.0
Limit (Chapter 4)			89.8	95.7	99.5	275.0
Limit (Chapter 14)			88.8	94.7	98.5	268.0
Limit (FAA NPRM)			89.8	95.7	99.5	271.5

* Caution: additional rules are applicable according to the ICAO Noise Chapters/NPRM, e.g., no trading Refs. [1,37].

Certification Noise Levels

The predicted EPNL values are now compared to available noise limits specified according to the Noise Chapters of ICAO Annex 16 [37] and the proposed FAA regulations [4], see Table 4. At the departure measurement points, defined limits of the FAA NPRM are exceeded by the aircraft *TWO* along the standard departure (+1.8 EPNdB at the flyover point and +1.7 EPNdB at the lateral point). Applying the advanced procedure for this vehicle will reduce the noise levels but still exceed or barely met the limits (−1.1 EPNdB at the flyover point and +0.7 EPNdB at the lateral point). As already mentioned, exceeding the limits at the individual measurement points is no longer permitted, which is why certification according to the proposals of the NPRM or any other than ICAO Noise Chapter 3 would not be possible for the aircraft *TWO*.

Overall, noise level reductions along an advanced procedure can be confirmed for the aircraft *TWO* vehicle but are of reduced effectiveness compared to an over-wing aircraft as documented by NASA and the comparative study with the novel simulation process already shown for the 55t STCA. Compared to the 55t STCA, the limits are exceeded not only at the lateral point but also at the flyover measurement point, which results from similar EPNL and the reduced number of engines. An aircraft, like the aircraft *TWO*, with two engines requires more thrust for a similar take-off mass, which leads to higher nozzle exit velocities and thus more jet noise.

For a better comparability of the two procedures, the predicted EPNL contours for both departure procedures are depicted, see Figure 9. The upper half of the plot shows the EPNL for the advanced take-off procedure and the lower half the EPNL for the standard take-off procedure. The longer acceleration on the ground extends the area with increased

EPNL along the advanced procedure, as the aircraft stays longer on the ground. This is confirmed by a different lateral measurement position along the two procedures. After the pilot initiated cutback, the contour areas strongly decrease and become narrower for the advanced procedure compared to the results of the standard procedure. In summary, the advanced procedure reduces the EPNL countour areas of the aircraft *TWO* with the exception of the area shortly after take-off. Nevertheless, for the two certification measurement points lateral and flyover, smaller EPNLs result for the advanced procedure are observed.

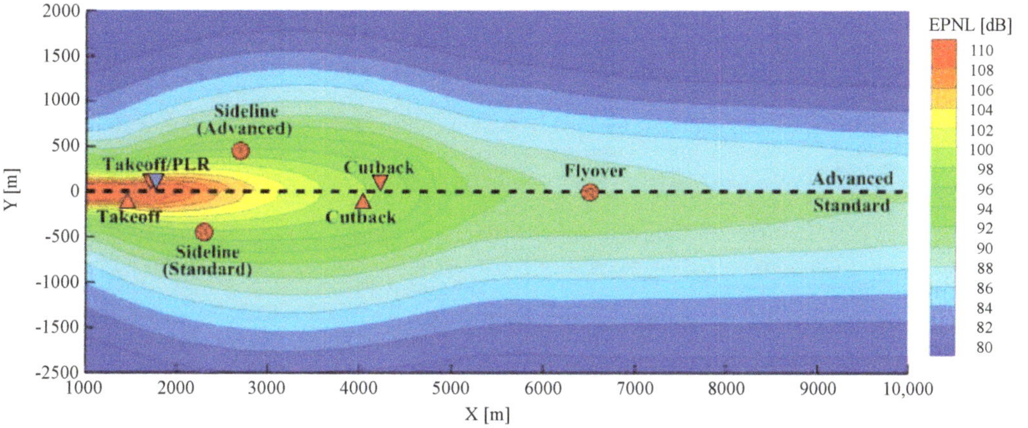

Figure 9. Comparison of EPNL contours for advanced (**top**) and standard procedure (**bottom**) of the aircraft *TWO*. Point of take-off, PLR and Cutback (triangles) as well as lateral and flyover reference noise measurement point (red dots) are marked with symbols. The x position is specified relative to the brake-release point.

6. Conclusions and Future Work

First results to estimate the landing and take-off noise of future SST aircraft are presented. Upgrades to an existing simulation process for the simulation of noise certification in aircraft preliminary design as well as interfaces to high fidelity simulation tools are presented.

Recalculations of the NASA studies on their 55t STCA configuration based on NASA trajectories were performed to verify the simulation process. Similar simulation methods have been applied compared to the NASA work. A satisfying agreement was observed between the PNLT and EPNL results, with an additional quantification of the effect of engine noise shielding for this vehicle. Detailed comparison of individual noise sources with NASA's results are yet still pending and are planned for the near future.

An application with complete simulation of noise certification is presented for a supersonic configuration developed at the Technical University of Braunschweig with two engines mounted under the wing which is referred to as aircraft *TWO*. The application includes the determination of optimal trajectories based on the noise certification regulations for the take-off procedure. The regulations of the certification process are established and the certification noise is predicted. NASA concepts for noise reduction via PLR and high speed climbout are confirmed. Overall, noise level reductions along an advanced procedure versus a standard procedure are predicted for the aircraft *TWO* in a lesser extend of what is known for the NASA 55t STCA. Inherent differences among under-wing and over-wing engine installation are demonstrated and disadvantages are experienced if engine noise is not shielded by the airframe. This can be attributed to a more dominant fan noise for any non-shielded engine concept if compared to a shielded variant. In addition along an advanced departure procedure, fan noise is generally reduced to a lesser extend compared to the jet noise. In conclusion, the proposed advanced departure procedure is more advantageous if jet noise is clearly dominating fan noise, i.e., the over-wing engine installation case. Current industry developments feature under-the-wing engine concepts and based on the presented findings of this study, a successful noise certification according to the proposed FAA regulations will be even more challenging. Novel technologies to reduce individual noise sources in addition to advantageous flight procedures become essential for these concepts.

In conclusion, the feasibility of the novel DLR simulation process and the applicability of the process toward SST vehicle assessments is confirmed. Results are promising and future updates will contribute to a better assessment of future SST air transportation. Yet, it should be noted that the simulation core process is based on semi-empirical methods and not validated against any experimental data for such SST vehicles. Validation of semi-empirical results with high-fidelity simulation is still ongoing but initial results look very promising.

Next to the planned NASA benchmark activities, future work will focus on a 3-engine configuration, referred to as THREE. For this vehicle, the third engine is integrated into the rear fuselage (tailplane). This aircraft is very similar in geometry compared with the two-engined aircraft studied here. In the upcoming study, certification regulations for different numbers of engines will be studied. In addition high fidelity simulations are currently carried out, which will be used to verify the results of the low to mid-fidelity simulations. The results will be processed directly for an overall assessment of the aircraft *TWO* and aircraft *THREE*. These simulations include CFD simulations of the aerodynamic performance with the DLR TAU Code [41] and CAA simulations of the engine noise shielding with DLR FMP [42] for characteristic operation conditions. Due to the relevance of fan noise for under-the-wing configurations, which is demonstrated in this work, the shielding of fan and jet noise sources has to be taken into account within these high-fidelity evaluations. At this point, especially the fan models currently used in PANAM and other similar noise prediction tools show a large uncertainty in predicting the noise of such fan architectures for SST aircraft. Therefore a further development of the fan noise prediction is pending and needed to improve predictive capabilities and noise assessment.

Author Contributions: Conceptualization, M.N., L.B. and M.S.; methodology, M.N., M.S., L.B., M.P.; software, M.N., L.B., M.P., M.K., R.J.; validation, M.N., L.B. and M.S.; formal analysis, M.N., M.S., L.B., M.P. and M.K.; investigation, M.N., M.S., L.B., M.P. and M.K.; resources, M.N., M.S., M.P., M.K. and J.J.B.; data curation, M.N., L.B.; writing—original draft preparation, M.N., M.S., L.B., M.P.; writing—review and editing, M.N., M.S., L.B., M.P., R.J., J.J.B.; visualization, M.N. and M.S.; supervision, M.N. and L.B.; project administration, L.B.; funding acquisition, L.B. All authors have read and agreed to the published version of the manuscript.

Funding: Internal funding supported the DLR work activities. Other contributors have not received any funding.

Acknowledgments: Members of the ELTON_SST Project, namely Jan Delfs, Lars Enghardt, Roland Ewert, Jochen Kirz, and Markus Lummer, provided a lot of helpful comments and discussions to different subdivisions of this work. Special thanks to NASA for providing the trajectories and an extensive technical exchange.

Conflicts of Interest: The authors declare no conflict of interest.

Abbreviations

The following abbreviations are used in this manuscript:

ATTCS	Automatic Take-off Thrust Control System
CFD	Computational Fluid Dynamics
DLR	Deutsches Zentrum für Luft- und Raumfahrt e.V.
ELTON_SST	Estimation of Landing and Take-off Noise for Supersonic Transport Aircraft
FAA	Federal Aviation Administration
FLIPNA	Flightpath for Noise Analysis
GTlab	Gas Turbine Laboratory
ICAO	International Civil Aviation Organization

JAXA	Japan Aerospace Exploration Agency
LTO	Landing and Take-off
NASA	National Aeronautics and Space Administration
PAA	Propulsion Airframe Aeroacoustic integration or interaction effects
PLR	Progammed Lapse Rate
PrADO	Preliminary Aircraft Design and Optimization Program
PANAM	Parametric Aircraft Noise Analysis Module
SENECA	(LTO) noiSe and EmissioNs of supErsoniC Aircraft
SSBJ	Supersonic Business Jet
SST	Supersonic Transport
STCA	Supersonic Technology Concept Aeroplane
TAU	DLR CFD software package
TsAGI	Central Aerohydrodynamic Institute of Russia
QueSST	Quiet SuperSonic Technology
VNRS	Variable Noise Reduction System
Variables/metrics	unit
A_W	Wing area, m^2
EPNL	Effective perceived noise level, dB
L_{Aeq}	Time-weighted, equivalent continuous sound pressure level, dB
l_{AC}	Aircraft length, m
LAMAX	Maximum noise level from a single noise event
Ma_C	Cruise Mach number, -
M_{MTOM}	Maximum take-off mass, -
N	Fresnel zone number, -
n_E	Number of engines, -
n_{PAX}	Number of passengers, -
PNLT	Tone corrected perceived noise level, dB
PNLTM	Maximum of tone corrected perceived noise level, dB
R_D	Range at design point, NM
s	Wing span, m
SEL	Sound exposure level, dB (also referred to as $L_{p,AE}$ or L_{AX})
SPL	Sound pressure level, dB (also referred to as L or L_p)
SPL(A)	A-weighted SPL, dB (also referred to as L_A or $L_{p,A}$)
t	Time, s
TAS	True Air Speed, m/s
$V_{fuel,max}$	Maximum Fuel Tank Volume, m^3
V_{Ref}	Reference Speed, m/s
V_2	Take-off safety speed, m/s
δ	Difference in shortest source-edge-receiver distance, m
Δ	Difference in sound pressure level, dB
Φ	Azimuthal/lateral directivity, +90° equals to starboard
Θ	Polar/longitudinal directivity, 0° equals flight direction
Plot legend	
airf	sum of t.e., l.e. and gear
eng	sum of combustion, fan and jet
com	combustion noise (all engines)
fan	sum of inlet and exhaust fan noise (all engines)
fan bb	fan: only broadband noise (all engines)
fan t	fan: only tonal noise (all engines)
jet	sum of jet noise (all engines)

Appendix A

Table A1. aircraft *TWO*: geometrical data from Ref. [14].

Parameter	Value	Unit
Aircraft length	36.80	m
Wing area	150.0	m^2
Wing span	18.50	m
Leading edge sweep angles	72.5, 52.0	deg
Dihedral angles	0	deg
Number of engines	2	–
Max. fuel volume	44.7	m^3

Table A2. Extract of the requirements from Ref. [14].

Parameter	Value	Unit	Remarks
Number of Passengers (design point)	8	-	-
Passenger mass (incl. luggage)	91	kg	-
Maximum Payload	1728	kg	-
Range at design point	4000 (7408)	nm (km)	-
Cruise Mach number	1.6	-	-
V_{AT}, speed at threshold	<140	kts	
Max. allowed runway length	2200	m	serve smaller airports.
Max. cruise altitude	FL600	-	avoid interference with normal air traffic. Same as for Concorde.
ETOPS	180	min	allow to serve North and Latin America

Table A3. *TWO*: PrADO results, Design Point Ma = 1.4 from Ref. [14].

Parameter	Value	Unit	Remark
m_{MTOM}	55,412	kg	Max. take-off mass
m_{OEW}	24,313	kg	Operational empty mass (weight)
$m_{fuel,max}$	28,015	kg	Max. fuel mass
$W_{S,max}$	348.84	$\frac{kg}{m^2}$	Max. wing loading
S_0	291.72	kN	Total static thrust
$ALT_{begin,cruise,D}$	14.341	km	Altitude at the begin of cruise at the design point
$ALT_{end,cruise,D}$	17.860	km	Altitude at the end of cruise at the design point
$L/D_{begin,cruise,D}$	6.30	m	Lift-to-drag ratio at the begin of cruise at the design point
V_{AT}	237, (128)	$\frac{km}{h}$, (kts)	Speed at threshold
$l_{RWY,T}$	2389	m	Runway length at take-off
$l_{RWY,LD}$	1369	m	Runway length at landing

Figure A1. Side view of the aircraft TWO.

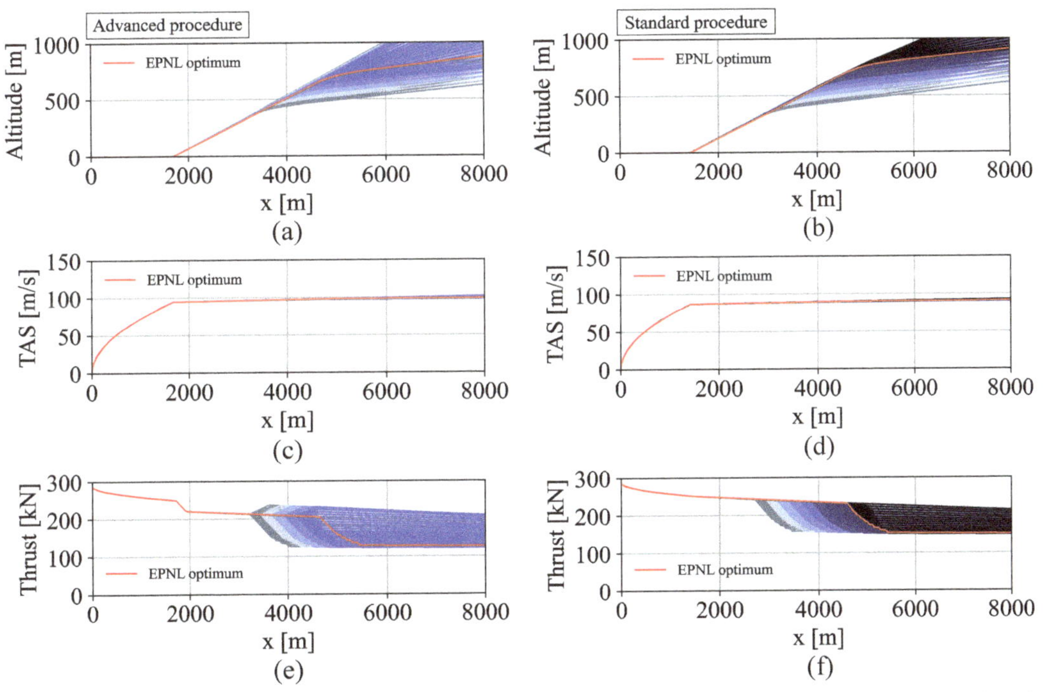

Figure A2. Flight altitude (**a**,**b**), True Air Speed (**c**,**d**) and Thrust (**e**,**f**) over X in the form of the distance to the brake release point is plotted for all trajectories considered in the flight path optimization. On the left side are all advanced and on the right side all standard departure procedures. The EPNL optimum, i.e., the trajectory with the lowest EPNL at flyover position is colored in red and trajectories with the same cutback height are assigned to one specific color (grey = lower, blue = middle, and black = higher cutback heights).

References

1. FAA. PART 36—Noise Standards: Aircraft Type and Airworthiness Certification, Subpart D—Noise Limits for Supersonic Transport Category Airplanes. Standard. 2020. Available online: https://www.law.cornell.edu/cfr/text/14/36.301 (accessed on 21 December 2020).
2. Japan Airlines (JAL). Japan Airlines Press Release. December 2017. Available online: https://press.jal.co.jp/en/2017/12/index_2.html (accessed on 21 December 2020).
3. Boom Technology, Inc. Available online: https://boomsupersonic.com/united (accessed on 17 August 2021).
4. Department of Transportation and Federal Aviation Administration. Noise Certification of Supersonic Airplanes. Technical Report. Federal Register, Notice of proposed rulemaking (NPRM). 2021. Volume 85, No. 71, pp. 20431–20447. Available online: https://www.govinfo.gov/content/pkg/FR-2020-04-13/pdf/2020-07039.pdf (accessed on 20 March 2021).
5. Grantham, W.D.; Smith, P.M. Development of SCR Aircraft Takeoff and Landing Procedures for Community Noise Abatment and their Impact on Flight Safety. *Supersonic Cruise Res.* **1980**, *1979*, 299–333.
6. NASA. NASA Tests 30-Mile-Long Microphone Array in Preparation for Quiet Supersonic X-59. 2019. Available online: https://www.nasa.gov/centers/armstrong/features/NASA-Tests-Microphone-Array-for-X-59.html (accessed on 21 December 2020).
7. NASA. Low-Boom Flight Demonstration. 2020. Available online: https://www.nasa.gov/mission_pages/lowboom/overview (accessed on 21 December 2020).
8. Berton, J.J.; Jones, S.M.; Seidel, J.A.; Huff, D.L. Advanced Noise Abatement Procedures for a Supersonic Business Jet. In Proceedings of the 23rd International Symposium on Air Breathing Engines (ISABE): Economy, Efficiency and Environment, Manchester UK, 3–8 September 2017; ISABE-2017-22538.
9. Berton, J.J.; Huff, D.L.; Geiselhart, K.; Seidel, J. Supersonic Technology Concept Aeroplanes for Environmental Studies. In Proceedings of the AIAA Scitech 2020 Forum, Orlando, FL, USA, 6–10 January 2020; p. 263.
10. Rizzi, S.A.; Berton, J.J.; Tuttle, B.C. Auralization of a Supersonic Business Jet Using Advanced Takeoff Procedures. In Proceedings of the AIAA Scitech 2020 Forum, Orlando, FL, USA, 6–10 January 2020; p. 266.
11. Akatsuka, J.; Ishii, T. System Noise Assessment of NASA Supersonic Technology Concept Aeroplane Using JAXA's Noise Prediction Tool. In Proceedings of the AIAA Scitech 2020 Forum, Orlando, FL, USA, 6–10 January 2020; p. 265.
12. Sergey Chernyshev (TsAGI). Supersonic Transport: From the Tu-144 to the New Generation. October 2019. Available online: https://ftfsweden.se/wp-content/uploads/2019/10/FT2019-Plenary_TsAGI-Sergey-Chernyshev.pdf (accessed on 7 October 2021).
13. Stephen Trimble. Russia Working on Quiet Supersonic Business Jet. 2012. Available online: flightglobal.com (accessed on 14 March 2012).
14. Schuermann, M. Supersonic Business Jets in Preliminary Aircraft Design. Ph.D. Thesis, TU-Braunschweig, Niedersächsisches Forschungszentrum für Luftfahrt (NFL), Braunschweig, Germany, 2016.
15. Nöding, M.; Bertsch, L. Application of noise certification regulations within conceptual design. *Aerospace* **2021**, *8*, 210. [CrossRef]
16. Heinze, W. *Beitrag Zur Quantitativen Analyse Der Technischen Und Wirtschaftlichen Auslegungsgrenzen Verschiedener Flugzeugkonzepte für Den Transport Grosser Nutzlasten*; Technical Report ZLR-Forschungsbericht; Inst. für Flugzeugbau und Leichtbau, Technische Universität Braunschweig: Braunschweig, Deutschland, 1994; ISBN 3-928628-14-3.
17. Schuermann, M.; Horst, P.; Gaffuri, M. Extensions to aircraft pre-design for supersonic aircraft. In Proceedings of the 52nd Aerospace Sciences Meeting, National Harbor, MD, USA, 13–17 January 2014; p. 0184.
18. Reitenbach, S.; Vieweg, M.; Becker, R.; Hollmann, C.; Wolters, F.; Schmeink, J.; Otten, T.; Siggel, M. Collaborative Aircraft Engine Preliminary Design Using a Virtual Engine Platform, Part A: Architecture and Methodology. In Proceedings of the AIAA Scitech 2020 Forum, Orlando, FL, USA, 6–10 January 2020; p. 0867.
19. Blinstrub, J. *Immission-Based Noise Reduction within Conceptual Aircraft Design*; Technical Report DLR-FB-2019-12, DLR; Deutsches Zentrum für Luft- und Raumfahrt (DLR): Göttingen, Germany, 2019. [CrossRef]
20. Maekawa, Z. Noise reduction by screens. *Appl. Acoust.* **1968**, *1*, 157–173. [CrossRef]
21. Fleischer, F. Zur Anwendung von Schallschirmen. *Laermbekaempfung* **1970**, *6*, 131–136.
22. Bertsch, L. *Noise Prediction within Conceptual Aircraft Design*; Technical Report DLR-FB-2013-20, DLR; Deutsches Zentrum für Luft- und Raumfahrt (DLR): Göttingen, Germany, 2013. [CrossRef]
23. Bertsch, L.; Sanders, L.; Thomas, R.H.; LeGriffon, I.; June, J.C.; Clark, I.A.; Lorteau, M. Comparative Assessment of Aircraft System Noise Simulation. *AIAA J. Aircr.* **2021**, *58*, 4. [CrossRef]
24. Stone, J.; Krejsa, E.; Clark, B.; Berton, J. *Jet Noise Modeling for Suppressed and Unsuppressed Aircraft in Simulated Flight*; Technical Report NASA/TM—2009-215524; NASA, Glenn Research Center: Cleveland, OH, USA, 2009.
25. Kazin, S.; Matta, R.; Bilwakesh, K.; Emmerling, J.; Latham, D. *Core Engine Noise Control Program. Volume III. Prediction Methods*; Technical Report; General Electric CO Cincinnati OH Aircraft Engine Business Group: Cincinnati, OH, USA, 1974.
26. Dobrzynski, W.; Chow, L.; Guion, P.; Shiells, D. A European Study on Landing Gear Airframe Noise Sources. In Proceedings of the 6th AIAA/CEAS Aeroacoustics Conference, Lahaina, HI, USA, 12–14 June 2000. [CrossRef]
27. Dobrzynski, W.; Pott-Pollenske, M. Slat Noise Source Studies for Farfield Noise Prediction. In Proceedings of the 7th AIAA/CEAS Aeroacoustics Conference, Maastricht, The Netherlands, 28–30 May 2001. [CrossRef]
28. Pott-Pollenske, M.; Dobrzynski, W.; Buchholz, H.; Gehlhar, B.; Walle, F. Validation of Semiempirical Airframe Noise prediction Method through Dedicated A319 Flyover Noise Measurements. In Proceedings of the 8th AIAA/CEAS Aeroacoustics Conference, Breckenridge, CO, USA, 17–19 June 2002. [CrossRef]

29. Rossignol, K.S. Development of an empirical prediction model for flap side-edge noise. In Proceedings of the 16th AIAA/CEAS Aeroacoustics Conference, Stockholm, Sweden, 7–9 June 2010. [CrossRef]
30. Rossignol, K.S. Empirical Prediction of Airfoil Tip Noise. In Proceedings of the 17th AIAA/CEAS Aeroacoustics Conference, Portland, OR, USA, 5–8 June 2011. [CrossRef]
31. Heidmann, M. *Interim Prediction Method for Fan and Compressor Source Noise;* Technical Report NASA TMX-71763; NASA: Washington, DC, USA, 1979.
32. Lummer, M. Maggi-Rubinowicz Diffraction Correction for Ray-Tracing Calculations of Engine Noise Shielding. In Proceedings of the 14th AIAA/CEAS Aeroacoustics Conference, Vancouver, BC, Canada, 5–7 May 2008. [CrossRef]
33. International Organization for Standardization (ISO). *Acoustics—Attenuation of Sound during Propagation Outdoors. Part 1: Calculation of the Absorption of Sound by the Atmosphere;* Technical Report ISO 9613-1:1993; ISO: Geneva, Switzerland, 1993.
34. Society of Automotive Engineers (SAE). *Prediction Method for Lateral Attenuation of Airplane Noise during Takeoff and Landing;* Technical Report Aerospace Information Report, AIR 1751; SAE: Warrendale, PA, USA, 1981.
35. Moreau, A. A Unified Analytical Approach for the Acoustic Conceptual Design of Fans of Modern Aero-Engines. Ph.D. Thesis, TU-Berlin, Berlin, Germany, 2017.
36. Airplanes, B.C. *High-Speed Civil Transport Study;* NASA CR-4233; NASA: Seattle, WA, USA, 1989.
37. International Civil Aviation Organization (ICAO). *Environmental Protection. Annex 16 to the Convention on International Civil Aviation. Vol. I, Aircraft Noise*, 5th ed.; Technical Report Annex 16; ICAO: Montreal, QC, Canada, 2008; Volume I.
38. Kontos, K.; Janardan, B.; Gliebe, P. *Improved NASA-ANOPP Noise Prediction Computer Code for Advanced Subsonic Propulsion Systems;* Technical Report NASA-CR-195480; NASA: Washington, DC, USA, 1996.
39. Bertsch, L.; Wolters, F.; Heinze, W.; Pott-Pollenske, M.; Blinstrub, J. System noise assessment of a tube-and-wing aircraft with geared turbofan engines. *J. Aircr.* **2019**, *56*, 1577–1596. [CrossRef]
40. Greco, G.F.; Bertsch, L.; Ring, T.P.; Langer, S.C. Sound quality assessment of a medium-range aircraft with enhanced fan-noise shielding design. *CEAS Aeronaut. J.* **2021**, *12*, 481–493. [CrossRef]
41. Schwamborn, D.; Gerhold, T.; Kessler, R. DLR-TAU Code—An Overview. In Proceedings of the 1st ONERA/DLR Aerospace Symposium, Paris, France, 21–24 June 1974.
42. Lummer, M. Calculation of Acoustic Shielding at Full-Scale Aircraft Configurations. In Proceedings of the 16th CEAS-ASC Workshop, 2nd Scientific Workshop of X-Noise, Braunschweig, Germany, 11–12 October 2012.

aerospace

Article

System Noise Assessment and Uncertainty Analysis of a Conceptual Supersonic Aircraft

Junichi Akatsuka * and Tatsuya Ishii

Aviation Environmental Sustainability Innovation Hub, Japan Aerospace Exploration Agency (JAXA), Tokyo 182-8522, Japan; tatsuya.ishii@jaxa.jp
* Correspondence: akatsuka.junichi@jaxa.jp

Abstract: This paper describes a system noise assessment of a conceptual supersonic aircraft called the NASA 55t Supersonic Technology Concept Aeroplane (STCA), its prediction uncertainty, and related validation tests. A landing and takeoff noise (LTO) standard for supersonic aircraft is needed to realize future supersonic aircraft, and the noise impact due to the introduction of future supersonic aircraft should be analyzed to develop the standard. System noise assessments and uncertainty analyses using Monte Carlo simulation (MCS) were performed. The predicted noise levels showed good agreement with the prior study for both the benchmark case and statistics of the predictions. The predicted cumulative noise level satisfied the ICAO Chapter 4 noise standard, and its standard deviation was approximately 2 EPNdB. Moreover, sensitivity analysis using the obtained datasets revealed strong correlations with the takeoff noise for jet noise, fan exhaust noise at the flyover measurement point, and airframe trailing edge noise. Further understanding of these extracted factors, which were estimated to have a significant impact on the LTO noise, will be beneficial for the development of LTO noise standards and the design of supersonic aircraft.

Keywords: aircraft noise prediction; noise certification; supersonic aircraft; NASA STCA; ICAO Annex 16 Vol.1

Citation: Akatsuka, J.; Ishii, T. System Noise Assessment and Uncertainty Analysis of a Conceptual Supersonic Aircraft. *Aerospace* **2022**, *9*, 212. https://doi.org/10.3390/aerospace9040212

Academic Editors: Lothar Bertsch and Adrian Sescu

Received: 25 February 2022
Accepted: 7 April 2022
Published: 12 April 2022

1. Introduction

In recent years, the development of supersonic business jets and supersonic transport has gained interest due to their potential for high-speed transportation. Several manufacturers have already announced the development of these aircraft; however, the development of such new aircraft requires appropriate standards to obtain type certifications. Concorde, a commercial supersonic aircraft, was launched more than 50 years ago, and the standards for environmental compatibility at that time are now outdated. One of the most important standards for environmental compatibility is the standard for landing and takeoff (LTO) noise, because a trade-off relationship between LTO noise and cruise performance is expected in the design of supersonic aircraft, which are different from conventional subsonic airliners. In general, supersonic aircraft are characterized by the use of high-specific-thrust engines and lower aerodynamic performance during takeoff and landing, compared with conventional subsonic airliners, to maintain supersonic cruise performance.

A new LTO noise standard for supersonic aircraft is anticipated to be added to the International Civil Aviation Organization (ICAO) Annex 16 Vol.1 Chapter 12 [1]. The ICAO has been discussing the development of the LTO noise standard for supersonic aircraft; however, there is currently no specific description. According to the published documents described in the following paragraphs, considering the noise of supersonic aircraft, the important points are as follows: (1) the necessity of predicting the noise of aircraft which does not exist at this time; (2) the necessity of clarifying the difference between supersonic and subsonic aircraft, because the LTO noise trend is expected to be different from that of subsonic aircraft; and (3) the necessity of assessing the environmental impact of supersonic aircraft, and identifying the uncertainty of the prediction and the sensitive factors that need to be taken into account. In this respect, system noise prediction and assessment are important. This effort can be seen in the published literature. Piccirillo et al. [2] predicted the noise of Concorde, the only supersonic aircraft for which data are available, and showed that the difference between the prediction and published data was ±2.19% for LAmax and SEL; guidelines for supersonic aircraft prediction were clarified. Stone et al. [3] performed a preliminary assessment of airport noise for supersonic business jets using several noise reduction concepts. They demonstrated that the most critical noise component that satisfies the noise standards specified in ICAO Chapter 3 is the jet at the lateral noise measurement point, and the mixed exhaust velocity should be less than approximately 400 m/s, and should be shock-free jets. Many novel noise reduction methods for supersonic jets have also been studied. Chevron [4] and fluidic injection [5,6] have been shown to be effective in reducing shock noise, and their mechanisms have been investigated. To extend the range of supersonic cruise, a high jet Mach number and low bypass ratio during takeoff must be maintained as much as possible. From this point of view, noise reduction devices for supersonic jets have promising potential. However, because the difference between the sideline take-off noise of Concorde [2] and the noise limit for subsonic aircraft (ICAO Chapters 3, 4 and 14) exceeds 20 EPNdB, it is still challenging to fill this gap only with jet noise reduction devices, and noise reduction techniques for the entire system, including engine design and multi-purpose optimization, should be studied. Berton et al. [7] performed a comparative study on the engine cycle for future commercial supersonic aircraft and examined the effects of the takeoff procedure using automatically derated engine thrust, called the programmed lapse rate (PLR), to reduce the lateral noise levels. Huff et al. [8] and Henderson et al. [9] assessed several noise-reduction techniques to satisfy the latest LTO standard (Chapter 14). They reported that future commercial supersonic aircraft have the potential to satisfy Chapter 4 or 14 noise standards using noise-reduction techniques and PLR. However, they also suggested that detailed mission studies are required to investigate the impact of the range resulting from the incorporation of noise-reduction techniques. In 2017, Berton et al. [10] extended their study and proposed an advanced noise abatement takeoff procedure for a supersonic aircraft. In their study, they created a 55-ton conceptual reference airplane with three engines called the NASA 55t Supersonic Technology Concept Aeroplane (STCA). This conceptual reference airplane was also used to discuss future noise standards for supersonic aircraft in ICAO, and the dataset of the STCA was published by NASA [11]. The Japan Aerospace Exploration Agency (JAXA) has conducted a silent supersonic research program (S4) [12]. In this program, the authors studied LTO noise based on a 70-ton class conceptual supersonic airplane [13] and developed a system noise estimation tool called AiNEST [14]. They also predicted the NASA 55t STCA [15] and discussed technical issues through the ICAO working group. The German Aerospace Center (DLR) is also conducting a project to investigate the LTO noise of supersonic aircraft, and recently reported results of their system noise assessments for supersonic aircraft, including the NASA 55t STCA [16].

Scope of This Article

These previous studies have revealed many LTO noise levels of supersonic aircraft that will be developed in the near future, and the differences from those of conventional subsonic airliners. However, further analyses and discussions are required to clarify the uncertainty of the prediction. In addition, it is beneficial to identify the key factors of the LTO noise of supersonic aircraft obtained through prediction to support the discussion for future LTO noise standards and aircraft design.

The objective of this study is to investigate the prediction uncertainty of LTO noise and extract key factors to perform a system noise assessment for future supersonic aircraft. To determine key uncertainty models, component-based validation studies were conducted. Then, a system noise assessment of the NASA 55t STCA and an uncertainty analysis using a Monte Carlo simulation (MCS) were performed as an aircraft representing a supersonic aircraft for which noise standards are currently under consideration. In addition, based on the sensitivity analysis, important factors to be considered in the system noise assessment of future supersonic aircraft were discussed.

2. Technical Approach for System Noise Prediction of Supersonic Aircraft

2.1. Standard for LTO Noise

While discussing the LTO noise of supersonic aircraft, the LTO noise standard for subsonic aircraft can be used as a reference, and the standards are briefly introduced in this section.

The noise standards for a subsonic aircraft mentioned in ICAO Chapters 3, 4, and 14 [1] are defined by the margins for limits at three reference noise measurement points: lateral full-power, flyover, and approach. These limits are related to the maximum takeoff mass and the number of engines.

According to the latest standard (Chapter 14), the cumulative margin must be 17 EPNdB or more, and the margins at each measurement point must be 1 EPNdB or more. The position of each reference noise measurement point is shown in Figure 1.

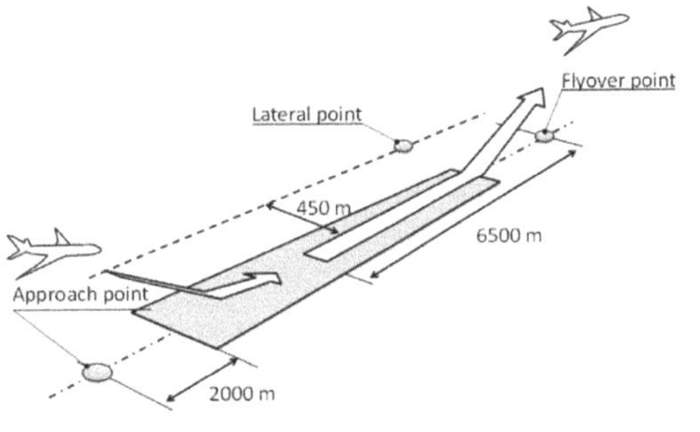

Figure 1. Schematic of reference noise measurement points.

The lateral reference noise measurement point is on a line parallel to and 450 m from the runway center line, where the noise level is at its maximum during takeoff. The flyover reference noise measurement point is on the extended center line of the runway and at a distance of 6.5 km from the start of the roll. The approach reference noise measurement point is on the ground, on the extended center line of the runway 2000 m from the threshold.

The reference takeoff and approach procedures are defined in the standard. The average takeoff thrust is maintained at the lateral measurement point. For the flyover measurement point, the thrust can be reduced at a height above 260 m from the runway for an aircraft with three engines, but the thrust should not be reduced below that required to maintain a climb gradient of 4%, or level flight with one engine inoperative. The takeoff climb speed is between V2 + 10 kt and V2 + 20 kt. The approach glide path is 3°, and Vref + 10 kt is maintained over the reference measurement point. Further information is available in Ref. [1].

2.2. NASA 55t Supersonic Technology Concept Aeroplane

The NASA 55t STCA is a conceptual airplane defined by NASA [10,11] to perform a gap analysis on the LTO noise between the conventional subsonic airliner and the supersonic aircraft currently being planned. The maximum takeoff weight, cruise Mach number and designed range of the STCA are 55 tons, 1.4 and 4243 nautical miles, respectively. The engine is a commercial off-the-shelf CFM56-7B. Using the core engine, it is assumed that the fan, low-pressure system, and nacelle are redesigned for supersonic aircraft. The engine type is a fixed-cycle mixed-flow turbofan. A summary of the engine performance provided by NASA [11] is presented in Table 1. The bypass ratio is 3, and the fan pressure ratio is 1.9 at sea level static. As indicated by previous studies, jet noise is the dominant noise component, and a shock-free jet at takeoff is necessary for the noise levels at takeoff and landing to be similar to those of a subsonic aircraft. Therefore, the exhaust jet Mach number of the NASA 55t STCA at takeoff is approximately 1. It is not necessary to consider the broadband shock-associated noise and screech tones generated by supersonic jets in LTO noise prediction, as represented by the STCA.

Table 1. Engine performance summary provided by NASA, reprinted from Ref. [11].

	M1.4, 50 kft, ISA	M0.25, Sea Level, ISA + 27 °F	Sea Level Static, ISA + 27 °F
Net thrust, lb/engine	3330	14,140	16,620
Specific fuel consumption, lb/hr/lb	0.943	0.588	0.479
Bypass ratio	2.9	2.9	3
Burner temperature, °R	3300	3150	3130
Turbine inlet temperature, °R	3180	3040	3020
Compressor exit temperature, °R	1450	1440	1430
Overall pressure ratio	22	21	21
Fan pressure ratio	2	1.9	1.9
Compressor pressure ratio	11.2	11.1	11.2
Extraction ratio	1.1	1.1	1.1
Nozzle pressure ratio	5.9	1.9	1.8

2.3. LTO Noise Prediction Tools and Input Data

AiNEST [14], developed by JAXA, is a semi-empirical prediction tool for LTO noise. An outline of this tool is shown in Figure 2. It is a software package that predicts the noise level of an entire aircraft from each noise source component and propagation effect based on semi-empirical or theoretical models. The inputs are flight trajectories, aircraft specifications, and engine operating conditions, and the outputs are the time histories of the tone-corrected perceived noise level (PNLT) and effective perceived noise level (EPNL) at the locations specified in the input.

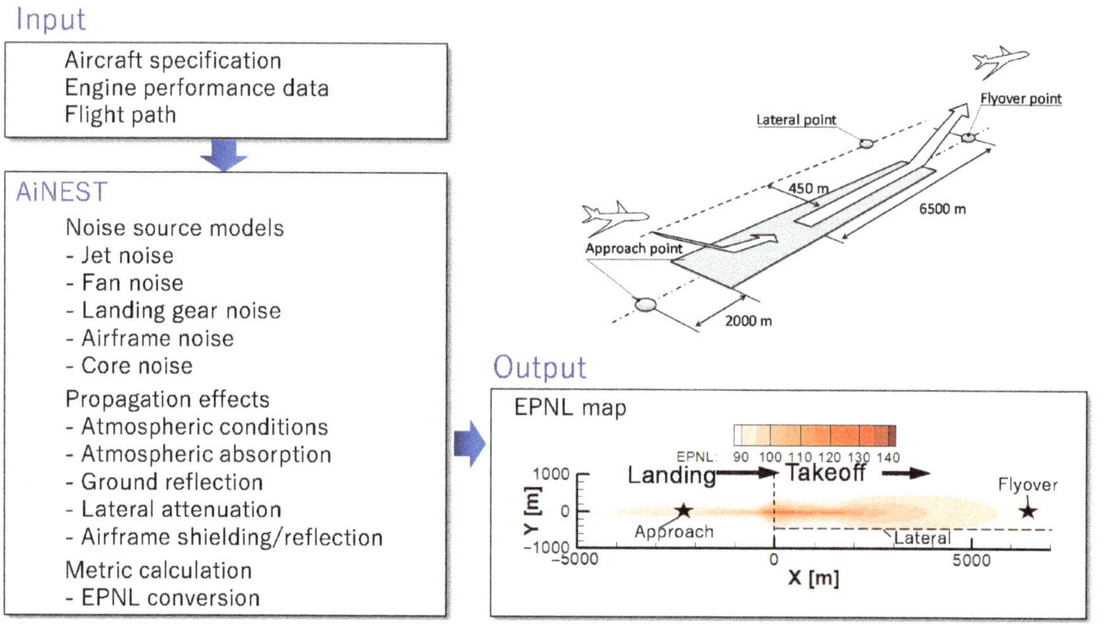

Figure 2. JAXA's LTO noise prediction tool, AiNEST [14].

The noise sources and propagation models used are listed in Table 2. To clarify the difference between the results of the prior noise predictions made by NASA, a configuration similar to that of NASA's ANOPP [17], used in Ref. [11], was selected for our tool, and the differences are clarified herein.

The first difference is in the jet noise prediction model. The LTO noise of a supersonic aircraft is dominated by jet noise; therefore, verifying its sensitivity to the selection of the jet noise model is important. The jet noise model used in this study was based on SAE ARP876 [18]; however, several modifications were implemented in the original model. The model was adjusted to our rig and the engine test data previously conducted, and calculations of the forward flight effects of jet noise were performed based on Viswanathan's model [19]. Another difference was in the selection of the airframe noise model. We used the original Fink model [20] without any special correction for supersonic aircraft. The prediction method for the fan noise shielding effect was also different from that used in a previous study. The shielding effect was predicted by ray-tracing using the Maekawa method [21] for a given plane form. Only the shielding effect of fan inlet noise by the main wing was considered in the STCA study. Fan treatment was considered based on the GE model [22]; however, a few modifications were performed by NASA in the STCA study, and their results were used to predict the fan treatment effect. Based on the differences from prior NASA studies, the uncertainty models for jet noise and fan noise shielding effects were set independently in this study. These were determined based on the results of our validation tests, which are described in the following sections.

The specifications of the aircraft and engine operating conditions as input values were provided by NASA. The noise abatement takeoff procedure, called the advanced takeoff procedure, is examined in this study. An advanced takeoff procedure was proposed by Berton [10,11]. It uses a 10% PLR and high-speed climbout to reduce takeoff noise. Figure 3a–c shows a comparison of the trajectory, flight velocity, and engine thrust for the conventional standard takeoff procedure and advanced takeoff procedure. Notably, this takeoff procedure is not allowed for use in contemporary noise certifications for subsonic airliners; however, the newly announced notice of proposed rule-making (NPRM) from FAA [23] allows the use of this procedure. Prior studies [11,15,16] have clarified that this procedure is effective in reducing the noise impact of supersonic aircraft.

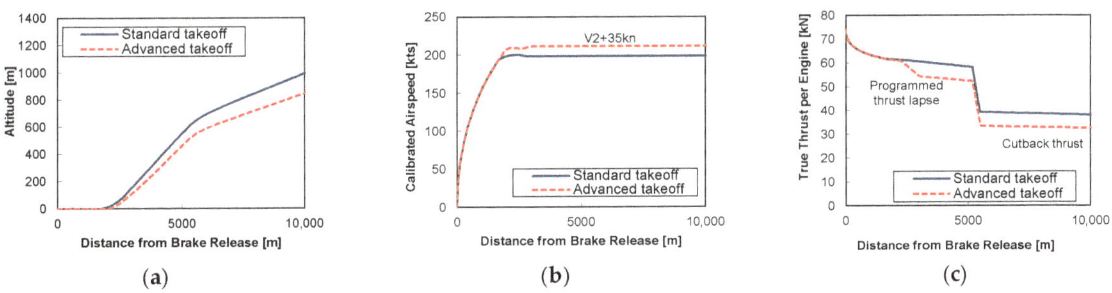

Figure 3. Flight path and thrust setting for the standard and the advanced takeoff procedure: (**a**) trajectory; (**b**) calibrated air speed; (**c**) thrust per engine.

Table 2. Noise source models and propagation models.

Component	JAXA AiNEST [14]	NASA ANOPP [11]
Jet noise	Modified SAE method	SAE method [20]
Fan noise	GE-Heidmann method [24]	GE-Heidmann method [24]
Treatment	Modified GE method [22]	Modified GE method [22]
Core noise	Emmerling method [25]	Emmerling method [25]
Airframe noise	Fink method [20]	Modified Fink method
Atmospheric absorption	ISO 9613-1:1993 [26] (ISA + 10 °C, 70%RH)	SAE ARP866 [27] (ISA + 10 °C, 70%RH)
Ground reflection	Chien-Soroka method [28] (Grass-covered ground)	Chien-Soroka method [28] (Grass-covered ground)
Lateral attenuation	SAE AIR 5662 [29] (Fuselage-mounted engines)	SAE AIR5662 [29] (Fuselage-mounted engines)
Shielding effect	Ray-tracing + Maekawa method [21]	Maekawa diffraction method [30]

3. Component-Based Validation

To investigate the difference between our predictions and previous studies, a validation test was conducted for the original acoustic model of jet noise and fan noise shielding, and the predictions were compared with the validation data obtained.

3.1. Scale Model Tests for a Single Heat-Simulated Jet

Acoustic tests using a scaled nozzle model were conducted to validate the jet noise model under the exhaust velocity conditions of the NASA 55t STCA, which are different from those of recent subsonic aircraft equipped with high bypass-ratio engines.

The JAXA jet noise test facility [31], the schematic of which is shown in Figure 4a, was used. The test nozzle was mounted vertically in a 4.1 m wide, 5.7 m deep, and 3.3 m high anechoic room equipped with a settling chamber on the floor. The methodology was based on a prior study performed by Doty [32] and validated using the published jet noise database [33].

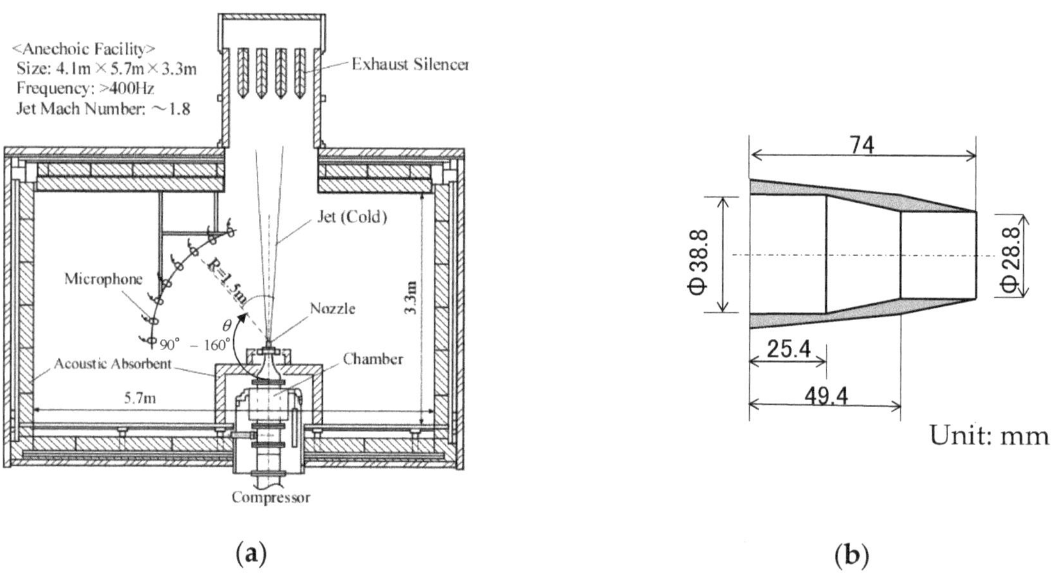

Figure 4. Experimental setup of the scale model tests for a single heat-simulated jet: (**a**) schematic diagram of JAXA's jet noise test facility; (**b**) schematic diagram of the test nozzle.

The test nozzle is shown in Figure 4b. The nozzle had an exit diameter, D, of 28.8 mm, and it had a convergent configuration of a convergent–divergent nozzle used to simulate a nozzle for a typical supersonic aircraft.

Far-field noise measurements were conducted with an array of eight Brüel & Kjær 4939 1/4-inch microphones arranged in an arc 1.5 m from the nozzle exit at polar angles, θ, from 90° to 160° at 10° intervals. The lower-limit frequency of the anechoic room was 400 Hz. The acoustic data were sampled at 200 kHz for 5 s and obtained through an 80 kHz low-pass filter using a TEAC DS160R data logger. Fast Fourier transform analysis was performed for each 4096-data point and averaged over the entire sampling time. The estimated uncertainties in the 1/3-octave band SPL, including uncertainties associated with the instruments, jet conditions, duration, and sampling rate, were ±0.47 dB for the heat-simulated jets at 400 Hz, which was the minimum frequency. The SPL data were corrected for atmospheric absorption and each microphone spectral response, and subsequently normalized as SPL per unit St for a reference distance of 100 D.

Figure 5 shows a typical result of the comparisons between the measurements and predictions of the four types of jet noise prediction models, namely, Stone2 [34], SAE ARP876, modified SAE ARP876, and Viswanathan's scaling law [35]. The jet condition had a jet velocity of 392.8 m/s and a jet total temperature ratio (TTR) of 1.75. The condition was set to simulate the takeoff. The results of the validation tests showed that the modified SAE ARP876 model used in this study and the original SAE ARP876 model used by NASA in previous studies showed good agreement between the predictions and experimental values under the jet conditions covering the STCA takeoff and cutback ratings.

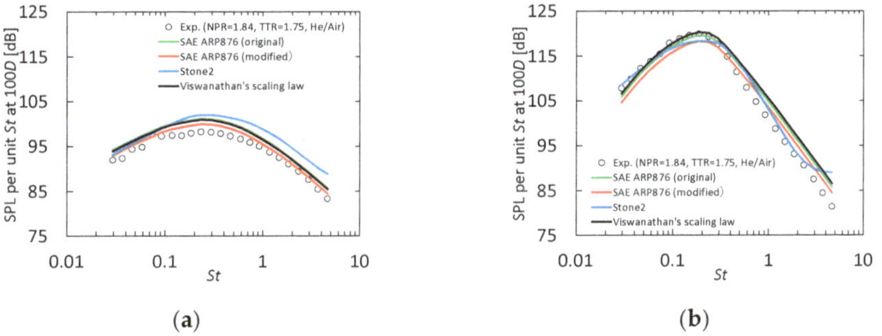

Figure 5. Comparison between the predicted and measured SPLs for a single heat-simulated jet operated at the jet velocity of 392.8 m/s and the jet total temperature ratio of 1.75: (**a**) θ = 90°, (**b**) θ = 150°.

The statistical differences between the predicted and measured SPLs were examined for the jet noise model used in this study. The average difference was approximately 1 dB (over-prediction) at θ = 90°, and less than the measurement uncertainty at θ = 150°. The standard deviation was more sensitive to changes in the jet noise spectral shape, due to the polar angle. The values were 0.6 dB at θ = 90°, and 2 dB at θ = 150°. Further details of the validation test, including other results not described here, can be found in Ref. [36].

3.2. Ground Noise Measurement Tests Using JAXA's Experimental Aircraft

Ground noise measurement tests using a JAXA experimental aircraft (Hisho [37]) were conducted at Nagoya Airport Flight Research Center to validate the prediction of the jet noise generated by the engine, similar to that anticipated for use in the NASA 55t STCA. The original aircraft, a Cessna 680, was equipped with two PW306C engines on its rear fuselage. The engine was a two-spool mixed-flow turbofan. The bypass ratio under static sea-level conditions was 4.5.

The layout of the aircraft and measurement system, and a photograph of the test site, are shown in Figure 6. The noise was measured using eight microphones arranged in an arc, which was 40 times the nozzle exit diameter from the nozzle exit of the left engine, at polar angles θ ranging from 90° to 160° at 10° intervals. The microphones used for the acoustic measurements were 1/4-inch pressure-field microphones (B&K4938-L-002 and GRAS40BD). The sampling frequency and time were 102.4 kHz and 20 s, respectively. A 20 Hz high-pass filter was passed through the preamplifier. The estimated precision limit for the averaged SPL based on the results of the repeated tests was less than 0.6 dB.

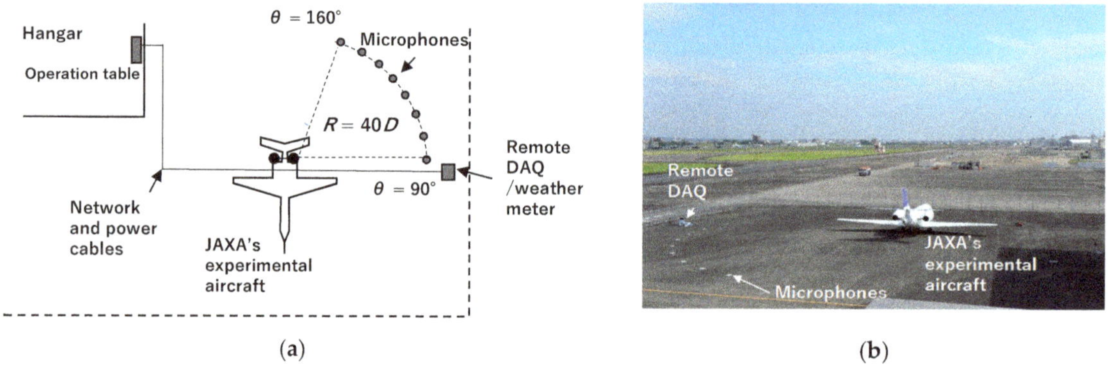

Figure 6. Experimental setup of ground noise measurement tests using JAXA's experimental aircraft: (**a**) schematic diagram of the layout of the instruments; (**b**) photograph of the test field.

For safety reasons, the maximum engine rating was restricted to low-pressure spool speed N1 = 95% for single-engine operation. The estimated jet velocity and TTR based on our engine cycle calculation system using the measured FADEC data were 322 m/s and 1.5, respectively, at N1 = 95%. This jet condition was similar to that of the cutback phase of NASA 55t STCA.

Figure 7 shows a typical result of the comparisons between the measurements and predictions of the four types of jet noise prediction models at N1 = 95%. The results showed that the modified and original SAE ARP876 models showed good agreement between the predictions and experimental values at θ = 90°, whereas the models underpredicted the jet noise at θ = 150°. The averaged differences of the modified SAE ARP876 used in this study were 1 dB (under-prediction) at θ = 90° and 2.7 dB (under-prediction) at θ = 150° The standard deviation was 0.8 dB at θ = 90° and approximately 1.8 dB at θ = 150°. Further information on this test can be found in Ref. [36].

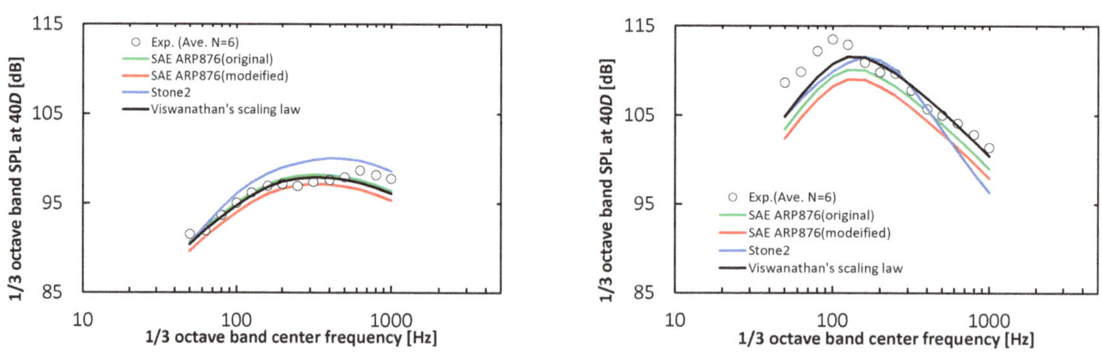

Figure 7. Comparison between the predicted and measured jet noises for single-engine operation at N1 = 95%, jet velocity =322 m/s, and TTR = 1.5: (**a**) θ = 90°; (**b**) θ = 150°.

3.3. Fan Noise Shielding Effect

Validation tests of the fan noise shielding effect were performed using a 10% scale delta wing model, which represents the initial planform of the STCA defined by NASA, and a simulated point sound source with high-pressure air. Schematics of the experimental setup are shown in Figure 8. The tests were conducted in an anechoic room (12 m × 12 m × 9 m) at Kawasaki Heavy Industries. The cutoff frequency of the room was 160 Hz. The delta-wing model was made of an aluminum plate with a thickness of 3 mm, and it was supported by ropes from both the ceiling and the floor. Acoustic data were obtained using 21 GRAS 46BF microphones and Brüel & Kjaer LAN-XI 3052-A-030 systems. The data were sampled at 204.8 kHz and evaluated for differences with and without the wing model over the frequency range of 1 kHz to 80 kHz at 1/3-octave-band frequency.

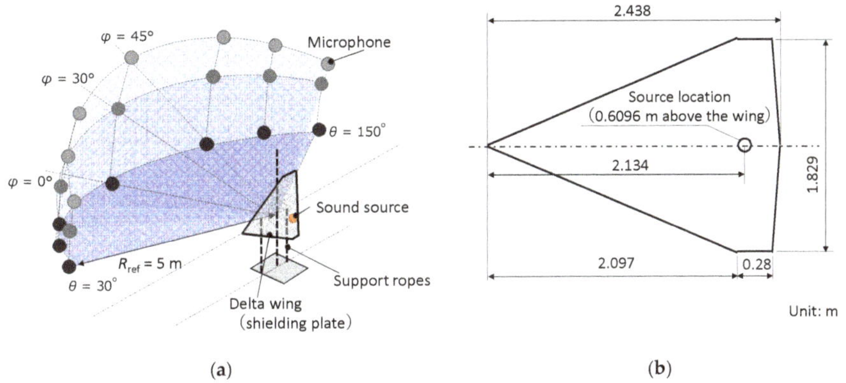

Figure 8. Schematics of scale model tests of the fan noise shielding effect: (**a**) experimental setup; (**b**) test model and the source location.

Comparisons between the predicted and experimental results are shown in Figure 9a,b. Prediction using ray-tracing with the Maekawa method can predict the trend of the shielding effect. However, the prediction model overestimated the effect over almost the entire frequency range. The mean difference between the prediction and the measured values over the frequency range was 8.9 dB for $\theta = 60°$, 5.2 dB for $\theta = 90°$, and 3.8 dB for $\theta = 120°$ at the direction below the aircraft ($\varphi = 0°$). Figure 9b shows a comparison of the shielding effects at $\varphi = 45°$. A trend similar to that of $\varphi = 0°$ can be seen at $\varphi = 45°$, and the mean difference between the predicted and measured values was slightly smaller than that for $\varphi = 0°$.

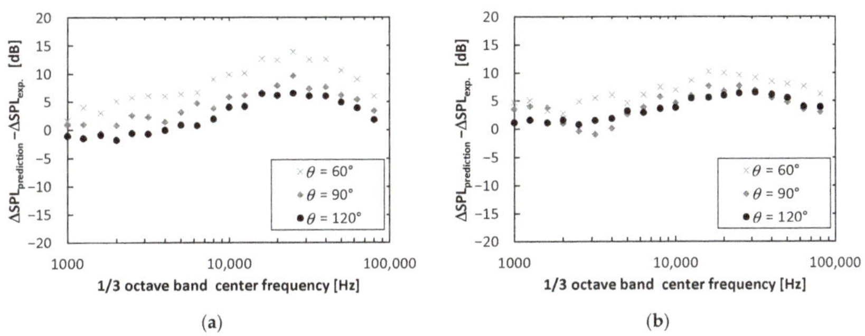

Figure 9. Results of validation tests for fan noise shielding effect: (**a**) $\varphi = 0°$, (**b**) $\varphi = 45°$.

3.4. Uncertainty Analysis of the System Noise Assessment

An uncertainty analysis was performed using a Monte Carlo simulation (MCS) to clarify how much variation could be expected when estimating the LTO noise of a conceptual supersonic aircraft with the best possible effort.

The uncertainty models required to perform the MCS are listed in Table 3. The uncertainty models used were almost the same as those used by NASA in a previous study [11]. The major differences between the previous study and the present study are the uncertainty models for jet noise and fan noise shielding effects, which were determined based on the validation tests described above.

Table 3. Uncertainty models used in the MCS.

No.	Item	Mode	Model	Min	Max	Std.dev.	Offset
1	Approach indicated airspeed, kt	Benchmark case	Triangular	−10	+10	-	-
2	Lateral indicated airspeed, kt	Benchmark case	Triangular	−10	+10	-	-
3	Flyover indicated airspeed, kt	Benchmark case	Triangular	−10	+10	-	-
4	Approach angle of attack, °	Benchmark case	Triangular	−1	+1	-	-
5	Lateral angle of attack, °	Benchmark case	Triangular	−1	+1	-	-
6	Flyover angle of attack, °	Benchmark case	Triangular	−1	+1	-	-
7	Flyover altitude, ft	Benchmark case	Triangular	−140	+140	-	-
8	Fan inlet noise adjustment (lateral), dB	-	Normal	-	-	1	1
9	Fan inlet noise adjustment (flyover), dB	-	Normal	-	-	1	1
10	Fan inlet noise adjustment (approach), dB	-	Normal	-	-	4	−4
11	Fan exit noise adjustment (lateral), dB	-	Normal	-	-	2	−2
12	Fan exit noise adjustment (flyover), dB	-	Normal	-	-	3	−3
13	Fan exit noise adjustment (approach), dB	-	Normal	-	-	3	−3
14	Core noise adjustment (lateral), dB	-	Normal	-	-	3	0
15	Core noise adjustment (flyover), dB	-	Normal	-	-	1	0
16	Core noise adjustment (approach), dB	-	Normal	-	-	1	0
17	Gear noise adjustment, dB	-	Normal	-	-	5	0
18	Flap noise adjustment, dB	-	Normal	-	-	5	0
19	Airframe trailing edge noise adjustment, dB	-	Normal	-	-	5	0
20	Inlet treatment effectiveness, dB	0	Triangular	−2	+2	-	-
21	Exhaust treatment effectiveness, dB	0	Triangular	−2	+2	-	-
22	Ground specific flow resistance, sl/s-ft3	291	Triangular	262	320	-	-
23	Shielding effect, dB	-	Normal	-	-	2.6	5
24	Jet noise adjustment (20°–90°), dB	-	Normal	-	-	1.1	0.1
25	Jet noise adjustment (100°), dB	-	Normal	-	-	1.0	0.2
26	Jet noise adjustment (110°), dB	-	Normal	-	-	1.3	0.8
27	Jet noise adjustment (120°), dB	-	Normal	-	-	1.1	0.6
28	Jet noise adjustment (130°), dB	-	Normal	-	-	1.2	1.3
29	Jet noise adjustment (140°), dB	-	Normal	-	-	1.7	1.9
30	Jet noise adjustment (150°), dB	-	Normal	-	-	1.6	1.4
31	Jet noise adjustment (160°), dB	-	Normal	-	-	2.1	1.5

For the jet noise uncertainty model, the results of the tests described in Sections 3.1 and 3.2 were summarized, and the uncertainty was modeled for each polar angle. For fan noise shielding, the uncertainty was modeled as independent of the polar and azimuthal angles based on the test results described in Section 3.3. Other uncertainty models were based on the values provided by NASA. Note that the analysis focused on the uncertainty of the acoustic prediction model. Uncertainties in aircraft design variables and specifications are outside the scope of this analysis.

The calculations were performed using JAXA's general-purpose parallel computing system utilizing Monte Carlo simulation [38]. The system can execute the AiNEST in parallel to accelerate the simulation. A total of 10,000 datasets were sampled and used to calculate statistical values for comparison with NASA's results.

4. Results and Discussions

4.1. Results of System Noise Assessment of NASA 55t STCA

The results of the system noise assessment of the NASA 55t STCA in this study are shown in Table 4, in comparison with the NASA results. The predictions for the benchmark case agreed well with those of the NASA. The ranges of the samples were slightly larger than those of NASA for all measurement points, whereas the standard deviations were in good agreement. They were 1.58 EPNdB, 0.56 EPNdB, and 1.11 EPNdB for the approach, lateral, and flyover, respectively. Although there is a difference in the selection of the noise and uncertainty models between the prior study performed by NASA and the present study, there is no significant difference between these two predictions and the statistics. The results of this cross-validation indicate that predictions and their statistics can be reproduced if the prediction model is appropriately selected for the target aircraft.

Table 4. Summary of the results and statistics of the system noise assessment of NASA 55t STCA.

Source of Data	NASA Ref. [11]				Present Study			
Statistic in EPNL	**Approach**	**Lateral**	**Flyover**	**Cumulative**	**Approach**	**Lateral**	**Flyover**	**Cumulative**
Benchmark case	96.4	93.0	87.0	276.4	97.5	93.0	86.1	276.6
Min. of samples	91.6	90.3	83.9	270.2	91.5	86.7	83.2	268.6
Max. of samples	102.6	97.5	92.9	286.8	103.7	96.1	97.1	286.1
Range of samples	11.0	7.2	9.0	16.6	12.2	9.4	13.9	17.6
Mean of samples	95.4	94.0	88.1	277.5	96.4	93.8	86.0	276.1
Standard deviation	1.33	0.96	1.32	2.27	1.58	0.56	1.11	2.13
LTO Noise Standard for Subsonic Aircraft					**Approach**	**Lateral**	**Flyover**	**Cumulative**
Noise limits in EPNL (Chapter4)					99.5	95.7	92.8	278.0
Noise limits in EPNL (Chapter14)					98.5	94.7	91.8	271.0

Figure 10 shows the jet, fan, and airframe noise levels in EPNL at each measurement point. Even with the advanced takeoff procedure, the contribution of jet noise is still large for the lateral measurement point, whereas for the flyover and approach measurement points, other components also have a large impact. This indicates that the small standard deviation of the lateral measurement point in Table 4 is because of the large influence of a single component: jet noise. In terms of the magnitude of LTO noise for supersonic aircraft, lateral measurement points are often considered to be an issue; however, flyover and approach measurement points are more difficult for noise prediction, and many factors need to be considered.

Figure 11 shows a histogram of the predicted cumulative noise. The shape of the histogram is close to a normal distribution, and there is no large degree of skewness or any outliers.

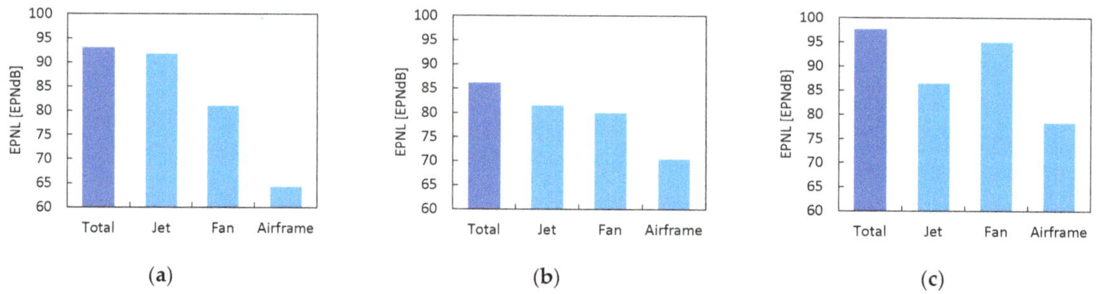

Figure 10. Predicted component noise: (**a**) lateral measurement point; (**b**) flyover measurement point; and (**c**) approach measurement point.

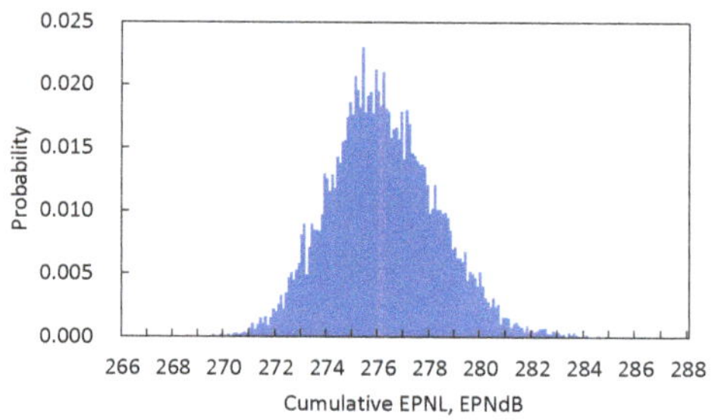

Figure 11. Histogram of the predicted cumulative noise.

4.2. Sensitivity Analysis and Discussion

A sensitivity analysis was performed based on the noise dataset obtained using MCS. In the analysis, the sum of the lateral and flyover noise levels was used as the explained variable, and a multiple regression analysis was repeated to extract explanatory variables with large correlations. The explanatory variables were each of the factors listed in Table 3, standardized by standard deviation and mean value.

The analysis revealed that the uncertainty of the total jet noise, fan exit noise during the flyover, and airframe trailing edge noise were identified in the order of the strongest correlation. In addition, when the correlation was analyzed for each polar angle of jet noise, a larger correlation was observed at a polar angle of 140°.

Figure 12a shows the values of the top three explanatory variables for the ten cases with the highest takeoff noise, and Figure 12b shows those for the ten cases with the lowest takeoff noise. Both figures clearly show that these three factors are correlated with takeoff noise levels.

Jet noise is a significant factor in the takeoff noise of supersonic aircraft, which have lower bypass ratio engines than recent subsonic airliners, which also support this feature. The strong correlation with jet noise at a polar angle of 140° can be explained by the directivity of the jet noise and the change in distance to the observer. This point should be considered in the design of jet-noise reduction devices.

The strong correlation with the fan exit noise at the flyover measurement point might be because the effect of the jet noise was relatively weakened by the cutback thrust over the flyover measurement point shown in Figure 10b, and shielding was assumed only for the fan inlet noise for the aircraft configuration of the NASA 55t STCA. This also implies that it is important to consider the shielding effect when shielding is assumed. Even if there is a relatively large uncertainty due to the low fidelity of the prediction method, it is recommended to employ the shielding model in the system noise assessment because it has a large impact on the fan noise.

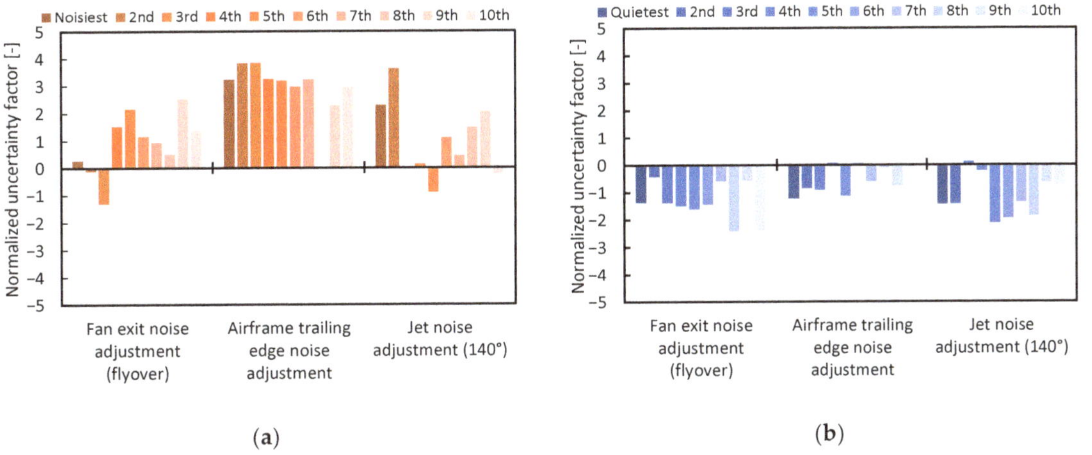

Figure 12. Normalized values of top three explanatory variables for: (**a**) the 10 cases with the loudest takeoff noise; (**b**) the 10 cases with the quietest takeoff noise.

The reason for the strong correlation with the airframe trailing-edge noise is that its directivity does not overlap with jet noise and shielded fan exit noise, and the aircraft is assumed to climb at a high speed due to the advanced takeoff procedure, as shown in Figure 10b. As for the prediction modeling of airframe trailing-edge noise of supersonic aircraft, there is a difference in the correction between the model used in this study and the model used by NASA, although both models originate from the same model. The airframe trailing-edge noise generated by the delta wing during high-speed climbout requires further discussion because there are not enough data to validate this.

5. Summary and Future Work

A system noise assessment of a conceptual supersonic aircraft called NASA 55t STCA and an uncertainty analysis using MCS were performed to investigate the prediction uncertainty of LTO noise and extract key factors to perform a system noise assessment of future supersonic aircraft.

In comparison with the prior study, the predicted noise levels showed good agreement both for the benchmark case and for the statistics of the sampled 10,000 predictions, although there was a difference in the selection of the noise models and uncertainty models between the prior study and the present study. The results of this cross-validation indicate that predictions and their statistics can be reproduced if the prediction model is appropriately selected for the target aircraft. The predicted cumulative noise level satisfied ICAO Chapter 4 noise standard, and its standard deviation was approximately 2 EPNdB.

The sensitivity analysis revealed that strong correlations with the takeoff noise were found for jet noise, fan exhaust noise at the flyover measurement point, and airframe trailing edge noise. Regarding the jet noise, there was a strong correlation between the noise at a polar angle of 140° and the takeoff noise. This should be considered in the design of jet-noise reduction devices. As for the fan exit noise at the flyover measurement point, it was suggested that the effect of fan noise was relatively large at the flyover point because of the cutback thrust, and that only the fan inlet noise was shielded, due to the STCA airframe configuration. This result indicates the importance of fan noise shielding. As for the airframe trailing edge noise, the relationship between takeoff noise and high-speed climbout is indicated; however, further study is needed.

In future, the LTO noise of different aircraft configurations will be investigated; the effect of aircraft configuration has been suggested in this study. Further validation and updating of the prediction model will be considered for the factors where uncertainty significantly affects the LTO noise estimation results.

Author Contributions: Conceptualization, J.A.; methodology, J.A. and T.I.; software, J.A.; validation, J.A.; formal analysis, J.A.; investigation, J.A.; resources, J.A.; data curation, J.A. and T.I.; writing—original draft preparation, J.A.; writing—review and editing, J.A. and T.I.; visualization, J.A.; supervision, T.I.; project administration, J.A.; funding acquisition, J.A. and T.I. All authors have read and agreed to the published version of the manuscript.

Funding: This research received no external funding.

Data Availability Statement: Not applicable.

Acknowledgments: The authors would like to thank H. Tomita, M. Naruoka, R. Hirai, H. Sawada, I. Shirota, K. Mizobe, T. Kumazawa, and T. Kumagai for their technical support with ground noise tests. We also thank H. Oinuma, K. Nagai, K. Wada, Y. Naka, S. Koganezawa, and O. Hamamura for their contributions to the experiments and useful discussions. T. Motoda developed and supported the MCS system used in this study. Special acknowledgement is extended to J. Berton for developing and providing the NASA 55t STCA model.

Conflicts of Interest: The authors declare no conflict of interest.

References

1. International Civil Aviation Organization. Chaps. 3, 4, 12 and 14. In *Annex 16 Environmental Protection Volume I Aircraft Noise*, 7th ed.; International Civil Aviation Organization: Montreal, QC, Canada, 2014.
2. Piccirillo, G.; Viola, N.; Fusaro, R.; Federico, L. Guidelines for the LTO Noise Assessment of Future Civil Supersonic Aircraft in Conceptual Design. *Aerospace* **2022**, *9*, 27. [CrossRef]
3. Stone, J.R.; Krejsa, E.A.; Halliwell, l.; Clark, B.J. Noise Suppression Nozzles for a Supersonic Business Jet. In Proceedings of the 36th AIAA/ASME/SAE/ASEE Joint Propulsion Conference and Exhibit, Las Vegas, NV, USA, 24–28 July 2000. AIAA 2000-3194.
4. Rask, O.; Kastner, J.; Gutmark, E.J. Understanding How Chevrons Modify Noise in a Supersonic Jet with Flight Effects. *AIAA J.* **2011**, *49*, 8. [CrossRef]
5. Semlitsch, B.; Cuppoletti, D.R.; Gutmark, E.J.; Mihăescu, M. Transforming the Shock Pattern of Supersonic Jets Using Fluidic Injection. *AIAA J.* **2019**, *57*, 5. [CrossRef]
6. Scupski, N.; Akatsuka, J.; McLaughlin, D.K.; Morris, P.J. Experiments with Rectangular Supersonic Jets with Potential Noise Reduction Technology. *J. Acoust. Soc. Am.* **2022**, *151*, 56–66. [CrossRef]
7. Berton, J.J.; Haller, W.J.; Senick, P.F.; Jones, S.M.; Seidel, J.A. *A Comparative Propulsion System Analysis for the High-Speed Civil Transport*; NASA/TM-2005-213414; NASA: Washington, DC, USA, 2005.
8. Huff, D.L.; Henderson, B.S.; Berton Jeff, J.; Seidel, J.A. Perceived Noise Analysis for Offset Jets Applied to Commercial Supersonic Aircraft. In Proceedings of the 54th AIAA Aerospace Sciences Meeting, San Diego, CA, USA, 4–8 January 2016. AIAA 2016-1635.
9. Henderson, B.; Huff, D.L. The Aeroacoustics of Offset Three-Stream Jets for Future Commercial Supersonic Aircraft. In Proceedings of the 22nd AIAA/CEAS Aeroacoustics Conference, Lyon, France, 30 May–1 June 2016. AIAA 2016-2992.
10. Berton, J.J.; Jones, S.M.; Seidel, J.A.; Huff, D.L. Advanced Noise Abatement Procedures for a Supersonic Business Jet. In Proceedings of the 23rd International Symposium on Air Breathing Engines (ISABE): Economy, Efficiency and Environment, Manchester UK, 3–8 September 2017. ISABE-2017-22538.
11. Berton, J.J.; Huff, D.L.; Geiselhart, K.; Seidel, J. Supersonic Technology Concept Aeroplanes for Environmental Studies. In Proceedings of the AIAA Scitech 2020 Forum, Orlando, FL, USA, 6–10 January 2020. AIAA2020-0263.
12. Yoshida, K. Supersonic drag reduction technology in the scaled supersonic experimental airplane project by JAXA. *Prog. Aerosp. Sci.* **2009**, *45*, 124–146. [CrossRef]
13. Akatsuka, J.; Ishii, T. Experimental and Numerical Study of Jet Noise Reduction for Supersonic Aircraft Using Variable Folding Nozzle Concept. In Proceedings of the 2018 AIAA/CEAS Aeroacoustics Conference, Atlanta, GA, USA, 25–29 June 2018. AIAA 2018-3612.
14. Akatsuka, J. *Development of Aircraft Noise Estimation Tool*; JAXA RR 2016-0005; JAXA: Tokyo, Japan, 2017. (In Japanese)
15. Akatsuka, J.; Ishii, T. System Noise Assessment of NASA Supersonic Technology Concept Aeroplane Using JAXA's Noise Prediction Tool. In Proceedings of the AIAA Scitech 2020 Forum, Orlando, FL, USA, 6–10 January 2020. AIAA202-0265.
16. Nöding, M.; Schuermann, M.; Bertsch, L.; Koch, M.; Plohr, M.; Jaron, R.; Berton, J.J. Simulation of Landing and Take-Off Noise for Supersonic Transport Aircraft at a Conceptual Design Fidelity Level. *Aerospace* **2022**, *9*, 9. [CrossRef]
17. Zorumski, W.E. *Aircraft Noise Prediction Program Theoretical Manual, Parts 1 and 2*; NASA TM-83199; NASA: Washington, DC, USA, 1982.
18. *ARP 876E*; Gas Turbine Jet Exhaust Noise Prediction. Society of Automotive Engineers (SAE): Warrendale, PA, USA, 2012.
19. Viswanathan, K.; Czech, M.J. Measurement and Modeling of Effect of Forward Flight on Jet Noise. *AIAA J.* **2011**, *49*, 216–234. [CrossRef]
20. Fink, M.R. *Airframe Noise Prediction Method*; FAA RD-77-29; FAA: Washington, DC, USA, 1977.

21. Ogino, S.; Kanazaki, M.; Ito, Y.; Murayama, M.; Yamamoto, K. Noise Evaluation Based on Geometrical Acoustics Approach and Its Application to Fan Noise Shielding Effect of Nacelle Over Wide-Body Aircraft. In Proceedings of the 2018 JSASS North Branch Annual Meeting, Sendai, Miyagi, Japan, 3–5 March 2018. (In Japanese)
22. Kontos, K.B.; Krafta, R.E.; Gliebe, P.R. *Improved NASA-ANOPP Noise Prediction Computer Code for Advanced Subsonic Propulsion Systems Volume 2: Fan Suppression Model Development*; Technical Report NASA CR 202309; NASA: Washington, DC, USA, 1997.
23. Department of Transportation and Federal Aviation Administration. Noise Certification of Supersonic Airplanes. Technical Report. Federal Register, Notice of Proposed Rulemaking (NPRM). 2021; Volume 85, No. 71; pp. 20431–20447. Available online: https://www.govinfo.gov/content/pkg/FR-2020-04-13/pdf/2020-07039.pdf (accessed on 21 February 2022).
24. Kontos, K.; Janardan, B.; Gliebe, P. *Improved NASA-ANOPP Noise Prediction Computer Code for Advanced Subsonic Propulsion Systems*; Technical Report NASA-CR-195480; NASA: Washington, DC, USA, 1996.
25. Emmerling, J.J.; Kazin, S.B.; Matta, R.K. *Core Engine Noise Control Program. Vol. III, Supplement 1-Prediction Methods*; FAA RD-74-125, III-I; FAA: Washington, DC, USA, 1976.
26. International Organization for Standardization (ISO). *Acoustics—Attenuation of Sound during Propagation Outdoors. Part 1: Calculation of the Absorption of Sound by the Atmosphere*; Technical Report ISO 9613-1:1993; ISO: Geneva, Switzerland, 1993.
27. *ARP 866B*; Standard Values of Atmospheric Absorption as a Function of Temperature and Humidity. Society of Automotive Engineers (SAE): Warrendale, PA, USA, 2012.
28. Chien, C.F.; Soroka, W.W. Sound Propagation Along an Impedance Plane. *J. Sound Vib.* **1975**, *43*, 9–20. [CrossRef]
29. Society of Automotive Engineers (SAE). *Method for Predicting Lateral Attenuation of Airplane Noise*; Aerospace Information Report, AIR5662; Society of Automotive Engineers (SAE): Warrendale, PA, USA, 1981.
30. Maekawa, Z. Noise reduction by screens. *Appl. Acoust.* **1968**, *1*, 157–173. [CrossRef]
31. Ishii, T.; Yamamoto, K.; Nagai, K.; Ishii, Y. Overview of Jet Noise Measurement at JAXA. *Trans. Jpn. Soc. Aeronaut. Space Sci. Aerosp. Technol. Jpn.* **2014**, *12*, 123–132. [CrossRef]
32. Doty, M.J.; McLaughlin, D.K. Acoustic and mean flow measurements of high-speed, helium-air mixture jets. *Int. J. Aeroacoustics* **2003**, *2*, 293–333. [CrossRef]
33. Tanna, H.K.; Dean, P.D.; Burin, R.H. *The Generation and Radiation of Supersonic Jet Noise: Vol. I11 Turbulent Mixing Noise Data*; U.S.A.F. Aero Propulsion Lab; Technical Report AFAPLTR-76-65; Air Force Aero Propulsion Laboratory: Dayton, OH, USA, 1976.
34. Stone, J.; Krejsa, E.; Clark, B.; Berton, J. *Jet Noise Modeling for Suppressed and Unsuppressed Aircraft in Simulated Flight*; Technical Report NASA/TM—2009-215524; NASA, Glenn Research Center: Cleveland, OH, USA, 2009.
35. Viswanathan, K. Improved Method for Prediction of Noise from Single Jets. *AIAA J.* **2007**, *5*, 151–161. [CrossRef]
36. Akatsuka, J.; Ishii, T. Comparative Study of Semi-empirical Jet Noise Prediction Models for Future Commercial Supersonic Aircraft. In Proceedings of the AIAA Aviation 2021 Forum, Virtual Event, 2–6 August 2021. AIAA2021-2219.
37. Tomita, H.; Naruoka, M. JAXA Flying Test Bed "Hisho". *Aeronaut. Space Sci. Jpn.* **2014**, *62*, 195–201. (In Japanese)
38. Motoda, T.; Stengel, R.F.; Miyazawa, Y. Robust control system design using simulated annealing. *J. Guid. Control. Dyn.* **2002**, *25*, 267–274. [CrossRef]

aerospace

MDPI

Article

On the Fundamental Possibility of a Supersonic Civil Aircraft to Comply with ICAO Noise Requirements Using Existing Technologies

Victor F. Kopiev [1,2,*], Ivan V. Belyaev [1,2], Andrey I. Dunaevsky [1], Andrey A. Poukhov [3] and Igor L. Trofimovsky [3]

1 Central Aerohydrodynamic Institute (TsAGI), 105005 Moscow, Russia; ivan.belyaev@tsagi.ru (I.V.B.); dunaevsky@tsagi.ru (A.I.D.)
2 WCRC "Supersonic", Zhukovsky, 140180 Moscow, Russia
3 National Research Center "Zhukovsky Institute", 125167 Moscow, Russia; poukhovaa@nrczh.ru (A.A.P.); itrofimovskii@mail.ru (I.L.T.)
* Correspondence: vkopiev@mktsagi.ru

Abstract: Compliance with environmental protection regulations, in particular, community noise requirements, constitutes one of the major obstacles for designing future supersonic civil aircraft. Although there are several noise sources that contribute to the total noise level of supersonic aircraft, it is the turbulent jet that appears most problematic; jet noise is a dominant noise source for low-to moderate- bypass-ratio engines, and at present there are no effective methods of jet noise reduction other than decreasing jet speed by increasing bypass ratio, which, in turn, is constrained by aerodynamic requirements for supersonic flight. The present study considers a concept of supersonic civil aircraft under the assumption that its total noise is determined by turbulent jets; it is shown that compliance of the supersonic aircraft with the current regulations for subsonic aircraft noise (Chapter 14 Volume I Annex 16 ICAO) would require the decreased jet speed that corresponds to a prohibitively high bypass ratio of aircraft engines. To enable jet noise reduction without necessarily increasing bypass ratio, a novel configuration of supersonic aircraft is proposed that meets the requirements of Chapter 14, thereby demonstrating that the norms of Chapter 14 are achievable for future supersonic civil aircraft with the use of existing technologies.

Keywords: aircraft noise; supersonic aircraft; noise certification; jet noise

Citation: Kopiev, V.F.; Belyaev, I.V.; Dunaevsky, A.I.; Poukhov, A.A.; Trofimovsky, I.L. On the Fundamental Possibility of a Supersonic Civil Aircraft to Comply with ICAO Noise Requirements Using Existing Technologies. *Aerospace* **2022**, *9*, 187. https://doi.org/10.3390/aerospace9040187

Academic Editors: Adrian Sescu and Lothar Bertsch

Received: 25 February 2022
Accepted: 27 March 2022
Published: 1 April 2022

1. Introduction

The era of supersonic civil aviation began on 31 December 1968 with the maiden flight of a Soviet Tupolev-144 from Zhukovsky, Russia and ended on 26 November 2003 with the final flight of a British–French Concorde to Filton, UK and the following retirement of the aircraft. Although there are no supersonic civil aircraft in operation at the moment, the vision of resumed supersonic passenger flights continues to encourage scientists and engineers to develop novel supersonic flight vehicles. However, future supersonic civil aircraft have to meet new challenges that were not so prominent for Tupolev-144 and Concorde at their design stage; one of most important among them is compliance with environmental protection regulations, in particular, community noise requirements, which became impressively more stringent from the time of 1960s–1970s. Another environmental issue specific to supersonic aircraft is their sonic boom; the sonic boom levels that the general population agrees to tolerate have decreased markedly from the time of development of the first-generation supersonic civil aircraft. Although low levels of sonic boom are imperative for supersonic civil aircraft, in this paper we address only the former problem, i.e., community noise.

Community noise is quantified by noise certification regulations that require aircraft to have noise levels at three certification points (takeoff, flyover, and approach) lower than some specified values in the EPNL metric. For modern subsonic aircraft, the noise level values are described in Chapter 14 of Volume I Annex 16 ICAO, and although next-generation supersonic aircraft are not yet subject to these requirements, the population expects them to be as quiet as their modern subsonic counterparts, i.e., complying with the norms of Chapter 14.

As a result, community noise turns out to be a strong constraint for the novel concepts of supersonic civil aircraft, because the advances in noise reduction for subsonic aircraft and their engines originated from their remarkable evolution over the last half-century; for instance, modern subsonic aircraft are powered by the turbofan engines with dramatically higher bypass ratios (BPR = 8.5 for PD-14, 12.5 for PW1100G, etc.) than those of the engines from 1960s (BPR = 0.3 for Rolls-Royce Conway 508, 1.0 for NK-8, etc.). Higher bypass ratios mean lower jet speeds and, therefore, significantly lower jet noise levels, which are proportional to the eighth power of the jet speed in accordance with Lighthill's theory [1]. Consequently, jet noise, which has been a dominant noise source for engines with low bypass ratios, is so reduced for high-bypass-ratio engines that for modern airliners it is often overcome by other noise sources such as fan noise, airframe noise, and interaction noise.

However, this evolution path of increasing bypass ratios looks like a dead end for supersonic vehicles, because high-bypass-ratio engines would greatly deteriorate aerodynamic performance of an aircraft during cruise at supersonic flight speeds, making such an aircraft configuration economically unviable. Indeed, there are estimates that bypass ratios beyond 3.0 would hinder the aircraft to efficiently pass the transonic regime and would lead to unacceptably poor performance at supersonic cruise [2]. As a result, although concepts of future supersonic civil aircraft tend to use engines with the maximum bypass ratios allowed by the aerodynamic considerations, the bypass ratios are still low to moderate.

This makes jet noise a dominant noise source for supersonic aircraft, and although there are several other noise sources that contribute to the total noise levels, it is the turbulent jet that appears most problematic for supersonic vehicles, because at present there are no effective methods of jet noise reduction other than decreasing jet speed, which implies increasing the bypass ratio of the engines to compensate for the thrust and is constrained by aerodynamic requirements, as discussed above. There are other available methods of jet noise reduction, which are reviewed in [3,4] and include such technologies as nozzle shape modifications [5,6] or active noise control [7–9], but they are not technologically mature or efficient enough.

Unlike the turbulent jet, the other noise sources of supersonic aircraft (fan noise, airframe noise, etc.) can be significantly mitigated with the existing methods of noise reduction. For example, concepts of next-generation supersonic aircraft often have engines with longer inlet ducts that are located above the airframe; these features allow fan noise to be reduced with acoustic liners inside engine ducts [10] and with noise shielding by the airframe [11]. During approach and landing, the engines are throttled down so that airframe noise, i.e., noise due to flow around landing gears, high-lift devices, and other airframe elements, can become an important noise source. Nevertheless, because of its strong dependence on the flight speed, airframe noise can be effectively reduced by decreasing the approach speed of the aircraft. Moreover, there are different methods of airframe noise reduction, both passive and active [12], that can be used to further reduce airframe noise. Therefore, although mitigation of these noise sources constitutes a formidable scientific and technical problem, major advances in its solution can be expected, so that jet noise will continue as a dominant noise source for supersonic civil aircraft with low-to-moderate-bypass-ratio engines.

As a result, jet noise sets the lower limit and a reference point for the noise levels of a given supersonic aircraft. On the one hand, lacking effective methods of jet noise reduction, it is unlikely that the aircraft will have noise levels below the limit. In other words, if the noise levels based on aircraft engine jets exceed the noise requirements, then the aircraft will definitely fail to comply with the noise regulations when the other noise sources are accounted for. On the other hand, having effective methods of noise reduction for the other noise sources, it is likely that their contribution to the total aircraft noise will be reduced by applying appropriate practices, so that the total noise will be close to the estimations based on the jet noise source alone. Therefore, estimations of the allowable jet speed limit do not just constitute a theoretical minimum for aircraft noise; they are realistic enough when the modern methods of noise reduction are applied to all noise sources for supersonic civil aircraft with low-to-moderate-bypass-ratio engines. Such assessments are particularly useful at the early stages of the aircraft design process, enabling selection of the aircraft configuration that can, at least in principle, comply with the noise regulations. The present study applies this approach to a concept of next-generation supersonic civil aircraft and estimates its noise levels, as well as the jet speed limit that allows the aircraft to comply with the norms of Chapter 14 while meeting aerodynamic requirements on thrust, flight speed, and trajectory. In addition, a novel concept of supersonic aircraft is proposed and studied that allows decreasing jet speed without necessarily increasing bypass ratio of the engines. The missing part of the thrust is obtained from another source. The paper is organized as follows. In the next Section, the approach to noise level prediction based on jet alone is validated by application to Tupolev-144, where the prediction results are compared with the available experimental noise data. In Section 3, the approach is applied to a concept of next-generation supersonic civil aircraft; its noise levels and the required jet speeds for the aircraft are determined. Based on these results, a novel evolution path for future supersonic aircraft is proposed in Section 4, which describes a concept of supersonic aircraft that meets the requirements of Chapter 14.

2. Jet Contribution to Tupolev-144 Noise Levels

Let us consider the method of estimation of aircraft noise based on jet noise source alone in application to the supersonic civil aircraft for which experimental data on its noise levels are available, such as for Tupolev-144. For Tupolev-144, the noise levels at the certification points can be found in [13].

The location of certification points at takeoff, flyover, and approach are schematically shown in Figure 1a. It should be noted that they correspond to the requirements of Chapter 2 of Volume I Annex 16 ICAO, because it is this Chapter that was then in effect. As a result, sideline microphones that measure noise at takeoff were installed at the distance of 650 m from the trajectory; this distance was reduced to 450 m in the later Chapters (Chapter 3, Chapter 4, and current Chapter 14). Noise at takeoff is determined as the maximum noise level at these sideline microphones; noise at flyover and approach is measured by single microphones in the corresponding points as shown in Figure 1a.

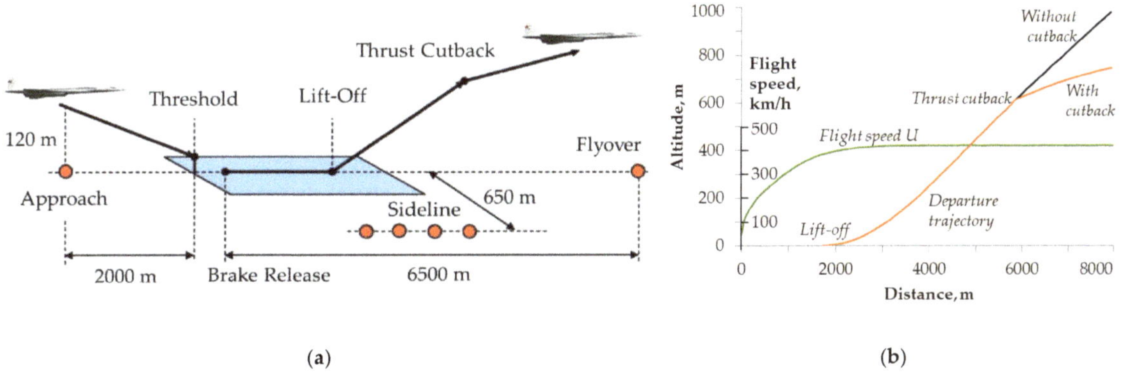

Figure 1. Parameters for Tupolev-144: (**a**) a scheme of the location of noise certification points; (**b**) the aircraft departure trajectory and flight speed.

As discussed in Introduction, we assume that the noise of the aircraft is determined by its jets; their parameters during departure and approach are provided in Table 1. Departure trajectory is shown in Figure 1b; at approach, the standard trajectory with 3° slope is used.

Table 1. Parameters of the jet during departure and approach for Tupolev-144.

Jet Parameter	Takeoff	Flyover	Approach
Jet speed, m/s	819	575	460
Total jet temperature, °K	1575	737	636
Jet Mach number	1.16	1.20	1.01
Nozzle area, m^2	1.32	0.834	0.834
Flight speed, m/s	118	118	80.5

Jet noise is calculated according to SAE ARP 876D. The accuracy of SAE ARP 876D for jet noise prediction was analyzed recently in [14] and exhibited robust and appropriate prediction results for jet noise. Since the jets are supersonic during departure and approach (their Mach number (M) > 1), the additional noise source associated with shocks (so-called BBSAN—Broadband Shocks Associated Noise [15]) is also included in the jet noise calculations according to SAE ARP 876D.

The EPNdB levels in the certification points are obtained with SOPRANO software [16], which was developed within the framework of EU SILENCE(R) project as a common tool to compare different technologies in terms of their effect on community noise. TsAGI performed a validation study of SOPRANO software through comparisons with the results of its own in-house code and experimental data, which showed a good agreement. In application to supersonic aircraft, TsAGI used SOPRANO for calculating noise of a Supersonic Technology Concept Aeroplane (STCA) developed by NASA ([17], p. 40); the obtained noise estimates agreed with those provided by NASA and JAXA.

Effects of atmospheric absorption (SAE ARP 866A), lateral attenuation (SAE AIR 1751), and ground reflection [18] are included in the calculation scheme. The results of EPNL calculations for Tupolev-144 at the certification points are provided in Table 2 and compared with the experimental data and the requirements of Chapter 2.

Table 2. Calculations of noise levels for Tupolev-144 at the certification points.

Noise Level	Takeoff	Flyover	Approach
Jet speed, m/s	819	575	460
Experiment, EPNdB	116.9	115.6	114.6
Calculations, EPNdB	116.1	114.6	114.0
Chapter 2, EPNdB	107.0	107.0	105.6

It can be seen that the results of calculations are quite close to the experimental data (within the range of 1 EPNdB); this encouraging agreement between the results shows that it is the turbulent jet that determines the noise of this aircraft at all three certification points. This allows us to use this approach of noise calculation for other supersonic aircraft as well, with the same assumptions. Comparison with the requirements of Chapter 2 shows that Tupolev-144 exceeded the noise levels for subsonic aircraft at all certification points with a large margin.

3. Noise Levels of Next-Generation Supersonic Aircraft Concept

As noted above, the experience in designing the first-generation supersonic civil aircraft (Tupolev-144 and Concorde) showed that their main sources of community noise are the high-speed jets. For the next-generation aircraft concepts, engines with higher bypass ratios up to BPR = 3.0 are considered. This allows jet speed to be reduced significantly, from 800 m/s down to 350–400 m/s, with the corresponding reduction of the noise levels.

We will discuss later whether such a reduction is sufficient and what can be done if this is not enough.

Let us consider as an example a concept of next-generation supersonic aircraft developed by TsAGI [19] to assess the allowable jet speed that makes it possible for the aircraft to comply with noise requirements of Chapter 14. A general view of the supersonic civil aircraft concept is shown in Figure 2a. It is a business jet with the maximum takeoff mass (MTOM) of 56 tons, powered by two engines with bypass ratio (BPR) of 2.5 at takeoff. The aircraft has a seating capacity of eight passengers and is designed for the cruise speed of $M = 1.8$ and flight range of 7400 km. The departure trajectory of the aircraft is shown in Figure 2b.

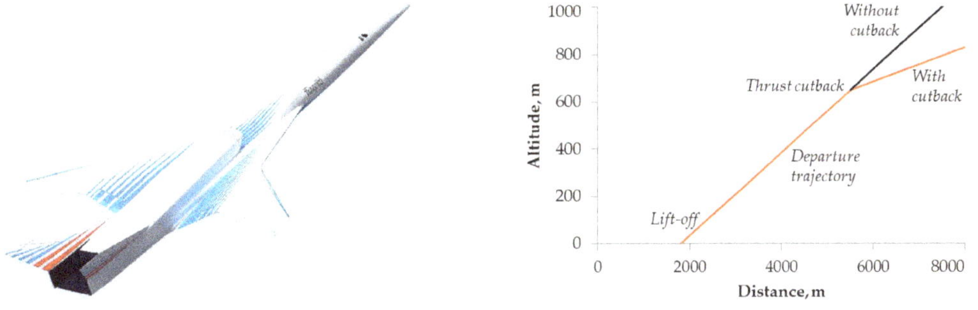

(a) (b)

Figure 2. A concept of next-generation supersonic civil aircraft developed by TsAGI: (**a**) a general view of the aircraft; (**b**) departure trajectory for the aircraft.

The turbulent jet will be considered as a dominant noise source of the supersonic aircraft in the same manner as has been done in Section 2 for Tupolev-144. Parameters of the turbulent jet for the aircraft during departure and approach are provided in Table 3. In should be noted that unlike Tupolev-144, the jets for this supersonic aircraft are subsonic during departure and approach (their Mach number (M) < 1).

Table 3. Jet parameters for the next-generation aircraft concept during departure and approach.

Jet Parameter	Takeoff	Flyover	Approach
Jet speed, m/s	395	345	214
Total jet temperature, °K	509	462	392
Jet Mach number	0.95	0.85	0.56
Nozzle area, m^2	1.02	1.02	1.02
Flight speed, m/s	100.6	100.6	83.3

The same calculation scheme as in Section 2 for Tupolev-144 is used here for determining noise levels at the certification points for the next-generation supersonic civil aircraft, with the difference that the distance to sideline microphones is set to 450 m. The results of these calculations are summarized in Table 4 and compared with the requirements of past Chapter 3 and current Chapter 14 for subsonic aircraft.

Table 4. Results of noise levels calculations for the next-generation supersonic civil aircraft.

Noise Level	Calculations	Chapter 3	Chapter 14
Takeoff, EPNdB	93.3	95.7	94.7
Flyover, EPNdB	87.4	89.9	88.9
Approach, EPNdB	86.4	99.6	98.6
Cumulative, EPNdB	267.1	285.2	268.2
Margin, EPNdB		18.1	1.1

As illustrated in Table 4, the limit of Chapter 14 for the maximum noise level at each of the three certification points is 1 EPNdB below the maximum noise level permitted by Chapter 3. Moreover, the limit of Chapter 14 for cumulative (total) noise, which is defined as an arithmetic sum of EPNdB noise levels at all three certification points, is 17 EPNdB below the cumulative noise permitted by Chapter 3.

From Table 4, it can be seen that for the next-generation supersonic civil aircraft the noise levels calculated by the jets alone fulfill the requirements of Chapter 14 at all certification points. Cumulative (total) noise, which is defined as an arithmetic sum of EPNdB noise levels at three certification points, meets the requirements of Chapter 14 with the margin of 1.1 EPNdB.

Nevertheless, one should note that the calculated noise level at approach is strikingly low; it is less than Chapter 14 requirements by a margin of 12.2 EPNdB. For comparison, one can mention that the quietest modern subsonic aircraft with MTOM = 56 tons (Embraer E190-E2, powered by two turbofan engines with BPR = 12) has the noise level 91.4 EPNdB at the approach certification point. It appears extremely unlikely that the noise of the supersonic aircraft at approach will be 5 EPNdB less than the noise of the quietest subsonic aircraft with the same MTOM. A reasonable explanation for this low noise level is that at approach, the jet of the considered aircraft has a rather low speed, so that the other noise sources (fan noise, airframe noise, etc.) produce noise levels similar or even exceeding the noise levels of the low-speed turbulent jet. Since the contributions of these noise sources are not accounted for in the calculations in Table 3, this leads to the underprediction of noise level specifically at approach.

Therefore, calculations of noise levels at approach have to include all noise sources of supersonic aircraft, which significantly complicates the calculation scheme, especially at the early design stage when a number of required parameters of the powerplant and airframe can still be undefined. A simpler method would be to use the noise level of an analogous subsonic aircraft at approach as a reference value for the noise level for the supersonic aircraft; however, it is unclear how to select the analogous subsonic aircraft. The choice of the reference subsonic aircraft can have major implications for cumulative noise assessment, since noise levels at approach for the subsonic aircraft with MTOM = 56 tons lie within a wide range from 91.4 EPNdB for Embraer E190-E2 to 100.1 EPNdB for Boeing 737-300.

For the purpose of the present study, the approach noise level of the supersonic aircraft is set equal to the requirements of Chapter 14 at the approach certification point. The idea behind such a choice is that if the approach noise level is higher than the Chapter 14 requirements, the aircraft will definitely fail to comply with Chapter 14. Therefore, the Chapter 14 requirements provide the upper limit for the approach noise. When these considerations are taken into account, total noise of the considered aircraft (Table 4) will exceed requirements of Chapter 14 by more than 10 EPNdB.

Let us now determine the jet speed of the supersonic aircraft that allows it to meet the requirements of Chapter 14. To ensure that the aircraft has the same thrust when the jet speed changes, the nozzle area will be changed simultaneously with the jet speed to compensate for the thrust losses. From the basic one-dimensional equations of fluid dynamics, we obtain a relationship between thrust F of the jet flow from a nozzle with area A_j and the jet speed V_j:

$$F = \rho V_j A_j (V_j - U) \tag{1}$$

where ρ is the fluid density and U is the flight speed. The density, ρ, and the jet speed, V_j, are related with the nozzle pressure ratio, π_s, by equations

$$\rho = m_r \frac{P_{st}}{T^*} \pi_s^{-\frac{\gamma-1}{\gamma}}, \; V_j = \sqrt{\frac{2\gamma}{\gamma-1}\left(1 - \pi_s^{-\frac{\gamma-1}{\gamma}}\right)\frac{T^*}{m_r}} \tag{2}$$

Here, P_{st} stands for static pressure, T^* for jet total temperature, γ for the adiabatic constant; m_r is the relative molar mass, i.e., the molar mass divided by the molar mass constant. Expressing the nozzle pressure ratio, π_s, through V_j and substituting it into the equation for the density, ρ, leads to the following expression:

$$\rho = m_r \frac{P_{st}}{T^*}\left(1 - \frac{\gamma - 1}{2\gamma} m_r \frac{V_j^2}{T^*}\right) \tag{3}$$

Substituting Equation (3) into Equation (1) results in a general relationship between the thrust F, the nozzle area A_j, and the jet speed V_j.

$$F = \frac{P_{st}}{T^*} m_r V_j A_j \left(1 - \frac{\gamma - 1}{2\gamma} m_r \frac{V_j^2}{T^*}\right)(V_j - U) \tag{4}$$

From Equation (4), we can obtain the value of nozzle area, A_j, for the decreased value of jet speed, V_j, that results in the same thrust as the baseline nozzle. For example, according to Table 3, the supersonic aircraft at takeoff has T^* = 509 °K and U = 100.6 m/s. Assuming the standard atmospheric conditions and air flow, we obtain P_{st} = 101350 Pa, γ = 1.4, and m_r = 0.029 / 8.31 = 0.00349. If the jet speed is decreased from the baseline value 395 m/s to, for example, 360 m/s, then to preserve the thrust F, the nozzle area A_j has to be increased from 1.02 to 1.31 m^2 according to Equation (4). Similar calculations can be performed for other jet speeds at takeoff. In calculations at the flyover point, it is assumed that the nozzle area, A_j, does not change and is known from the takeoff part, so that Equation (4) is used to determine the jet speed, V_j, after the cutback. The implications of decreasing jet speed for bypass ratio will be discussed later.

Since the engine thrust does not change, noise calculations assume that the aircraft trajectories do not change as well. Figure 3 demonstrates the cumulative noise levels of the supersonic aircraft for different jet speeds at takeoff obtained from the calculation procedure described above. It can be seen that a decrease in jet speed from 395 m/s to 335 m/s allows the supersonic aircraft to comply with Chapter 14 under the made assumptions.

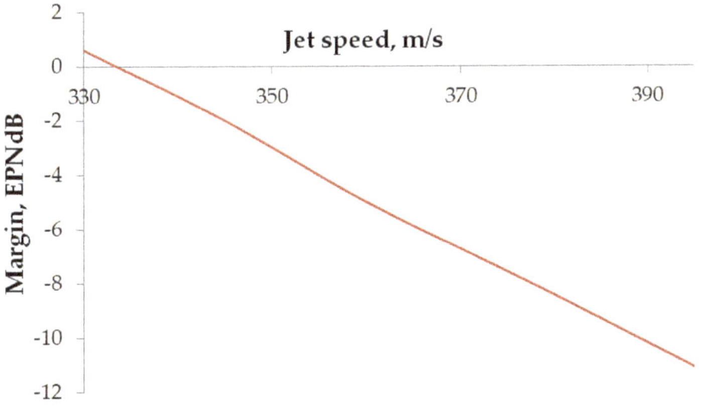

Figure 3. Noise margin with respect to Chapter 14 for different jet speeds at takeoff for the next-generation supersonic civil aircraft.

If the parameters of the turbofan engines are known, the jet speed can be related to its bypass ratio with engine-modeling software, such as NPSS [20] or pyCycle [21]. A simpler, although less accurate, approach would be to make use of the empirical relationship between the jet speed and bypass ratio obtained in [22]. According to it, the bypass ratio depends on jet speed, V_j, as $V_j^{-2.4}$. Applying this relationship for the baseline engine configuration, which has jet speed (V_j) of 395 m/s at bypass ratio 2.5, we obtain the jet speed (V_j) of 335 m/s for the engine with bypass ratio 3.7, which is larger than the allowable limit of 3.0 for bypass ratios of supersonic aircraft [2].

It should be noted that the value of 335 m/s is specific for the considered concept of supersonic aircraft; for other concepts the value may differ.

4. Novel Concept for Supersonic Civil Aircraft

The previous Section demonstrates that the concept of next-generation supersonic civil aircraft needs modifications to reach the objective of complying with Chapter 14 for subsonic aircraft. Some additional benefits can be obtained by utilizing advanced departure procedures such as "programmed lapse rate"; estimates by NASA for the STCA concept for supersonic civil aircraft demonstrated that implementation of such procedures reduced the cumulative noise level by 4 EPNdB [23]. Still, achieving this objective is a formidable technical problem, which is reflected in the proposed U.S. norms for supersonic civil aircraft by FAA [24]. In these norms, the requirements for cumulative noise levels are less than those of Chapter 14 by 3.5 EPNdB. For such an approach, in the probable case of making the ICAO norms for subsonic aircraft even more stringent, the possibility for supersonic civil aircraft to comply with the norms for subsonic aircraft seems virtually precluded.

It appears that to further reduce supersonic aircraft noise, novel aircraft concepts are required. To develop such a concept, let us note that the noise requirements are applied only to the landing and takeoff (LTO) part of aircraft flight envelope. The contradiction between aerodynamic and acoustic requirements lies in the fact that during LTO, the aircraft should have engines with low-speed jets (and therefore, as it seems, with high bypass ratios), whereas during the cruise, the bypass ratio of the engines should be minimized. This can be achieved, for instance, with variable-cycle engines that are under active development but still have low Technology Readiness Level. Another option is to use active control of jet noise based on, e.g., plasma actuators or water injection. The actuators operate during LTO and can be turned off during the cruise; however, their efficiency at the moment is quite low. One more option is to use buster turbofan engines that allow cruising engines to either operate at lower thrusts or be turned off altogether during LTO [25,26]. If the buster engines have a sufficiently high bypass ratio, such a configuration will produce the same noise levels as those of subsonic aircraft. During the cruise, the buster engines are either turned off or retracted inside the aircraft airframe so as to not deteriorate its aerodynamic performance at supersonic speeds. Such an approach has the drawbacks of additional mass associated with the buster turbofan engines, deterioration of aerodynamic performance during the cruise if the engines are not retracted, and high complexity of the retraction system for the engines if they are retracted.

However, the concept of aircraft with buster engines can be fundamentally rethought when turboprop engines are used instead of turbofan engines during LTO (Figure 4). This concept leverages the increasing usage of electrical systems and hybrid power plants in the quest for more electric aircraft, both subsonic and supersonic.

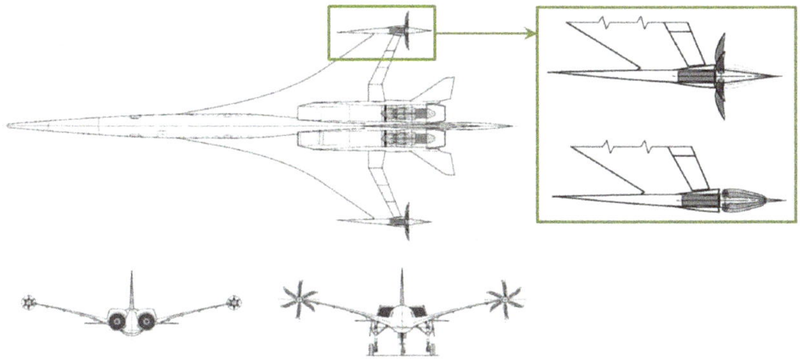

Figure 4. A scheme of notional supersonic aircraft with two cruise engines and two turboprop engines in pusher configuration.

The aircraft concept implies that during LTO, the thrust of the cruising engines (and their jet speed) is reduced, and the thrust is compensated with the turboprop electric engines. After reaching a certain Mach number, the propeller blades of the turboprop engines fold (Figure 4) so as to not deteriorate aerodynamic characteristics of the aircraft at the high-speed regimes, while the thrust of the cruise engines is increased. The concept will be effective if the propeller noise of buster engines is significantly lower than the noise due to higher-speed jets that in this case is removed. Taking into account the eighth-power law of noise dependency on jet speed, such a possibility seems realistic.

Let us assess the noise levels at the certification points for the supersonic civil aircraft, considered in Section 3, which is equipped with two additional turboprop engines with folded blades. The maximum takeoff weight (MTOM) of the modified aircraft with propellers is supposed to be the same as that for the baseline configuration. At departure, we assume that its cruising engines operate at 50% thrust of the baseline configuration, the corresponding jet speed, V_j, can be obtained from Equation (1). The other half of the required thrust is provided by turboprop engines. For the purpose of this study, we assume that the turboprop engines are identical; their propellers have six blades with the diameter of 3.6 m and rotate with speed 1200 RPM. The noise of the aircraft was also calculated with SOPRANO software with the additional propeller noise source included; it was modeled according to SAE AIR 1407. The departure trajectory of the aircraft with propellers is assumed to be the same as for the baseline aircraft.

The calculations of noise levels for the supersonic aircraft with propellers indicate that it has lower noise levels than those of the baseline configuration; noise reduction equals 9.0 EPNdB in the takeoff (sideline) certification point, to 6.3 EPNdB in the flyover certification point. This leads to a comfortable cumulative margin of 4.2 EPNdB with respect to Chapter 14, indicating that such an aircraft is a viable concept for ensuring compliance with the current and future noise requirements.

It should be noted that the value of 50% thrust in the calculations above was chosen rather arbitrarily. It is of interest to compare the cumulative noise levels of the supersonic aircraft for different ratios of thrust produced by turbofan and turboprop engines. Such a comparison is shown in Figure 5.

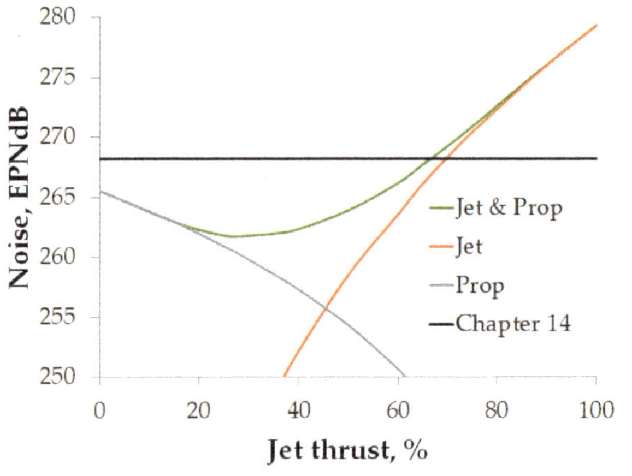

Figure 5. Cumulative noise levels for the supersonic aircraft with turboprop engines for different ratios of thrust produced by turbofan engines and turboprop engines.

From Figure 5, it can be seen that the supersonic aircraft meets the requirement of Chapter 14 when the thrust of its cruising engines (thrust produced by the jet) is 65% or less of the total required thrust. For large jet thrusts, the cumulative noise level is determined by the jet, whereas for small jet thrusts, the cumulative noise level is produced by propellers, as expected. It is also interesting to note that there appears a minimum of cumulative noise level at the jet thrust 30%, which indicates that parameters of turboprop engines can be optimized to ensure minimum noise during LTO cycle.

5. Discussion

Jet noise is a dominant noise source for supersonic aircraft and the most problematic one, since there are no efficient methods for its reduction at the moment, except for a decrease in jet speed.

Indeed, prediction methods based on jet noise alone succeed in predicting noise levels that are very similar to the experimental data for supersonic civil aircraft of the first generation, such as Tupolev-144. The same approach was applied to the next-generation aircraft for obtaining its noise levels at the certification points at the early stages of aircraft design. These assessments of noise levels allow determining whether the aircraft can in principle meet the noise requirements, because if the noise levels exceed the allowable limits, the aircraft will definitely fail to comply with the noise regulations when the other noise sources are accounted for. The calculation of noise levels for the concept of next-generation supersonic civil aircraft developed by TsAGI has shown that it does not meet the requirements of Chapter 14 and thus needs a decrease in jet speed (i.e., increasing the bypass ratio of the engines). This result implies that for this supersonic aircraft configuration, compliance with the noise requirements of subsonic aircraft is likely to be problematic.

At the first glance, this assessment seems to validate the idea that the next-generation supersonic civil aircraft cannot meet the requirements for noise levels of subsonic aircraft. The problem can be addressed either by setting different norms for supersonic aircraft than for subsonic aircraft (with the expected backlash from the population regarding the noise levels) or by considering novel concepts that can lead to a large decrease in community noise. One such a novel concept is proposed in this paper, and the first estimations for the supersonic aircraft with buster turboprop engines indicate that it leads to a significant noise reduction of 15.3 EPNdB with respect to the baseline case. Therefore, a more detailed study of the aircraft concept seems reasonable, where the other noise sources are accounted for. Moreover, based on aerodynamic or strength requirements or sonic boom reduction considerations, the aircraft concept can be modified to account for these considerations (Figure 6).

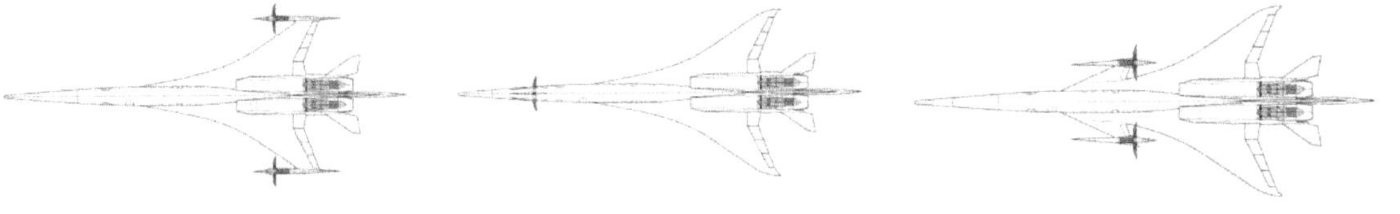

(a) (b) (c)

Figure 6. Different variants of supersonic aircraft with buster turboprop engines: (**a**) propellers in tractor configuration; (**b**) a propeller at the aircraft nose; (**c**) propellers at forward winglets.

Overall, the authors believe that these results support the idea that the current technological level allows designing supersonic civil aircraft that can meet the requirements on noise for their subsonic counterparts while ensuring viable aerodynamic performance during cruise, so the norms for supersonic aircraft noise levels should reflect this technological feasibility.

6. Patents

This work resulted in a patent application for invention No. 079634 "Supersonic aircraft" on 21 December 2021. Registration number RU2021137975.

Author Contributions: Conceptualization, V.F.K. and A.A.P.; methodology, V.F.K. and I.V.B.; validation, V.F.K. and A.A.P.; formal analysis, A.I.D. and I.L.T.; investigation, I.V.B. and A.I.D.; writing—original draft preparation, V.F.K. and I.V.B.; visualization, I.L.T.; supervision, V.F.K. All authors have read and agreed to the published version of the manuscript.

Funding: This research was funded by the Ministry of Science and Higher Education of the Russian Federation (grant no. 075-11-2020-023) within the program for the creation and development of the World-Class Research Center "Supersonic" for 2020–2025.

Data Availability Statement: Not applicable.

Conflicts of Interest: The authors declare no conflict of interest. The funders had no role in the design of the study; in the collection, analyses, or interpretation of data; in the writing of the manuscript, or in the decision to publish the results.

Abbreviations

The following abbreviations are used in this manuscript:

BPR	bypass ratio
EPNL	effective perceived noise level
FAA	U.S. Federal Aviation Administration
ICAO	International Civil Aviation Organization
JAXA	Japan Aerospace Exploration Agency
LTO	landing and takeoff
MTOM	maximum takeoff mass
NASA	U.S. National Aeronautics and Space Administration
SAE	Society of Automotive Engineers
STCA	Supersonic Technology Concept Aeroplane

Variables:

A_j	nozzle area
F	thrust
P_{st}	static pressure
M	Mach number
m_r	relative molar mass
T^*	jet total temperature
U	flight speed
V_j	jet speed
γ	adiabatic constant
π_s	nozzle pressure density
ρ	fluid density

References

1. Lighthill, M.J. On sound generated aerodynamically I. General theory. *Proc. R. Soc. Lond. A* **1952**, *211*, 564–587. [CrossRef]
2. Papamoschou, D.; Debiasi, M. Conceptual development of quiet turbofan engines for supersonic aircraft. *J. Propuls. Power* **2003**, *19*, 161–169. [CrossRef]
3. Nesbitt, E. Current engine noise and reduction technology. *CEAS Aeronaut. J.* **2019**, *10*, 93–100. [CrossRef]
4. Liu, X.; Zhao, D.; Guan, D.; Becker, S.; Sun, D.; Sun, X. Development and progress in aeroacoustic noise reduction on turbofan aeroengines. *Prog. Aerosp. Sci.* **2022**, *130*, 100796. [CrossRef]
5. Bridges, J.; Brown, C. Parametric testing of chevrons on single flow hot jets. In Proceedings of the 10th AIAA/CEAS Aeroacoustics Conference, Manchester, UK, 10–12 May 2004. [CrossRef]
6. Kopiev, V.; Zaytsev, M.; Ostrikov, N. Subsonic jet noise suppression by a corrugated nozzle. *Acoust. Phys.* **2013**, *59*, 207–209. [CrossRef]
7. Samimy, M.; Adamovich, I.; Webb, B.; Kastner, J.; Hileman, J.; Keshav, S.; Palm, P. Development and characterization of plasma actuators for high-speed jet control. *Exp. Fluids* **2004**, *37*, 577–588. [CrossRef]
8. Kopiev, V.F.; Bychkov, O.P.; Kopiev, V.A.; Faranosov, G.A.; Moralev, I.A.; Kazansky, P.N. Control of Instability Waves in an Unexcited Turbulent Jet Using Plasma Actuators in a Narrow Frequency Band. *Acoust. Phys.* **2021**, *67*, 413–420. [CrossRef]
9. Kopiev, V.F.; Akishev, Y.S.; Belyaev, I.V.; Berezhetskaya, N.K.; Bityurin, V.A.; Faranosov, G.A.; Grushin, M.E.; Klimov, A.I.; Kopiev, V.A.; Kossyi, I.A.; et al. Instability wave control in turbulent jet by plasma actuators. *J. Phys. D Appl. Phys.* **2014**, *47*, 505201. [CrossRef]
10. Envia, E. Fan noise reduction: An overview. *Int. J. Aeroacoustics* **2002**, *1*, 43–64. [CrossRef]
11. Denisov, S.L.; Ostrikov, N.N.; Granich, V.Y. Problems of Aviation Power Plants Noise Reduction by Means of Shielding Effect. *Acoust. Phys.* **2021**, *67*, 293–297. [CrossRef]
12. Zhao, K.; Okolo, P.; Neri, E.; Chen, P.; Kennedy, J.; Bennett, G.J. Noise reduction technologies for aircraft landing gear-A bibliographic review. *Prog. Aerosp. Sci.* **2020**, *112*, 100589. [CrossRef]
13. Zhmulin, E.M.; Munin, A.G.; Tupolev, A.A.; Cheremukhin, G.A. Local ground noise generated by supersonic transport planes. *Sov. Phys.-Acoust.* **1979**, *25*, 297–301.
14. Akatsuka, J.; Ishii, T. Comparative Study of Semi-empirical Jet Noise Prediction Models for Future Commercial Supersonic Aircraft. In Proceedings of the AIAA AVIATION 2021 FORUM, Chicago, IL, USA, 2–6 August 2021. [CrossRef]
15. Wong, M.H.; Jordan, P.; Maia, I.A.; Cavalieri, A.V.G.; Kirby, R.; Fava, T.C.L.; Edgington-Mitchell, D. Wavepacket modelling of broadband shock-associated noise in supersonic jets. *J. Fluid Mech.* **2021**, *918*, A9. [CrossRef]
16. Van Oosten, N. *SOPRANO Presentation (PDF)*; SOPRANO Workshop: Madrid, Spain, 21–22 June 2007.
17. European Union Aviation Safety Agency. Notice of Proposed Amendment 2020-06. 2020. Available online: https://www.easa.europa.eu/sites/default/files/dfu/NPA%202020-06.pdf (accessed on 14 February 2022).
18. Zaporozhets, O.I.; Tokarev, V.I.; Golembievskiy, G.G. Engine installation and sound propagation effects on aircraft noise levels in noise control points. In Proceedings of the 8th Aeroacoustics Workshop of the CEAS and 3rd Workshop of the X2-Noise EU Thematic Network on: Aeroacoustics of New Aircraft & Engine Configurations Impact of Aircraft Noise on Future Designs, Budapest, Hungary, 11–12 November 2004. Available online: https://www.win.tue.nl/ceas-asc/Workshop8/Documents/CEAS_Budapest_2004_Zaporozhets.pdf (accessed on 14 February 2022).
19. Chernyshev, S.L.; Lyapunov, S.V.; Wolkov, A.V. Modern problems of aircraft aerodynamics. *Adv. Aerodyn.* **2019**, *1*, 7. [CrossRef]
20. Lytle, J.K. *The Numerical Propulsion System Simulation: An Overview*; NASA/TM-2000-209915; Technical Report for NASA: Cleveland, OH, USA, 2000.
21. Hendricks, E.S.; Gray, J.S. pyCycle: A tool for efficient optimization of gas turbine engine cycles. *Aerospace* **2019**, *6*, 87. [CrossRef]
22. Michel, U. Correlation of aircraft certification noise levels EPNL with controlling physical parameters. In Proceedings of the 19th AIAA/CEAS Aeroacoustics Conference, Berlin, Germany, 27–29 May 2013. [CrossRef]
23. Berton, J.J.; Jones, S.M.; Seidel, J.A.; Huff, D.L. Noise predictions for a supersonic business jet using advanced take-off procedures. *Aeronaut. J.* **2018**, *122*, 556–571. [CrossRef]

24. Department of Transportation and Federal Aviation Administration. *Noise Certification of Supersonic Airplanes*; Technical Report; Federal Register, Notice of Proposed Rulemaking (NPRM): Washington, DC, USA, 2021; Volume 85, pp. 20431–20447. Available online: https://www.govinfo.gov/content/pkg/FR-2020-04-13/pdf/2020-07039.pdf (accessed on 14 February 2022).
25. Kopiev, V.F.; Ostrikov, N.N.; Chernychev, S.A.; Maslov, A.A.; Fournier, G.F. Are New Supersonic-Transport Configurations Insuring Engine Noise Reduction? In Proceedings of the 8th Aeroacoustics workshop of the CEAS and 3rd Workshop of the X2-Noise EU Thematic Network on: Aeroacoustics of New Aircraft & Engine Configurations Impact of Aircraft Noise on Future Designs (2004), Budapest, Hungary, 11–12 November 2004. Available online: https://www.win.tue.nl/ceas-asc/Workshop8/Documents/CEAS_Budapest_2004_Viktor_Kopiev.pdf (accessed on 14 February 2022).
26. Fournier, G.F. Supersonic-transport takeoff silencing. *Int. J. Aeroacoust.* **2004**, *3*, 249–258. [CrossRef]

aerospace MDPI

Article

Guidelines for the LTO Noise Assessment of Future Civil Supersonic Aircraft in Conceptual Design

Grazia Piccirillo [1,*], Nicole Viola [1], Roberta Fusaro [1] and Luigi Federico [2]

[1] Department of Mechanical and Aerospace Engineering, Politecnico di Torino, Corso Duca Degli Abruzzi 24, 10129 Torino, Italy; nicole.viola@polito.it (N.V.); roberta.fusaro@polito.it (R.F.)
[2] Centro Italiano Ricerche Aerospaziali, Via Maiorise, 81043 Capua, Italy; l.federico@cira.it
* Correspondence: grazia.piccirillo@studenti.polito.it

Abstract: One of the most critical regulatory issues related to supersonic flight arises from limitations imposed by community noise acceptability. The most efficient way to ensure that future supersonic aircraft will meet low-noise requirements is the verification of noise emissions from the early stages of the design process. Therefore, this paper suggests guidelines for the Landing and Take-Off (LTO) noise assessment of future civil supersonic aircraft in conceptual design. The supersonic aircraft noise model is based on the semi-empirical equations employed in the early versions of the Aircraft NOise Prediction Program (ANOPP) developed by NASA, whereas sound attenuation due to atmospheric absorption has been considered in accordance with SAE ARP 866 B. The simulation of the trajectory leads to the prediction of the aircraft noise level on ground in terms of several acoustic metrics (LAmax, SEL, PNLTM and EPNL). Therefore, a dedicated validation has been performed, selecting the only available supersonic aircraft of the Aircraft Noise and Performance database (ANP), that is, the Concorde, through the matching with Noise Power Distance (NPD) curves for LAmax and SEL, obtaining a maximum prediction error of ±2.19%. At least, an application to departure and approach procedures is reported to verify the first noise estimations with current noise requirements defined by ICAO at the three certification measurement points (sideline, flyover, approach) and to draw preliminary considerations for future low-noise supersonic aircraft design.

Keywords: LTO noise; SuperSonic Transport (SST); aircraft conceptual design

Citation: Piccirillo, G.; Viola, N.; Fusaro, R.; Federico, L. Guidelines for the LTO Noise Assessment of Future Civil Supersonic Aircraft in Conceptual Design. *Aerospace* **2022**, *9*, 27. https://doi.org/10.3390/aerospace9010027

Academic Editor: Adrian Sescu

Received: 25 October 2021
Accepted: 28 December 2021
Published: 4 January 2022

1. Introduction: Background and Motivation

The recent rise in environmental concern and renewed interest in supersonic flight has involved intense scientific activity that aims to realize a new generation of sustainable supersonic aircraft [1–7].

More than two decades ago, the Concorde project brought about a heated debate on the environmental impact of SST [8], which has led to the need for new Standards and Recommended Practices (SARPs) to ensure social acceptability for the next generation of supersonic aircraft [9]. Currently, the International Civil Aviation Organization (ICAO) Committee on Aviation Environmental Protection (CAEP) is cooperating with industries and research institutes to define a specific regulation for SST, allowing the certification of supersonic aircraft in the 2020–2025 time-frame [10].

Specifically, one of the most controversial and least accepted features of Concorde was the high community noise level around the airports, due to the higher thrust, jet speed and lift-off speed required for taking-off [11]. For this reason, one of the indispensable premises for the design of low-noise future supersonic aircraft is the integration of breakthrough technologies and flight procedures aimed at reducing noise, especially during LTO operations. To ensure that future supersonic aircraft will meet low-noise requirements, it is essential to move LTO noise evaluations up to the early stage of the design process. This will imply a paradigm shift in conceptual design towards a design-to-noise approach, including the integration of noise reduction measures together with preliminary evaluations of their impact on the overall aircraft configuration. Therefore, it will guarantee a significant saving in resources (time and money) and will avoid the generation of new aircraft concepts, which might not be socially acceptable.

To support this approach, this paper focuses on the application of a methodology aiming at predicting noise levels emitted by a supersonic aircraft during LTO operations since pre-conceptual studies (pre-phase A [12]). As a result, this approach will provide guidelines for the LTO noise assessment of future supersonic aircraft in conceptual design.

The supersonic aircraft noise model developed relies on the already existing extensive research about semi-empirical and parametric noise source models to assess aircraft noise, enabling a fast noise prediction within the design process. Since the 1970s, NASA Langley Research Center has started the development of the Aircraft NOise Prediction Program (ANOPP) [13], the first computer program with noise prediction capabilities integrable into the preliminary design process. Nowadays, many comprehensive similar tools have been developed by research institutes, such as the Parametric Aircraft Noise Analysis Module (PANAM) [14] and CARMEN [15]. A brief description for each of the already available tools is given below:

- One of the major applications of ANOPP has been to support the Supersonic Cruise Research (SCR) project at Langley, while the next application has been in conjunction with the Federal Aviation Administration (FAA) study to determine the economic and technological feasibility of noise limits for future supersonic transport [16]. The purpose is a high-fidelity system noise prediction along arbitrary flight paths, ANOPP embeds models for sound propagation, including the effects of the atmosphere and terrain, the installation effect, scattering and shielding. Over the years, the tool has been maintained and updated, including the ongoing development of new methods, essential for a more accurate physics-based prediction. Hence, in 2011, NASA announced the ANOPP2 release, which provides a modern prediction environment with a flexible framework meeting the needs of the future unconventional aircraft noise studies [17].
- PANAM has been developed by the German Aerospace Laboratory (DLR) to integrate noise prediction within the aircraft conceptual design and to support decision-making processes towards low-noise designs. Differently from ANOPP, the current version is only applicable to conventional tube-and-wing aircraft concepts [18]. However, PANAM uses proprietary source models for airframe noise, which are based on real modern aircraft [19]. Major aircraft noise components are modelled stand-alone, neglecting interactions. Sound propagation and convection effects are directly applied to the emitting noise source, to be more representative of the actual flight operating conditions [14].
- CARMEN is a tool developed by French aerospace laboratory ONERA and is connected with IESTA [20], a proprietary modular distributed simulation platform for the evaluation of air transport systems. The model dedicated to aircraft noise can predict the noise footprint around the airport on existing and future aircraft and is composed of three modules: the acoustic source models, the installation effects and the atmospheric propagation [15].

A comparison between these simulation tools has been carried out in [21]. However, all these methods are very similar, with the remaining differences in the individual code implementation. Indeed, Table 1 lists the different semi-empirical models employed within the three tools, considering the following noise contributions:

- Airframe noise, comprised of: clean/trailing edge noise and flap side edge noise (abbreviation: t.e.), leading edge noise (abbreviation: l.e.), main landing gear noise (abbreviation: m.g.), nose landing gear noise (abbreviation: n.g.);
- Engine tonal and broadband noise, comprised of: fan broadband noise (abbreviation: fan bb), fan tonal noise (abbreviation: fan t), jet noise (abbreviation: jet).

Table 1. Summary of simulation models used within PANAM, ANOPP, and CARMEN [21].

	PANAM	ANOPP	CARMEN
Airframe noise models			
t.e.	DLR Airframe	Boeing: Flap, Fink: Trailing edge	Boeing: Flap, Fink: Trailing edge
l.e.	DLR Airframe	Boeing Airframe	DLR Airframe
m.g.	DLR Airframe	Boeing Airframe	DLR Airframe
n.g.	DLR Airframe	Boeing Airframe	DLR Airframe
Engine noise models			
fan bb	mod. Heidemann	Heidmann Fan, GE Large Fan Option	mod. Heidmann Fan and Kontos
fan t	mod. Heidemann	Heidmann Fan, GE Large Fan Option	mod. Heidmann Fan and Kontos
jet	mod. Stone	Stone2	mod. Stone
Propagation effects	ISO9613	ISO9613	ISO9613

PANAM, ANOPP and CARMEN involve models that represent the state-of-the-art in the field of aircraft noise prediction and are a benchmark for the development of further methods to assess the aircraft system noise.

Unfortunately, none of these tools are open source, thus limiting the exploitation at universities or for research purposes. Although many of the underlying equations used are shown in reports for PANAM [22,23] and ANOPP [24,25], the full and most updated methods are generally not open access. In other cases, the noise models require too many input variables, which might be unknown at the beginning of the design process. In addition, all semi-empirical models developed for conceptual design have been applied to predict the overall aircraft noise of subsonic aircraft only. In some cases, the applicability of these models to supersonic case studies is only theoretically discussed, and no real applications are reported. As far as ANOPP is concerned, the available literature confirms that the model was developed by NASA to support the SCR project, even if real applications are missing as well.

To verify the applicability of already existing models and to guarantee their exploitation at the very beginning of the SST aircraft design process, this paper discloses the application of a methodology that includes a supersonic aircraft noise model in line with the noise models implemented in early versions of ANOPP. This paper confirms the adequacy of this supersonic aircraft noise model to Concorde-like configuration by means of a dedicated validation. In order to overcome the lack of experimental data for modern civil supersonic aircraft required by consolidated validation procedures [26], an alternative approach, more suitable for the conceptual design stage, is addressed in this paper. The methodology is employed to predict NPDs data, thus the accuracy assessment for flyover trajectories is carried out by the comparison with NPD curves provided by the ANP database [27]. ANP is an open-source database provided by Eurocontrol Experimental Centre, which contains noise data for any specific certified aircraft. To support the computation of noise contours around airports, the database is also comprised of NPDs and spectral classes data. NPD curves define received sound event levels directly beneath the aircraft as a function of distance, for steady straight flight at a reference speed and atmospheric conditions in a specified flight configuration.

Notwithstanding the simplified approach, the validation with Concorde data shows that a good accuracy is reached for flyover trajectories. In addition, quantitative estimations for noise increments resulting from the afterburner are derived.

The approach suggested in this paper will also support the ongoing development and update of certification guidelines for the future SST. Considering default flight procedure, the overall noise level produced by traditional supersonic aircraft is assessed at the three certification measurements. Then, the results obtained are compared with the current LTO noise limitations reported in ICAO Annex 16, Volume I, Chapter 12 [28].

Precisely, ICAO recommends taking SARPs defined for subsonic jet aeroplanes as guidelines for new generations of SST aircraft. However, significant differences exist between supersonic and subsonic aircraft, so up-to-date noise standards could not be appropriate for the supersonic case. Therefore, dedicated studies are being undertaken towards the identification of advanced departure and approach flight procedures aimed at reducing noise especially for take-off conditions [29–31].

In this context, the EU is supporting ICAO efforts to fund research activities on environmentally sustainable supersonic aviation through the MORE&LESS (MDO and REgulations for Low-boom and Environmentally Sustainable Supersonic aviation) project, a Horizon 2020 initiative which aims to develop a holistic framework that is able to assess the environmental impact of supersonic aviation through a multidisciplinary approach [32].

In conclusion, the current study deals with the application of a simplified LTO noise prediction methodology tailored to civil supersonic aircraft that aims at assessing noise emission from the beginning of the design process. The roadmap followed to achieve this objective is summed up in Figure 1. The present Section specifies the motivations and the background on which the intention of this work lies, also providing an overview of the state-of-the-art in aircraft noise prediction methods. Thereupon, Section 2 describes the overall methodology framework, focusing on the most relevant parts concerning aircraft noise evaluations. Subsequently, Section 3 comprises the validation with ANP experimental data and the application of the proposed methodology to departure and approach procedures, considering the Concorde as a case study. Lastly, conclusions and possible further improvements are drawn in Section 4.

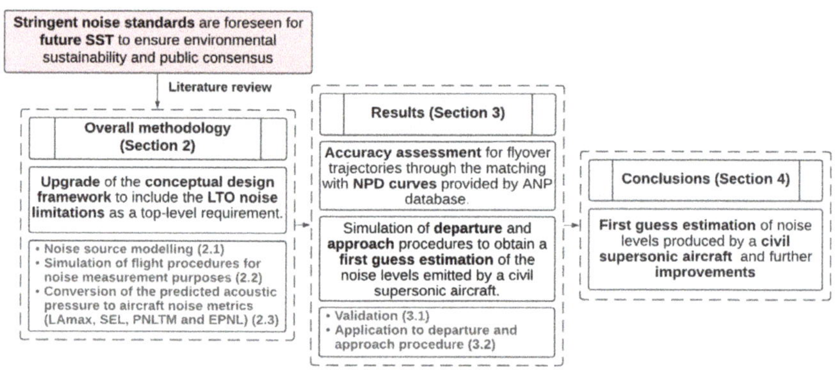

Figure 1. Roadmap of the performed activities.

2. Overall Methodology

Integrated conceptual design approaches for future advanced aircraft should be drawn on a specific set of top-level requirements, as reported in the literature, for example, by Raymer [33] and Torenbeek [34]. Hence, the first step towards a design-to-noise approach for future SST shall be the upgrade of the conceptual design framework to include LTO noise limitations within this set. Specifically, in order to increase the public consensus towards future SST and to foster their environmental sustainability, the maximum noise levels defined as requirements shall be in line with current standards developed for subsonic civil aviation. Relying on this list of requirements, a first guess of weights and all relevant geometrical, aerodynamics and performance parameters is given and details about the aircraft configuration and engine type are specified, together with the concept of operations. It is essential to incorporate the aircraft noise prediction at this early stage, when all other initial performance analyses are performed [35–39].

The determination of noise levels received on the ground is a complex multidisciplinary problem that must be simplified to be included at a conceptual design level. Figure 2 suggests a possible way to embed aircraft noise prediction into the traditional conceptual design process framework. In detail, in Figure 2, aircraft noise prediction analyses are added to enrich the original activity flow as reported in textbooks [33,34] and are marked in red.

A noise source model is needed to provide a mathematical formulation which correlates aircraft noise generation with design and operational parameters. Since there are still some uncertainties about the configuration in this phase of the project, interaction and installation effects are neglected and each noise source is modelled separately. As take-off and approach segments are fully characterized, the aircraft noise analysis can be performed, obtaining a first guess estimation of noise levels at the three certification measurement points defined by ICAO [28]. Thus, noise levels emitted by each source are predicted in terms of mean-square acoustic pressure and are then easily converted to the corresponding Sound Pressure Level (SPL), as a function of frequency expressed in a 1/3 octave centre frequency band. To evaluate the noise received by the observer on the ground, propagation effects shall be considered. In line with SAE ARP 866 B [40], sound losses due to atmospheric absorption are estimated as a function of frequency and temperature. Furthermore, the methodology suggested in this paper includes the possibility to convert the SPL into a set of well-established noise metrics (LAmax, SEL, PNLTM and EPNL). Ultimately, each design loop shall end with requirements verification, including LTO noise related requirements. If the aircraft exceeds the noise requirements, the iterative design-to-noise approach suggests the introduction of noise mitigation technologies or the evaluation of alternative flight procedures.

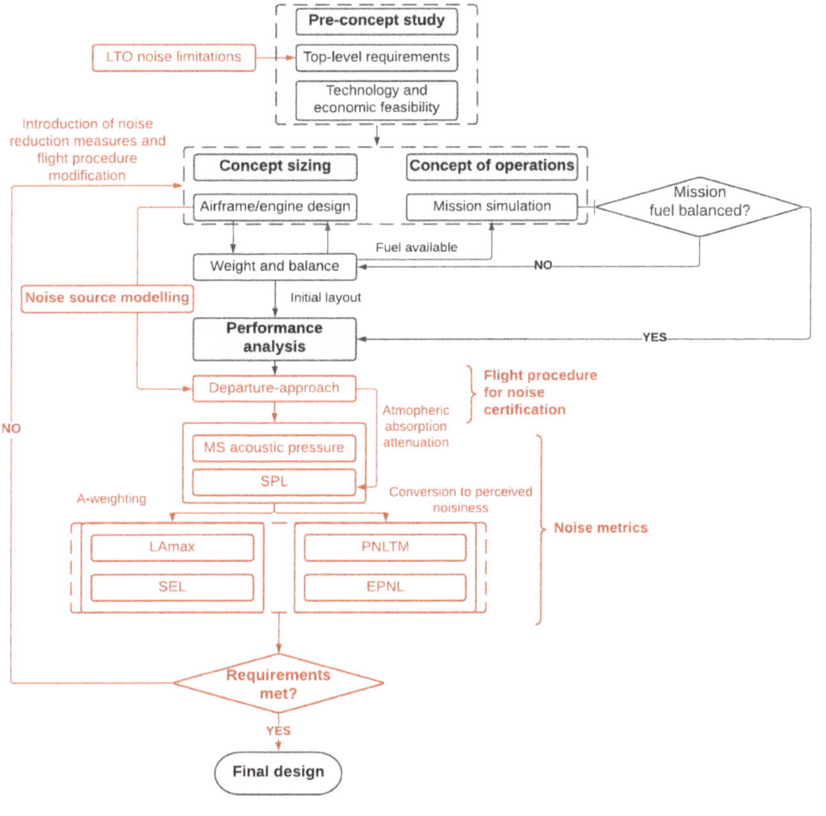

Figure 2. Overall methodology framework supporting a design-to-noise approach.

2.1. Noise Source Modelling

A theoretical and experimental review about noise generation mechanisms for each aircraft component is available in [41]. Complementary noise source modelling techniques are described in [42,43], classified on the basis of their fidelity-levels and suggested for specific applications accordingly. Among these model techniques, the scientific methods have prevailed to estimate the aircraft noise [43].

This paper follows this approach, which starts with the identification of the major noise sources on board the aircraft. Specifically, a clear distinction between non-propulsive noise (airframe) and propulsive noise (engine) is made. Then, both non-propulsive and propulsive noise sources are further broken-down. Figure 3 suggests a break-down specifically tailored on supersonic aircraft.

Differently from the noise sources identified and described in [44] for a subsonic aeroplane, SSTs usually do not have high lift devices and horizontal stabilizer. Furthermore, interaction and installation effects are neglected, as well as turbo-machinery and combustion noise, to keep the approach simple and make it applicable at the conceptual design level. Hence, the airframe noise can be computed by spectrally summing the contributions of clean delta wing, vertical tail and landing gear noise. Similarly, engine noise can be computed by spectrally summing jet and fan noise contributions. Each sub-component identified is modelled using the equations suggested in ANOPP [24].

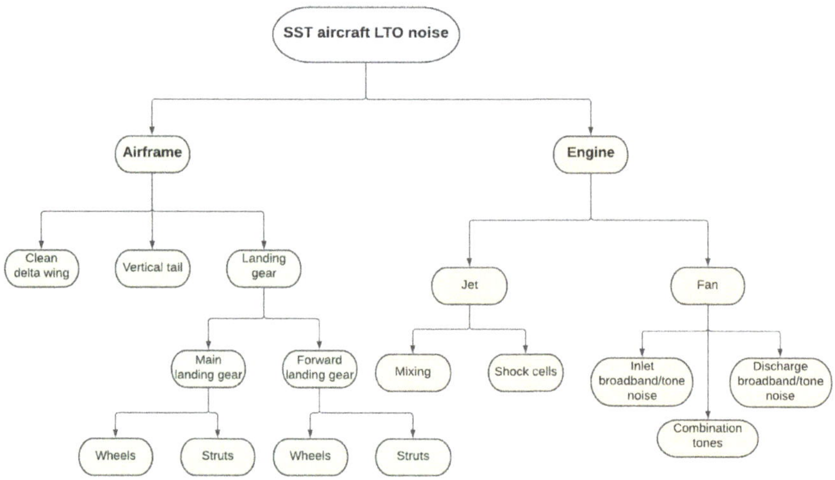

Figure 3. SST aircraft LTO noise sources break-down.

The ANOPP model correlates noise generation with operational and airframe/engine geometrical parameters, thus leading to the prediction of the mean-square acoustic pressure as a function of directivity angles and frequency. Once the mean-square acoustic pressure is computed for each sub-component, the total noise can be predicted by spectrally summing each acoustic pressure.

The procedure used to predict the overall aircraft noise level is illustrated in Equation (1), where the $< p^2 >^*$ is the dimensionless mean-square acoustic pressure, referred to as $\rho_\infty^2 c_\infty^4$, with ρ_∞ as the ambient density and c_∞ as the ambient speed of sound.

$$< p_{overall}^2 >^* = < p_{airframe}^2 >^* + < p_{engine}^2 >^* . \quad (1)$$

2.1.1. Airframe Noise

Despite the airframe noise not representing the predominant noise source of the aircraft, the introduction of high By-Pass Ratio (BPR) turbofans and the tightening of noise requirements led to a re-evaluation of noise produced by the airframe as a possible noise barrier [45]. A dissertation about the achievements of airframe noise research conducted in the last decades is presented in [46].

Several airframe noise prediction schemes can be employed. Specifically, this work benefits from the formulation suggested by Fink [47] and the mathematical formalism used in ANOPP. Fink's method is the first semi-empirical prediction method for airframe noise, based on a wide set of experimental data, which is still in use today. It is here applied to predicting the overall airframe noise of a generic supersonic aircraft as a combination of clean delta wing, vertical tail, and landing gear.

Noise radiation from a clean airframe, with all gear and high-lift devices retracted, is assumed to be entirely associated with turbulent boundary layer flow over the trailing edges of the wing and tail surfaces. The contribution of leading edges can be ignored as long as the airfoil chord remains large compared to the acoustic wavelength of the sound produced. Noise contributions from forward landing gear and main landing gear are calculated separately because the differences in architecture and size translate into different peak frequencies. In detail, each landing gear noise contribution is evaluated, considering only the most two dominant sources, which are struts and wheels.

In general, each airframe source can be mathematically modelled following Equation (2). Thereupon, the far-field mean-square acoustic pressure is calculated as:

$$< p^2 >^* = \frac{\Pi^*}{4\pi(r_s^*)} \frac{D(\theta,\phi)F(S)}{(1 - M_\infty \cos\theta)^4}, \tag{2}$$

where:

- Π^*: overall acoustic power, re $\rho_\infty c_\infty^3 b_w^2$;
- $D(\theta,\phi)$: directivity function;
- $F(S)$: spectrum function;
- S: Strouhal number;
- r_s^*: dimensionless distance from source to observer, re b_w;
- b_w: wingspan of the aerodynamic surface;
- $4\pi(r_s^*)$: spherical propagation factor;
- $(1 - M_\infty \cos\theta)^4$: Doppler factor accounting for the forward velocity effect;
- M_∞: aircraft Mach number;
- θ: polar directivity angle (deg);
- ϕ: azimuthal directivity angle (deg).

The acoustic power for the airframe Π^* can be expressed as:

$$\Pi^* = K(M_\infty)^a G, \tag{3}$$

where:

- K and a are constants determined from empirical data;
- G is a geometry function different for each airframe component and incorporated all geometrical effects on the acoustic power.

The values of K, a and G reported in Table 2 for each airframe noise source. Specifically, n, d and l are, respectively, the number of wheels per landing gear, the tire diameter and the strut length. The parameter δ^* is the dimensionless turbulent boundary-layer thickness, computed from the standard flat-plate turbulent boundary-layer model. Directivity functions and Spectrum function used for each airframe noise source are specified in Table 3.

Each described contribution is then summed over the 1/3 octave frequency band to predict the airframe noise (Equation (4)).

$$< p^2_{airframe} >^* = < p^2_{clean\ delta\ wing} >^* + < p^2_{vertical\ tail} >^* + < p^2_{landing\ gear} >^* . \tag{4}$$

Typically, landing gear noise is the most dominant airframe noise source during the LTO cycle, with the highest contribution during the approach phase [48]. This is true both for subsonic as well as supersonic aircraft. It is worth noting that, despite the differences in wing planform, configuration and landing gear size, the airframe noise of supersonic aircraft is expected to be comparable with the noise coming from a subsonic aircraft. However, a different conclusion could be drawn for unconventional designs. The methodology described in this section can be applied to supersonic aircraft, but it is tailored towards conventional wing-fuselage configuration (Concorde-like). Considering that future SSTs may be characterized by unconventional configurations, the inclusion of additional elements of the airframe noise break-down, such as canard or moving surfaces, or different architectures, such as Blended Wing Body (BWB), is needed to widen the applicability of the methodology.

Table 2. K, a and G for each airframe noise source [24].

	K	a	G
Clean delta wing (aerodinamically clean)	7.075×10^{-6}	5	δ_w^*
Vertical tail (aerodinamically clean)	7.075×10^{-6}	5	$\delta_v^* (\frac{b_v}{b_w})^2$
1-and-2 wheel landing gear wheel noise	4.349×10^{-4}	6	$n(\frac{d}{b_w})^2$
4 wheel landing gear wheel noise	3.414×10^{-4}	6	$n(\frac{d}{b_w})^2$
Landing gear strut noise	2.753×10^{-4}	6	$(\frac{d}{b_w})^2(\frac{l}{d})$

Table 3. Directivity function D and Spectrum function $F(S)$ for each airframe noise source [24].

	Directivity Function	Spectrum Function
Clean delta wing (aerodinamically clean)	$4\cos^2\phi\cos^2\frac{\theta}{2}$	$0.485(10S)^4[(10S)^{1.35}+0.5]^{-4}$
Vertical tail (aerodinamically clean)	$4\sin^2\phi\cos^2\frac{\theta}{2}$	$0.613(10S)^4[(10S)^{1.5}+0.5]^{-4}$
1-and-2-wheel landing gear wheel	$\frac{3}{2}\sin^2\theta$	$13.59S^2(12.5+S^2)^{-2.25}$
1-and-2-wheel landing gear strut	$\frac{3}{2}\sin^2\theta\sin^2\phi$	$5.32S^2(30+S^8)^{-1}$
4 wheel landing gear wheel	$\frac{3}{2}\sin^2\theta$	$0.0577S^2(1+0.25S^2)^{-1.5}$
4 wheel landing gear strut	$\frac{3}{2}\sin^2\theta\sin^2\phi$	$1.280S^3(1.06+S^2)^{-3}$

2.1.2. Engine Noise

Similar to the strategy adopted for airframe noise prediction, the contributions of engine noise can be further decomposed into several noise sources. Noise generated by the engine consists of several contributions, which in the literature are classified into fan noise, jet noise and engine core noise (compressor stages, combustor, turbine stages) [49]. However, considering the limited amount of data available during the early design phases, the engine noise model described in this paper considers only the two most predominant engine noise sources—fan and jet noise. This hypothesis is a well-established practice in conceptual design [19,20] and does not affect the engine noise prediction significantly for SSTs, due to the logarithmic nature of the noise levels and the prevalence of jet noise compared to other sources.

Among the aircraft noise sources, jet noise is the most widely studied and had its foundations in the work of Lighthill [50]. The most relevant finding of that work was the Lighthill's eighth power law, that states that the power of the sound created by a turbulent motion is proportional to the eighth power of the characteristic turbulent velocity.

In this work, jet noise is predicted using the Stone method [51], which is based on the Lighthill theory. The total far-field jet noise is typically computed as the sum of the jet mixing noise and shock noise, that occurs when $\sqrt{(M_1^2-1)}$ is greater than zero, with

M_1 the primary stream Mach number. The method uses empirical functions to provide the directivity and the spectral content of the field with the computed overall mean-square acoustic pressure at $\theta = 90^\circ$, that is $< p^2(\sqrt{A_e}, 90^\circ) >^*$, used to fix the amplitude throughout the field.

The equation used to calculate the jet mixing noise at a distance r_s from the nozzle exit is:

$$< p^2(r_s^*, \theta) >^* = \frac{< p^2(\sqrt{A_e}, 90^\circ) >^*}{(r_s^*)^2} \left[\frac{1 + (0.124V_1^*)^2}{(1 + 0.62V_1^* \cos\theta)^2 + (0.124V_1^*)^2} \right]^{\frac{3}{2}} \cdot D_m(\theta') F_m(S_m, \theta') H_m(M_\infty, \theta, V_1^*, \rho_1^*, T_1^*) G_c G_p \tag{5}$$

where $< p^2(\sqrt{A_e}, 90^\circ) >^*$ is the mean-square acoustic pressure for a stationary jet calculated at the reference distance $\sqrt{A_e}$ from the nozzle exit at $\theta = 90^\circ$, and is defined as:

$$< p^2(\sqrt{A_e}, 90^\circ) >^* = \frac{2.502 \times 10^{-6} A_{j,1}^* (\rho_1^*)^{\omega_\circ} (V_1^*)^{7.5}}{[1 + (0.124V_1^*)^2]^{\frac{3}{2}}}, \tag{6}$$

where the parameters are:

- r_s^*: dimensionless distance from the nozzle exit r_s, referred to as $\sqrt{A_e}$;
- $A_{j,1}^*, \rho_1^*, V_1^*$ and T_1^* : fully expanded jet area, density, velocity and total temperature respectively, with all three quantities evaluated for the primary stream, and normalized by $A_e, \rho_\infty, c_\infty$ and T_∞;
- θ': modified directivity angle, $\theta' = \theta(V_1^*)^{0.1}$;
- $D_m(\theta')$: directivity function;
- $F_m(S_m, \theta')$: spectral distribution function;
- $H_m(M_\infty, \theta, V_1^*, \rho_1^*, T_1^*)$: forward flight effects factor;
- G_c and G_p: configuration factors;
- S_m: jet mixing noise Strouhal number;
- $\omega_\circ$: empirical function of V_1^*.

The 1/3 octave band mean-square acoustic pressure due to shock turbulence interaction noise is calculated through the following equation:

$$< p^2 >^* = \frac{(3.15 \times 10^{-4}) A_{j,1}^*}{(r_s^*)^2} \frac{\beta^4}{1 - \beta^4} \frac{F_s(S_s) D_s(\theta, M_1) G_c}{1 - M_\infty \cos(\theta - \delta)}, \tag{7}$$

with β being the pressure ratio parameter, equal to $\beta = \sqrt{(M_1^2 - 1)}$, which must be greater than zero for shock cell noise to occur. The function $D_s(\theta, M_1)$ provides the dependence of the shock cell noise, for a stationary jet, on the directivity angle θ and the fully expanded primary stream Mach number M_1. This function is given by:

$$D_s(\theta, M_1) = \begin{cases} 1 & \theta \leq \theta_m \\ 1.189 & \theta > \theta_m, \end{cases} \tag{8}$$

where θ_m is the Mach angle defined by: $\theta_m = \arcsin \frac{1}{M_1}$. The total far-field jet noise is the sum of the shock noise and the jet mixing noise (Equation (9)) and its most influential parameters are the exhaust jet speed and the jet Mach number.

$$< p_{jet}^2 >^* = < p_{mixing}^2 >^* + < p_{shock}^2 >^* . \tag{9}$$

Fan noise dominates most flight conditions and can be higher than jet noise. As far as fan noise is concerned, efforts have been recently made in fan noise reduction and predictive models are available in the literature. These methods allow a first-order estimate of the acoustic pressures arising from any fan identified by a limited number of design parameters, such as diameter, tip chord, number of blades, rotational speed, fan-stator distance, pressure ratio, mass flow ratio, temperature rise across the fan [52]. The method proposed by Heidmann in the mid-1970s has come to dominate the arena of empirical fan and single-stage compressor noise prediction [43]. Heidmann prediction method is applicable to turbojet compressors and to single-and-two-stage turbofans with and without inlet guide vanes [53]. The total noise levels are obtained by spectrally summing the predicted levels of broadband, discrete-tone and combination-tone noise components. Precisely, the predicted free-field radiation patterns (neglecting the reflection of sound) consist of composite of the following separately predicted noise components:

- Noise emitted from the fan or compressor inlet duct (broadband noise, discrete-tone noise, combination-tone noise);
- Noise emitted from the fan discharge duct (broadband noise, discrete-tone noise).

Hence, the total fan noise has been predicted by summing the noise from six separate components: inlet broadband noise, inlet rotor–stator interaction tones, inlet flow distortion tones, combination tone noise, discharge broadband noise and discharge rotor–stator interaction tones. All noise sources are combined into single 1/3 octave band spectrum for each directivity angle.

The general approach is the same for each noise component and is based on the following equation for the computation of far-field mean-square acoustic pressure:

$$< p^2 >^* = \frac{A^* \Pi^*}{4\pi(r_s^*)} \frac{D(\theta)S(\eta)}{(1 - M_\infty \cos\theta)^4}, \tag{10}$$

where A is the fan inlet cross sectional area. The frequency parameter η is defined as:

$$\eta = (1 - M_\infty \cos\theta)\frac{f}{f_b}, \tag{11}$$

where f_b is the blade passing frequency depending on the rotational speed N. The acoustic power Π^* for the fan is expressed as:

$$\Pi^* = KG(i,j)(s^*)^{-a(k,l)} M_m^b (\frac{\dot{m}^*}{A^*})(\Delta T^*)^2 F(M_r, M_m), \tag{12}$$

with :

- $\dot{m}$: mass flow rate, re $\rho_\infty c_\infty A_e$;
- ΔT^*: total temperature rise across fan, re T_∞;
- M_r: relative tip Mach number;
- M_m: defined as $M_m = max(1, M_d)$, where M_d is the fan rotor relative tip Mach number at design point;
- s^*: rotor-stator spacing, re C (mean rotor blade chord);
- K, G, i, j, a, k, l: empirical constants and factors depending on geometry and configuration.

Equation (12) must be specialized for each noise component before computing the overall acoustic power:

- Inlet broadband noise

$$\Pi^* = (1.552 \times 10^{-4})(s^*)^{-a(k,l)} M_m^2 (\frac{\dot{m}^*}{A^*})(\Delta T^*)^2 F(M_r) \tag{13}$$

- Inlet rotor-stator interaction tones

$$\Pi^* = (2.683 \times 10^{-4})G(i,j)(s^*)^{-a(k,l)} M_m^{4.31} (\frac{\dot{m}^*}{A^*})(\Delta T^*)^2 F(M_r, M_m) \tag{14}$$

- Inlet flow distortion tones

$$\Pi^* = (1.488 \times 10^{-4})G(i,j)(s^*)^{-a(k,l)} M_m^{4.31} (\frac{\dot{m}^*}{A^*})(\Delta T^*)^2 F(M_r, M_m) \tag{15}$$

- Combination tone noise

$$\Pi^* = KG(i,j)(s^*)^{-a(k,l)} M_m^b (\frac{\dot{m}^*}{A^*})(\Delta T^*)^2 F(M_r, M_m) \tag{16}$$

with $K = 6.225 \times 10^{-4}$ for 1/8 fundamental combination tone, $K = 2.030 \times 10^{-3}$ for 1/4 fundamental combination tone and $K = 2.525 \times 10^{-3}$ for 1/2 fundamental combination tone.

- Discharge broadband noise:

$$\Pi^* = (3.206 \times 10^{-4}) G(i,j)(s^*)^{-a(k,l)} M_m^2 (\frac{\dot{m}^*}{A^*})(\Delta T^*)^2 F(M_r) \tag{17}$$

- Discharge rotor-stator interaction tones:

$$\Pi^* = (2.643 \times 10^{-4}) G(i,j)(s^*)^{-a(k,l)} M_m^2 (\frac{\dot{m}^*}{A^*})(\Delta T^*)^2 F(M_r). \tag{18}$$

The values of empirical constants and function $F(M_r, M_m)$ are reported in [24] for each fan noise component. Afterwards, the total fan noise is computed as the sum of the previously described contributions, obtaining the Equation (19) by appropriately summing broadband and tone noise components:

$$< p_{fan}^2 >^* = < p_{inlet}^2 >^* + < p_{combination\ tones}^2 >^* + < p_{discharge}^2 >^* . \tag{19}$$

Usually, the main broadband noise contribution is the discharge noise, whereas combination tone noise causes some peaks in the SPL that depend on the blade passing frequency. The parameters with a higher influence on fan noise generation are the air mass flow, the rotational speed, and the rise of temperature across the fan. Increasing the air mass flow and temperature produces an increment of SPL, whereas variations in rotational speed N can shift peak values along the frequencies band.

It is worth noting that the engine noise model described above perfectly fits the first generation of supersonic aircraft and related propulsive technologies. Indeed, the Olympus 593, which equipped the Concorde, can be described with this model. However, a different conclusion could be drawn for future supersonic propulsive technologies. The under-development of the future generation of SSTs might integrate more turbofan-oriented engines, which might be partially or completely embedded into the airframe. As observed in [54,55], at high power engine operation conditions, especially at take-off conditions, the noise levels observed from such future supersonic engines are very high. A major component of fan noise is expected to be the buzz-saw noise, produced by shocks at the fan blade tips at this high-power engine operation condition.

Ultimately, the total engine noise is computed as:

$$< p_{engine}^2 >^* = < p_{jet}^2 >^* + < p_{fan}^2 >^* . \tag{20}$$

Thus, by correctly summing the mean-square acoustic pressures for each frequency of the spectrum in 1/3 octave band between 50 Hz and 1000 Hz given from Equations (4) and (20), the overall aircraft mean-square acoustic pressure can be predicted, as outlined in Equation (1).

2.2. Flight Procedure for Noise Certification

LTO noise limitations and flight procedures for subsonic aircraft are specified by regulatory community by ICAO. Therefore, operational requirements can be directly elicited from these certification standards and used as a benchmark during the design process. Conversely, for supersonic aviation standards, flight procedures have not been defined yet. However, in order to guarantee an adequate public consensus towards the future SSTs, the scientific community is anticipating the possibility of extending the current subsonic flight procedures to civil supersonic aircraft [31]. In parallel, the regulatory community is fostering research activities aimed at investigating the technical and technological limitations to the application of already existing flight procedures to SSTs [29,30]. In this context, design methodologies, such as the one disclosed in this paper, are essential for suggesting possible updates of the flight procedures to fill in the gap between subsonic and supersonic aviation.

Therefore, it is necessary to couple traditional flight profile analysis with flight procedure simulations. During the first design loop, the reference ICAO flight procedures are adopted, while in the case of technical evidence of the impossibility of following such procedures, modifications to the main operational parameters are suggested.

Hereafter, the ICAO flight procedures for certification are considered as a reference. Specifically, the noise levels for certification are associated with three different operating conditions, physically represented by a ground reference measurement point (Figure 4):

- Sideline-maximum power condition: the measurement point is along the line parallel to the axis of the runaway centre line at 450 m, where the noise level is maximum during take-off. This operating condition corresponds to the so-called sideline measurement, which is the maximum sound level reached along the lateral full-power line;
- Flyover-intermediate power condition: the measurement point is along the extended runaway centre line at 6500 m from the start to roll;
- Approach-low power condition: the measurement point is 120 m vertically below the 3° descent path originating from a point 300 m beyond the threshold.

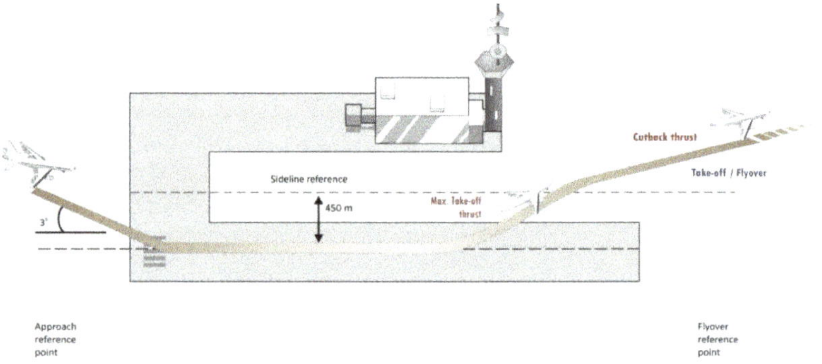

Figure 4. Aircraft noise certification reference measurement points [56].

The reference procedures are specified in the Environmental Technical Manual [57], and are specified for different classes of aircraft characterized by different Maximum Take-Off Mass (MTOM) and number of engines. Two take-off flight paths can be modelled: reduced power/thrust (or cutback) and full power/thrust. If a power/thrust reduction is applied, a slight decrease in the climb gradient may occur due to the power/thrust lapse that results from an increased air-craft height. The reference approach flight test configuration for noise certification is constrained by the ILS flight path; therefore the approach angle (steady glide angle) is $3^\circ \pm 0.5^\circ$ and the target aeroplane height vertically above the noise measurement point is 120 m (394 ft). A detailed description of the flight path is required for noise prediction. Examples of parameters needed include aircraft height, climb angle, airspeed, gross mass, aircraft configuration (flap position, landing gear position), engine thrust (power) setting parameters, and aircraft accessory conditions that may affect the measurement or adjustment of noise data and/or aircraft or engine performance.

For noise modelling purposes, the corresponding trajectory consists of straight flight segments with constant operational and configurational settings. This segmentation modelling is supported by the ANP database, which has been assembled through the years and updated by the aircraft manufacturers together with the noise certifying authorities. The ANP database contains a set of trajectories described by fixed-point data at constant the engine thrust setting, configuration, and acceleration/vertical speed.

In the frame of this work, the take-off and landing trajectories have been simulated as flight paths composed respectively by 5 and 3 flight segments constructed from an ANP set of fixed-point data. In addition, an engine model simulating the on-design and off-design operations is needed, since specific information about geometry, thermodynamic, performance and operational conditions are required to predict noise emitted by the aircraft. To keep the approach as simple as possible, in this paper a one-dimensional model for a two-spool turbojet engine with afterburner has been modelled as an example, taking the Olympus 593 MRK 610 as a reference [58], relying on the data and the results presented in [59]. The afterburner has been modelled by the addition of another component after the low-pressure turbine, which gives, in exit, a flow reaching a total temperature of 1700 K, considering an afterburner efficiency of 0.9. At least, an exhaust system composed by adiabatic, isentropic and variable-geometry convergent-divergent nozzle has been modelled to predict the exhaust jet parameters as a function of thrust.

During the simulation, the evolution of aircraft configuration is considered (e.g., retraction/extraction of landing gear, nozzle area variations) and operating parameters are continuously updated. Information about aircraft distance and noise sources directivity are recorded with a sampling time of 0.5 s over the 10 dB-down period for each aircraft noise certification points.

2.3. Noise Metrics

This sub-section includes the possibility of converting the mean-square acoustic pressure (measured following the method described in the paragraph Section 2.1) into a set of well-established noise metrics (LAmax, SEL, PNLTM and EPNL).

A noise metric is an expression used to assess the loudness or the annoyance of any quantity of noise on human hearing system. A variety of noise metrics are using among different aviation organization, countries, and airports [60]. However, they can be classified into three main groups [61]:

- Single event (or instantaneous) metrics: used to provide a description of noise occurring during one noise event, accounting for sound amplitude only;
- Exposure (or integral) metrics: used to provide a description of the type of noise exposure experienced over a given interval of time;
- Supplementary metrics: measurements often used in conjunction with the above, to provide a more meaningful depiction of the potential impact of noise exposure.

The most internationally accepted aircraft noise metrics in use belong to the first two categories, so they are typically single event or exposure metrics. Furthermore, aircraft noise metrics could be associated with frequency weighted or computed SPL distribution [62]. A-weighting is the most widely used type of weighting by federal, state, and local agencies for environmental noise analyses. In this case, the weighted SPL results from the adjustment of the spectral distribution to de-emphasize the low frequency portion of sounds to gain a better approximation of the human ear's response to sound. Noise metrics deriving from this correction are LAmax and SEL, which are respectively single event and exposure noise metric; both are expressed in dBA. Otherwise, computed SPL is obtained by the ICAO standardized procedure [63], which involves the conversion of SPL in noisiness levels by means of the correspondence with Noy tables [64] and the correction for spectral irregularities. Noise metrics deriving from this procedure are Maximum Tone Corrected Perceived Noise Level (PNLTM) and Effective Perceived Noise Level (EPNL), which are respectively a single event and exposure noise metric, expressed in PNdB and EPNdB. Based upon these different ways of processing the frequency distribution of energy, LAmax and SEL are considered loudness-based metrics, whereas PNLTM and EPNL are considered annoyance-based [61]. Annoyance-based metrics are more sensitive to the spectral shape and tonal content of the sound, therefore EPNL has been found to be more effective for the assessment of heavy jet-powered aircraft noise [65].

Even though some difference subsists, there is, in practice, a high correlation between the LAmax and PNLTM measures and, consequently, between SEL and EPNL [61]; thus, generally all the mentioned metrics are used for aircraft noise certification, depending on the specific procedure.

Anyway, a few preliminary considerations should be accounted for when predicting the noise levels received by the microphones on the ground. The aircraft noise model described in the previous paragraph gives as its output a dimensionless mean-square acoustic pressure that corresponds to the acoustic pressure of the sound signal at the noise source. Hence, the dimensionless mean-square acoustic pressure must be converted into a 1/3 octave band SPL by means of:

$$SPL = 10log_{10} < p^2 >^* +20log_{10}\frac{\rho_\infty c_\infty^2}{p_{ref}}, \tag{21}$$

where p_{ref} is the lowest sound pressure possible for the human ear to hear, which is approximately 20^{-5} Pa. After that, a number of factors which influence the propagation of noise should be considered to gain a higher fidelity level in prediction. However, the distances between the noise source and the observer on ground are such that sound attenuation in the atmosphere is the most significant phenomenon. Temperature and humidity are the parameters causing a major reduction in sound as distance increases. To determine the entity of these losses, the mathematical procedure suggested in SAE ARP 866 B [40] has been adopted. The SARP considers only the classical and molecular absorption of sound energy by the atmosphere. Precisely, the classical absorption results from energy dissipation through the effects of heat conduction and viscosity and it is a function of frequency and temperature, whereas molecular absorption results principally from rotational and vibrational relaxation process of oxygen and nitrogen molecules and is a function of frequency, temperature, and humidity. The classical absorption is relevant only at higher frequencies and varies slightly with temperature. By contrast, molecular absorption is the main contribution to sound attenuation, varying on a wide range of values, producing a more relevant sound reduction at highest frequencies. The total loss is expressed as the attenuation in dB/100 m and it is a function of frequency, temperature, and relative humidity. As a result, summing algebraically the losses for each centre frequency in 1/3 octave band of the spectrum, the approximated SPL received on ground is obtained. Once these two preliminary steps have been fulfilled, the resulting SPL distribution can be processed to quantify noise emitted by the aircraft. The A-weighting of the frequency spectrum and the ICAO procedure leads respectively to the SPL_A and PNL, that are function of frequency. Then, the Overall Sound Pressure Level (*OASPL*) and the Tone Corrected Perceived Noise Level (PNLT) as function of time can be computed with Equations (22) and (23), respectively.

$$OASPL(k) = 10log_{10}\sum_{i=1}^{24} 10^{(SPL_A(i,k)/10)} \tag{22}$$

$$PNL(k) = 40 + \frac{10}{log_{10}2}log_{10}N(k), \tag{23}$$

where *N (k)* is the total perceived noisiness, defined as:

$$N(k) = 0.85n(k) + 0.15\sum_{i=1}^{24} n(i,k), \tag{24}$$

where *n (k)* is the largest of the 24 values of *n (i, k)* of the Noy tables, *i* is the 1/3 frequency band considered between 50 Hz and 10,000 Hz and *k* the time instant.

Thus, the instantaneous noise metrics can be computed as the maximum sound level reached during the event over the 10 dB-down period, identified by kF and kL, which are the time instants when the sound level decreases up to 10 dB-down with respect to the peak:

$$LA_{max} = max(OASPL(k)) \tag{25}$$

$$PNLTM = max(PNL(k)). \quad (26)$$

Integral noise metrics are duration corrected single event metrics. Hence, *SEL* is an energy averaged A-weighted sound level over a specified period of time or single event, with a reference duration of 1 second. On the other hand, *EPNL* is an integration of the *PNLT* over a certain noise duration, normalized to a reference duration of 10 s. The relative formulas are:

$$SEL = 10log_{10} \sum_{k=kF}^{kL} 10^{(OASPL(k)/10)} \quad (27)$$

$$EPNL = 10log_{10} \sum_{k=kF}^{kL} 10^{(PNLT(k)/10)} - 13, \quad (28)$$

where 13 dB is a constant relating the one-half second values of *PNLT*(*k*) to the 10 s reference duration.

To sum up, the flow chart in Figure 5 schematically reports all the mentioned mathematical steps leading to the prediction of aircraft noise levels received on ground in terms of the most common aircraft noise metrics, starting from the normalized mean-square acoustic pressure.

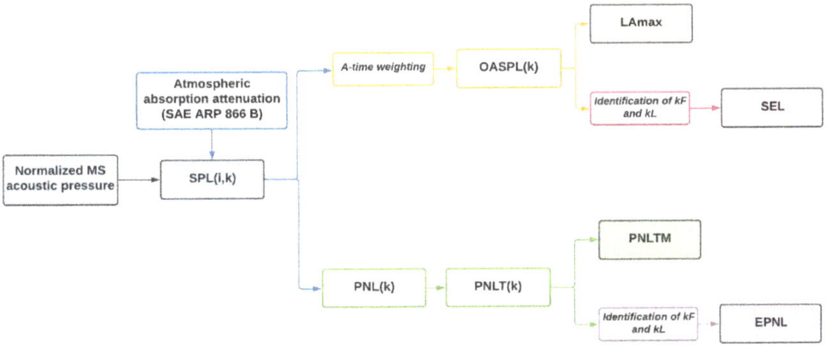

Figure 5. Procedure employed to assess aircraft noise level received on ground in terms of the most common aircraft noise metric.

3. Results

As outlined in the introduction, the results section is divided into two parts. First, the accuracy of the noise model in predicting overall aircraft noise level is assessed through a dedicated validation with ANP experimental data by the evaluation of the matching with LAmax and SEL NPDs. After that, reference departure and approach procedures reported in the ANP database are simulated and, using the methodology described above, a first guess of aircraft noise levels is evaluated. The only available civil supersonic aircraft of the ANP, that is, the Concorde, has been selected as a case study. Corresponding geometrical parameters affecting noise generation and their values are listed in Table 4. Engine on-design and off-design operations simulation are required, hence a two-spool turbojet with an afterburner is simulated through a one-dimensional model based on the results reported in [59] for the Olympus 593.

Table 4. Case study geometrical parameters affecting noise generation.

Airframe Parameters	Value
Wing span b_w	25.6 m
Wing surfrace S_w	358.25 m^2
Wing span (vertical tail) b_v	11.32 m
Wing surfrace (vertical tail) S_v	33.91 m^2
N struct main landing-gear n_{strut_m}	2
N wheels main landing-gear n_{wheels_m}	4
Tyre diameter main landing-gear d_{tyre_m}	1.2 m
Length strut main landing-gear l_{strut_m}	2.5 m
N struct forward landing-gear n_{strut_f}	1
N wheels forward landing-gear n_{wheels_f}	2
Tyre diameter forward landing-gear d_{tyre_f}	0.787 m
Length strut forward landing-gear l_{strut_f}	3 m
Engine parameters	**Value**
Number of engines N_e	4
Engine reference area A_e	1.15 m^2
Fan rotor diameter d_{rot}	1.21 m
Fan reference area A_{fan}	1.15 m^2
Number of stator vanes nV	32
Number of blades B	19
Mean rotor blade chord C	0.22 m
Rotor-stator spacing s	0.22 m
Fan rotor relative tip Mach number at design point M_d	1
Inlet guide vane index i	2

3.1. Validation

By the simulation of different flyover flight paths, ranging the altitude from 630 ft (192 m) to 10,000 ft (3048 m) and the thrust between 10,000 lb (44,482 N) and 32,000 lb (142,342 N), a total of 24 points have been obtained to produce the NPD curves.

The aircraft speed is fixed to 160 knots (82 m/s), in accordance with the reference airspeed used to derive the NPD from experimental measurements, whereas the ambient conditions have been set to the reference conditions suggested in [66] for noise contours modelling around the airports (ambient temperature equal to 15 °C and the relative humidity HR equal to 0.7). The matching with the NPD curves provided by the ANP database for LAmax and SEL are reported in Figures 6 and 7.

The degree of matching between predicted and experimental curves has been evaluated in a quantitative way through a numerical indicator for each validation point, respectively for LAmax and SEL:

$$E_{LAmax} = \left| \frac{LA_{max_p} - LA_{max_{ANP}}}{LA_{max_{ANP}}} \right| \tag{29}$$

$$E_{SEL} = \left| \frac{SEL_p - SEL_{ANP}}{SEL_{ANP}} \right|, \tag{30}$$

where LA_{max_p} and SEL_p are the predicted values, whereas $LA_{max_{ANP}}$ and SEL_{ANP} are the experimental ones.

The maximum prediction error is ±2.19% around the experimental value, attaining a good accuracy for application at a conceptual design level.

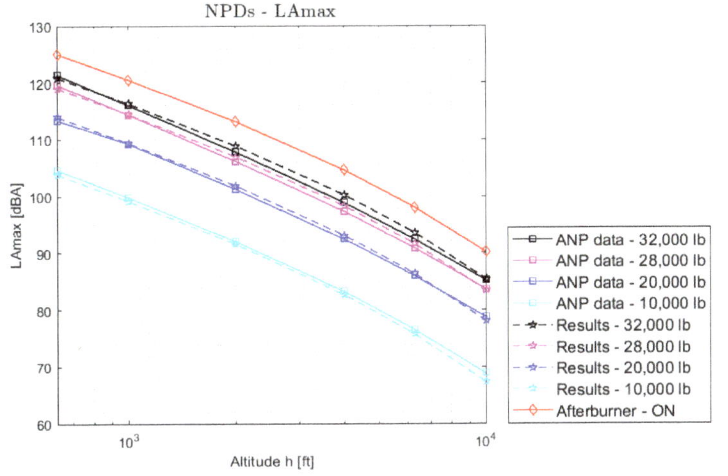

Figure 6. Matching between predicted and experimental NPDs-LAmax.

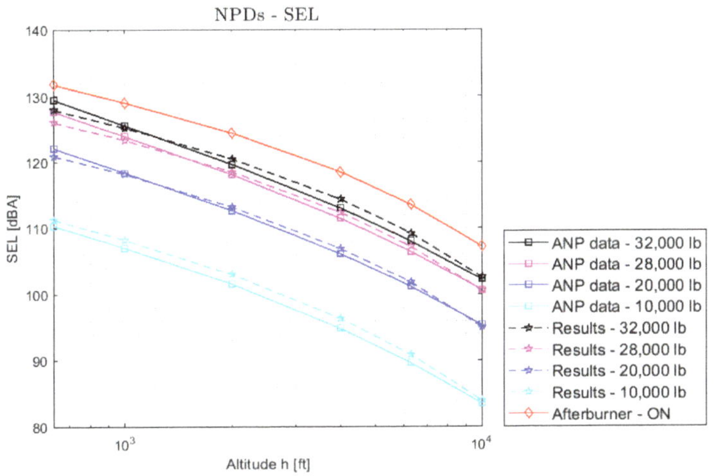

Figure 7. Matching between predicted and experimental NPDs-SEL.

Noise levels reached when the afterburner is turned on are not provided by the ANP database. However, knowing the increment of thrust due to the afterburner, it is possible to obtain a preliminary estimate of the resulting noise impact. The increase in jet exhaust speed and exhaust stream Mach number have been derived by the simulation of the engine model, considering the higher total temperature resulting from reheating and the adaptation of the throat section of the nozzle to avoid the choking of the duct. Hence, the corresponding NPD curve has been obtained setting a thrust level equal to 176,380 N, as it is the maximum thrust value reported in the ANP database for the Concorde take-off procedure (red line in Figures 6 and 7). The increment has been evaluated with respect to the noise level reached with the maximum dry thrust level with the following formulas, respectively for LAmax and SEL:

$$\Delta L_{max,ab} = \frac{L_{max,ab} - L_{max}}{L_{max}} \tag{31}$$

$$\Delta SEL_{max,ab} = \frac{SEL_{max,ab} - SEL_{max}}{SEL_{max}}, \tag{32}$$

where $\Delta L_{max,ab}$ and $\Delta SEL_{max,ab}$ represent the increment in noise levels when the afterburner is turned on. This parameter has been calculated for each point of the NPD curve, and then an arithmetic average has been calculated for both LA_{max} and SEL. The results expressed show that the afterburner causes an increase of 4.24% in instantaneous noise level and 3.52% in integrated noise level.

To assess the correlation between numerical and experimental results in a quantitative way, an adaptation of the Frequency Response Scale Factor (FRSF), usually employed in the field of frequency response analysis ([67]), has been used. Such index evaluates of how much the predicted results overestimate or underestimate the numerical ones, and acceptable values are included in the range 0.95–1.05. Precisely, if FRSF < 1, the predicted curve is on average at a higher level than the experimental one; whereas, if FRSF > 1, the predicted curve is on average at a lower level than the experimental one. The index has been readapted as:

$$FRSF = \frac{\{H_{exp(j,d)}\}\{H_{num(j,d)}\}}{\{H_{num(j,d)}\}\{H_{num(j,d)}\}}, \tag{33}$$

where j indicates the thrust level and d the distance related to the noise data H. The values obtained for LA_{max} and SEL are respectively 0.999 and 1.0033. Hence, this parameter also confirms the good accuracy of the methodology and, in addition, denotes a slight over-prediction for LA_{max} and an underprediction for SEL.

3.2. Application to Departure and Approach Procedures

A set of ANP departure fixed-point data (distance on ground, altitude, thrust, TAS/CAS) has been selected to simulate standard departure and approach procedures Figures 8 and 9.

Results are reported in Table 5. As expected, the most critical condition occurs during the departure procedure at the sideline noise measurement point. By contrast, flyover noise is reduced, as a cutback procedure has been simulated, reducing thrust level and, consequently, the climb angle. Ultimately, the approach condition is characterized by the lowest noise levels.

Table 5. Overall noise level at the three certification measurement point.

	Sideline	**Flyover**	**Approach**
LAmax [dBA]	113.22	106.30	104.92
SEL [dBA]	123.16	116.31	115.34
PNLTM [PNdB]	126.57	119.26	118.23
EPNL [EPNdB]	124.53	118.45	118.85

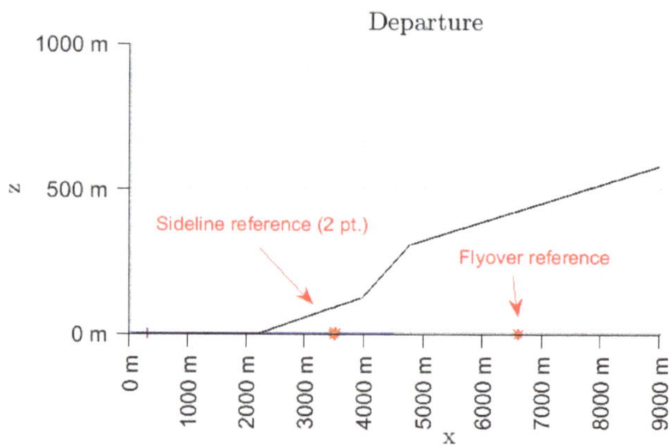

Figure 8. Simulated departure flight path.

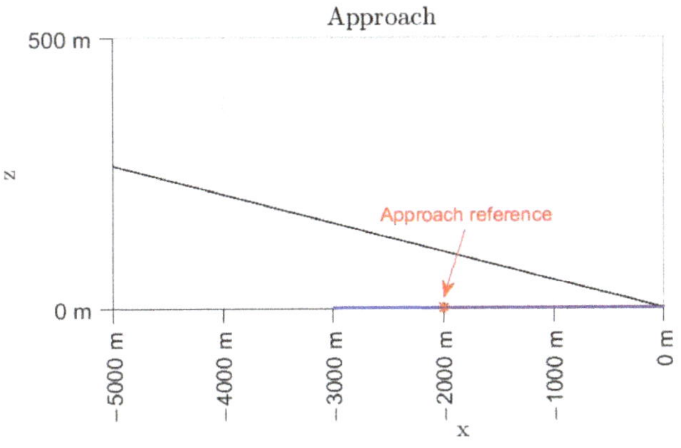

Figure 9. Simulated approach flight path.

In addition, noise contributions and overall noise have been reported for each measurement point in terms of LAmax obtained at the overflight point Figures 10–12. It is evident that the high thrust level required for taking-off and the use of the afterburner produces a high jet speed that greatly affect sideline noise. However, the impact of jet noise is significant at all three measuring points, as jet noise suppression measures have not been modelled in the frame of this work. A significant contribution of fan noise appears, especially in the approach condition. This result can be justified as the original method of Heidemann typically overestimates fan noise [68,69] .

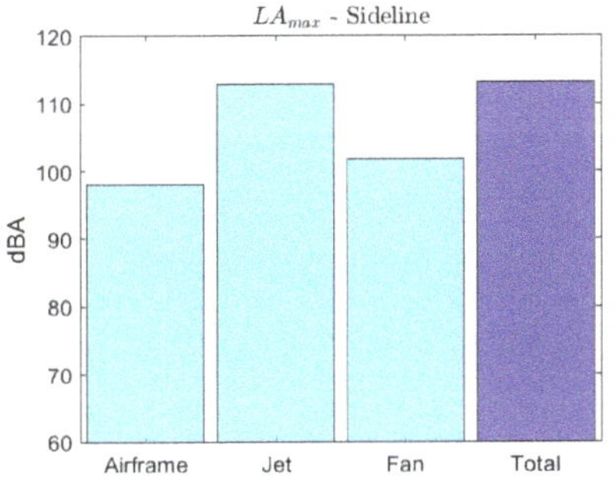

Figure 10. Noise contributions and overall noise for sideline measurement point.

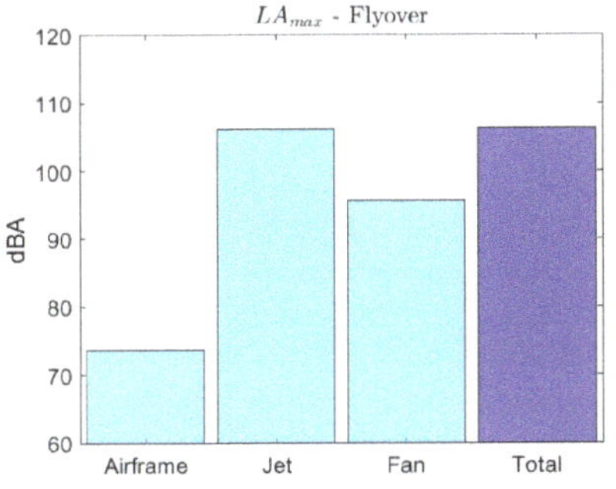

Figure 11. Noise contributions and overall noise for flyover measurement point.

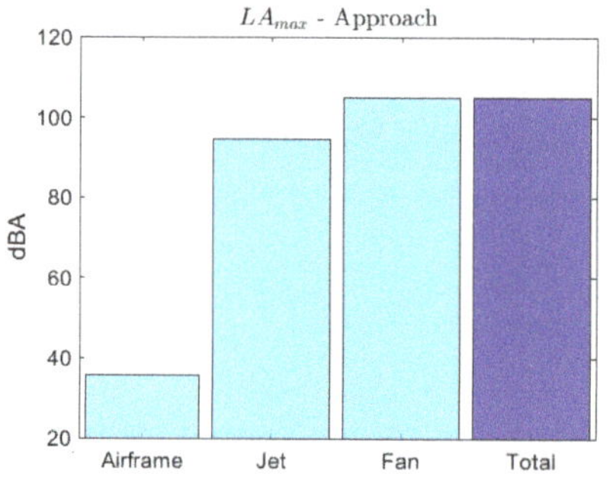

Figure 12. Noise contributions and overall noise for approach measurement point.

To enable a verification of the first noise estimations, the up-to-date LTO limitations imposed for Concorde-like sized subsonic aircraft equipped with four jet-powered engines have been considered as top-level requirements, as ICAO Annex 16, Volume 1, Chapter 12 [28] refers to subsonic aeroplanes noise standards for the certification of supersonic aeroplanes whose type certificate is issued after 1 January 1975. Therefore, in accordance with [70], the LTO noise requirements derived for M_{TOM} = 185,000 Kg are:

- Sideline-full power: $LIMIT_L = 100.15 EPNdB$
- Flyover-intermediate power: $LIMIT_F = 103.60 EPNdB$
- Approach-low power condition: $LIMIT_A = 101.86 EPNdB$

Consider the cumulative noise level, calculated as:

$$(LIMIT_L - EPNL_L) + (LIMIT_F - EPNL_F) + (LIMIT_A - EPNL_A) \geq 17. \quad (34)$$

An excess of 36.22 EPNdB has been estimated. Such a large discrepancy can be associated with the inadequacy of current regulation of the traditional civil supersonic aircraft case study. Therefore, it underlines the significance of developing new standards, which are not so penalizing for SST, and the review of traditional supersonic aircraft design with the introduction of innovative noise reduction concepts.

Chevron nozzles are one of the most studied noise reduction technologies, particularly efficient for medium and high by-pass ratio engines of subsonic aircraft [71]. The shaped edges serve to significantly reduce turbulence at the nozzle exit and thus jet noise, leading to over 2–3 EPNdB noise reduction [72]. However, this passive noise control might not be equally efficient for supersonic jets. Chevrons have been shown to be effective at reducing mixing noise for subsonic conditions, but effects on shock-cells associated noise are under investigation, as chevron nozzles reduce the shock cell spacing, causing the peak amplitude of the shock associated noise to shift to higher frequencies [73].

On the contrary, fluidic injection techniques are more efficient active noise reduction methods for supersonic jet, showing potential for over 4 EPNdB noise reduction [74]. By means of fluidic injection it is possible to rearrange the shock diamonds structure arising from the nozzle exit with respect to the throat shock diamonds, thus controlling shock associated noise [75]. However, further improvements in the engine design and performance, together with the adoption of low-noise flight procedures would still be required to ensure the Concorde-like configuration complies with current LTO limitations.

4. Conclusions

To ensure that future supersonic aircraft will meet low-noise requirements, this paper disclosed guidelines aiming at assessing LTO noise emission estimation since the very beginning of the design process. Therefore, a preliminary literature review has been conducted to identify the state-of-the-art methods currently adopted to complement the design process with aircraft noise prediction. Then, in line with the current well-established methods, a semi-empirical noise model has been applied to a civil supersonic aircraft case study. To confirm the adequacy of this noise model for Concorde-like configuration, a dedicated validation has been performed. The results attested a good accuracy for application at a conceptual design level, with a maximum relative prediction error of 2.19%. In addition, a noise level increment arising from the use of the afterburner has been determined, obtaining an increase of 4.24% in LAmax and 3.52% in SEL, which greatly discourage the use of the afterburner during take-off.

Afterwards, the suggested framework has been employed to evaluate LTO aircraft noise levels at the three certification measurement points defined by ICAO along standard departure and approach trajectories. As expected, the predicted LAmax for each contribution confirmed that the dominant noise source during take-off is the jet noise, and that the use of an afterburner greatly affects sideline noise level. By contrast, the approach condition is less critical for noise generation. However, it has been found that jet and fan noise become comparable at this condition. This result could be attributed to the original method of Heidmann, which has been seen in the literature to over-predict the intensity of fan tones. To enable first verification of the estimate, the predicted EPNLs have been compared with current LTO noise limitations for cumulative noise level, highlighting that the predicted values for supersonic aircraft significantly exceed those imposed for similar sized subsonic aircraft.

The outcome of this research activity led to several useful considerations in the field of supersonic aircraft LTO noise prediction. Firstly, the feasibility of introducing an initial noise assessment relying on information available in a conceptual design phase and using semi-empirical models found in the literature has been proven, gaining a good fidelity-level in the prediction of NPD relationships. Therefore, the procedure can theoretically be applied within the conceptual design process to produce NPD curves for different aircraft concepts by changing the design and operational parameters. After that, take-off, and precisely at the sideline noise measurement point, has been confirmed as the most critical noise condition, manly due to the engine contribution. In addition, the comparison with current cumulative noise requirement has remarked the need for both new specific noise standards for supersonic aircraft and a review of a conventional Concorde-like design by the incorporation of noise reduction measures and procedures from the earliest stage of the project.

Adjustments in noise modelling (especially for fan noise prediction) are required to gain a high-fidelity level in departure and approach procedures noise prediction as well as the introduction of simplified models to evaluate newer jet noise reduction concepts. Furthermore, since new supersonic aircraft designs could be very different from the past conventional design, the inclusion of additional elements (e.g., for the airframe noise breakdown) is a key improvement to widen the applicability of the methodology employed and enable the exploration of innovative low-noise configurations. Lastly, to obtain a better comprehension of supersonic aircraft noise, further advancements could also concern the evaluation of other annoyance-based or psychoacoustic noise metrics aimed at assessing the subjective response to aircraft noise, as suggested in [76].

Author Contributions: Conceptualization, N.V. and R.F.; methodology, G.P., N.V. and R.F.; software, G.P.; validation, G.P. and L.F.; investigation, G.P.; resources, N.V.; writing—original draft preparation, G.P. and R.F.; writing—review and editing, N.V. and L.F.; visualization, G.P.; supervision, N.V. and L.F.; project administration, N.V.; funding acquisition, N.V. All authors have read and agreed to the published version of the manuscript.

Funding: This research was funded by European Union's HORIZON 2020 research and innovation programme under grant agreement No.101006856—MDO and REgulations for Low boom and Environmentally Sustainable Supersonic aviation (MORE&LESS) Project.

Data Availability Statement: Not applicable.

Conflicts of Interest: The authors declare no conflict of interest.

Abbreviations

The following abbreviations are used in this manuscript:

ANOPP	Aircraft NOise Prediction Program
ANP	Aircraft Noise and Perfomance database
CAEP	Committee on Aviation Environmental Protection
DLR	German Aerospace Laboratory
EPNL	Effective Perceived Noise Level
EU	European Union
FAA	Federal Aviation Administration
FRSF	Frequency Response Scale Factor
ICAO	International Civil Aviation Organization
LAmax	A-weighted Sound Pressure Level
LTO	Landing and Take-Off
MORE&LESS	MDO and REgulations for Low-boom and Environmentally Sustainable Supersonic aviation
MTOM	Maximum Take-Off Mass
NASA	National Aeronautics and Space Administration
NPD	Noise Power Distance
OASPL	Overall Sound Pressure Level
ONERA	French Aeronautics and Space Research Center
PANAM	Parametric Aircraft Noise Analysis Module
PNL	Perceived Noise Level
PNLTM	Maximum Tone Corrected Perceived Noise Level
SAE	Society of Automotive Engineers
SARPs	Standard and Recommended Practices
SCR	Supersonic Cruise Research
SEL	Sound Exposure Level
SPL	Sound Pressure Level
SST	SuperSonic Transport
SAE	Society of Automotive Engineers
SARPs	Standard and Recommended Practices
SCR	Supersonic Cruise Research
SEL	Sound Exposure Level
SPL	Sound Pressure Level
SST	SuperSonic Transport

References

1. ICAO. Destination Green: The Next Chapter. In *Environmental Report*; ICAO: Montreal, QC, Canada, 2019.
2. Hardeman, A.; Maurice, L. Sustainability: Key to enable next generation supersonic passenger flight. *IOP Conf. Ser. Mater. Sci. Eng.* **2021**, *1024*, 012053. [CrossRef]
3. Antoine, N.E.; Kroo, I.M. Aircraft Optimization for Minimal Environmental Impact. *J. Aircr.* **2004**, *41*, 790–797. [CrossRef]
4. X-59 QueSST (Quiet SuperSonic Technology)-Silencing the Sonic Boom. Available online: https://www.lockheedmartin.com/en-us/products/quesst.html (accessed on 25 October 2021).
5. Boom Supersonic-Overture. Available online: https://boomsupersonic.com/overture (accessed on 25 October 2021).
6. SENECA-LTO noiSe and EmissioNs of supErsoniC Aircraft. Available online: https://seneca-project.eu/ (accessed on 25 October 2021).
7. RUMBLE-RegUlation and norM for Low Sonic Boom LEvels. Available online: https://rumble-project.eu/i/project/project-overview (accessed on 25 October 2021).
8. Drake, F.; Purvis, M. The effect of supersonic transports on the global environment: A debate revisited. *Sci. Technol. Hum. Values* **2001**, *26*, 501–528. Available online: http://www.jstor.org/stable/690166 (accessed on 25 October 2021). [CrossRef]
9. Tang, R.Y.; Elias, B.; Luther, L.; Morgan, D. Supersonic Passenger Flights. 2018. Available online: https://sgp.fas.org/crs/misc/R45404.pdf (accessed on 25 October 2021).
10. ICAO-Supersonic Aircraft Noise Standards Development. Available online: https://www.icao.int/environmental-protection/pages/Supersonic-Aircraft-Noise-Standards-Development.aspx (accessed on 25 October 2021).
11. Hay, J.A. Concorde-Community Noise. *Aerospace Engineering and Manufacturing Meeting*; Technical Report 0148-7191; SAE International: Warrendale, PA, USA, 1976.
12. Hirshorn, S.R. *NASA Systems Engineering Handbook*, 2nd ed.; NASA: Washington, DC, USA, 2016; pp. 17–23.

13. Raney, J. Development of a new computer system for aircraft noise prediction. In Proceedings of the AIAA 2nd Aeroacoustics Conference; Hampton, VA, USA, 24–26 March 1975. [CrossRef]
14. Bertsch, L.; Guerin, S.; Looye, G.; Pott-Pollenske, M. The Parametric Aircraft Noise Analysis Module-Status overview and recent applications. In Proceedings of the 17th AIAA/CEAS Aeroacoustics Conference (32nd AIAA Aeroacoustics Conference), Portland, OR, USA, 5–8 June 2011. [CrossRef]
15. Sanders, L.; Malbéqui, P.; LeGriffon, I. Capabilities of IESTA-CARMEN to predict aircraft noise. In Proceedings of the 23rd International Congress on Sound & Vibration (ICSV23): "From Ancient to Modern Acoustics", Athens, Greece, 10–14 July 2016. Available online: https://hal.archives-ouvertes.fr/hal-01385584 (accessed on 25 October 2021).
16. Raney, J.P.; Padula, S.L.; Zorumski, W.E. *NASA Progress in Aircraft Noise Prediction*; NASA Technical Memorandum 81915; National Aeronautics and Space Administration: Washington, DC, USA, 1981.
17. Lopes, L.V.; Burley, C.L. Design of the Next Generation Aircraft Noise Prediction Program: ANOPP2. In Proceedings of the 17th AIAA/CEAS Aeroacoustics Conference (32nd AIAA Aeroacoustics Conference), Portland, OR, USA, 5–8 June 2011. [CrossRef]
18. Bertsch, L.; Heinze, W.; Guerin, S.; Lummer, M.; Delfs, J. 10 years of joint research at DLR and TU Braunschweig toward low-noise aircraft design-what did we achieve? *Aeronaut. Aerosp. Open Access J.* **2019**, *3*, 89–104. [CrossRef]
19. Dobrzynski, W.; Ewert, R.; Pott-Pollenske, M.; Herr, M.; Delfs, J. Research at DLR towards airframe noise prediction and reduction. *Aerosp. Sci. Technol.* **2008**, *12*, 80–90. [CrossRef]
20. Elie, A.; Kervarc, R.; Dubot, T.; Bourrely, J. *IESTA: A Modular Distributed Simulation Platform for the Evaluation of Air Transport Systems*; MASCOT08-IMACS/ISGG Workshop; IAC-CNR: Roma, Italy, 2008.
21. Bertsch, L.; Clark, I.A.; Thomas, R.H.; Sanders, L.; Legriffon, I. The Aircraft Noise Simulation Working Group (ANSWr)-Tool Benchmark and Reference Aircraft Results. In Proceedings of the 25th AIAA/CEAS Aeroacoustics Conference, Delft, The Netherlands, 20–23 May 2019. Available online: https://hal.archives-ouvertes.fr/hal-02196662 (accessed on 30 September 2021).
22. Bertsch, L. Noise Prediction Within Conceptual Aircraft Design. Ph.D. Thesis, Technical University of Braunschweig, Braunschweig, Germany, 2012.
23. Bertsch, L.; Isermann, U. Noise prediction toolbox used by the DLR aircraft noise working group, INTER-NOISE 2013. In Proceedings of the 42nd International Congress and Exposition on Noise Control Engineering, Innsbruck, Austria, 15–18 September 2013.
24. Zorumski, W.E. *Aircraft Noise Prediction Program Theoretical Manual-Part 2*; NASA Technical Memorandum 83199; National Aeronautics and Space Administration: Washington, DC, USA1982.
25. ANOPP and ANOPP2. Available online: https://software.nasa.gov/software/LAR-19861-1 (accessed on 30 September 2021).
26. Filippone, A.; Bertsch, L.; Pott-Pollenske, M. Validation strategies for comprehensive aircraft noise prediction methods. In Proceedings of the 12th AIAA Aviation Technology, Integration, and Operations (ATIO) Conference and 14th AIAA/ISSM, Indianapolis, IN, USA, 17–19 September 2012. [CrossRef]
27. ANP-Eurocontrol Experimental Centre. Available online: https://www.aircraftnoisemodel.org/ (accessed on 30 September 2021).
28. ICAO. Environmental Protection. In *Annex 16 to the Convention on International Civil Aviation*, 8th ed.; Vol. I: Aircraft Noise, Chapter 12; ICAO: Montreal, QC, Canada, 2017.
29. Berton, J.J.; Huff, D.L.; Geiselhart, K.; Seidel, J. Supersonic Technology Concept Aeroplanes for Environmental Studies. In Proceedings of the AIAA SciTech Forum and Exposition, Orlando, FL, USA, 6–10 January 2020. [CrossRef]
30. Nöding, M.; Bertsch, L. Application of Noise Certification Regulations within Conceptual Aircraft Design. *Aerospace* **2021**, *8*, 210. [CrossRef]
31. Berton, J.J.; Jones, S.M.; Seidel, J.A.; Huff, D.L. Noise predictions for a supersonic business jet using advanced take-off procedures. *Aeronaut. J.* **2018**, *122*, 556–571. [CrossRef]
32. Project-MORE&LESS. Available online: https://www.h2020moreandless.eu/project/ (accessed on 25 October 2021).
33. Raymer, P.D. *Aircraft Design: A Conceptual Approach*, 5th ed.; Shetz, J.A., Ed.; AIAA Education Series; Wright-Patterson Air Force Base: Fairborn, OH, USA, 2013.
34. Torenbeek, E. *Advanced Aircraft Design: Conceptual Design, Analysis and Optimization of Subsonic Civil Airplanes*; Wiley: Hoboken, NJ, USA, 2013.
35. Ferretto, D.; Fusaro, R.; Viola, N. Innovative Multiple Matching Charts approach to support the conceptual design of hypersonic vehicles. *Proc. Inst. Mech. Eng. Part J. Aerosp. Eng.* **2020**, *234*, 1893–1912. [CrossRef]
36. Ferretto, D.; Fusaro, R.; Viola, N. A conceptual design tool to support high-speed vehicle design. In Proceedings of the AIAA AVIATION 2020 FORUM, Virtual Event, 15–19 June 2020; p. 22. [CrossRef]
37. Fusaro, R.; Ferretto, D.; Viola, N. MBSE approach to support and formalize mission alternatives generation and selection processes for hypersonic and suborbital transportation systems. In Proceedings of the 2017 IEEE International Systems Engineering Symposium (ISSE), Vienna, Austria, 11–13 October 2017; p. 8088275. [CrossRef]
38. Fusaro, R.; Viola, N.; Fenoglio, F.; Santoro, F. Conceptual design of a crewed reusable space transportation system aimed at parabolic flights: Stakeholder analysis, mission concept selection, and spacecraft architecture definition. *CEAS Space J.* **2017**, *9*, 5–34. [CrossRef]
39. Nijsse, J. Design and Noise Acceptability of Future Supersonic Transport Aircraft. Ph.D. Thesis, Delft University of Technology, Delft, The Netherlands, 2020.
40. *Aerospace Standard ARP866*; Standard Values of Atmospheric Absorption as a Function of Temperature and Humidity for Use in Evaluating Aircraft Flyover Noise. SAE International: Warrendale, PA, USA, 1964.

41. Hubbard, H.H. *Aeroacoustic of Flight Vehicles: Theory and Practice, Volume 1: Noise Sources*; NASA Reference Publication 1258, RDC Technical Report 90-3052; NASA Langely Center: Hampton, VA, USA, 1991.
42. Farassat, F.; Casper, J. Towards an airframe noise prediction methodology: Survey of current approaches. In Proceedings of the 44th AIAA Aerospace Sciences Meeting and Exhibit, Reno, NV, USA, 9–12 January, 2006. [CrossRef]
43. Filippone, A. Aircraft noise prediction. *Prog. Aerosp. Sci.* **2014**, *68*,27–63. [CrossRef]
44. Bertsch, L.; Simons, D.G.; Snellen, M. *Aircraft Noise: The Major Sources, Modelling Capabilities, and Reduction Possibilities*; DLR-Interner Berich, DLR-IB 224-2015 A 110, 29 S; 1st Joint DLR & TU Delft Aviation Noise Workshop; Institute of Aerodynamics and Flow Technology: Göttingen, Germany, 2015. [CrossRef]
45. Morgan, H.G.; Hardin, J.C. Airframe Noise-The Next Aircraft Noise Barrier. *J. Aircr.* **1975**, *12*, 622–624. [CrossRef]
46. Dobrzynski, W. Almost 40 Years of Airframe Noise Research: What Did We Achieve? *J. Aircr.* **2010**, *47*, 353–367. [CrossRef]
47. Fink, M.R. *Airframe Noise Prediction Method, FAA Research Report, FAA-RD-77-29*; Federal Aviation Administration RD-77-29: Springfield, VA, USA, 1977.
48. Merino-Martinez, R.; Bertsch, L.; Simons, D.; Snellen, M. Analysis of landing gear noise during approach. In Proceedings of the 22nd AIAA/CEAS Aeroacoustics Conference, Lyon, France, 30 May–1 June 2016. [CrossRef]
49. Ihme, M. Combustion and Engine-Core Noise. *Annu. Rev. Fluid Mech.* **2017**, *49*, 277–310. [CrossRef]
50. Lighthill, M.J. On sound generated aerodynamically I. General theory. *Proc. R. Soc. London. Ser. A. Math. Phys. Sci.* **1952**, *211*, 564–587.
51. Stone, J.R., Groesbeck, D.E.; Zola, C.L. Conventional profile coaxial jet noise prediction. *AIAA J.* **1983**, *21*, 336–342. [CrossRef]
52. Feiler, C.E.; Conrad, E.W. Fan Noise from Turbofan Engines. *J. Aircr.* **1976**, *13*, 128–134. [CrossRef]
53. Heidmann, M. *Interim Prediction Method for Fan and Compressor Source Noise*; Technical Memorandum 19750017876; NASA Lewis Research Center: Cleveland, OH, USA, 1975.
54. Chima, R.V. *Analysis of Buzz in a Supersonic Inlet*; NASA, Glenn Research Center: Cleveland, OH, USA, 2012.
55. Adetifa, O.E. Prediction of Supersonic Fan Noise Generated by Turbofan Aircraft Engines. Ph.D. Thesis, University of Southampton, Faculty of Engineering and the Environment, Southampton, UK, 2015.
56. ICAO-Reduction of Noise at Source. Available online: https://www.icao.int/environmental-protection/pages/reduction-of-noise-at-source.aspx (accessed on 30 September 2021).
57. ICAO. *Environmental Technical Manual, Vol. I, Procedures for the Noise Certification of Aircraft*; ICAO: Montreal, QC, Canada, 2018.
58. Talbot, J.E. Concorde Development-Powerplant Installation and Associated Systems, SAE Transactions. *J. Aerosp.* **1991**, *100*, 2681–2698. Available online: http://www.jstor.org/stable/44548124 (accessed on 30 September 2021).
59. Stiuso, G. Tecnica di Simulazione Numerica delle Prestazioni Stazionarie e Transitorie di Turbomotori. Ph.D. Thesis, Politecnico di Torino, Turin, Italy, 2019. Available online: http://webthesis.biblio.polito.it/id/eprint/11258 (accessed on 30 September 2021).
60. Vasov, L.; Stojiljkovic, B.; Cokorilo, O.; Mirosavljevic, P.; Gvozdenovic, S. Aircraft noise metrics. *Saf. Eng.* **2014**, *4*. [CrossRef]
61. Jones, K.; Cadoux, R. *Metrics for Aircraft Noise*; R ERCD Report 0904; Environmental Research and Consultancy Department: London, UK, 2009.
62. Bennett, R.L.; Pearsons, K.S. *Handbook of Aircraft Noise Metrics*; NASA Contractor Report; Bolt, Beranek, and Newman, Inc.: Canoga Park, CA, USA, 1981.
63. ICAO. Environmental Protection. In *Annex 16 to the Convention on International Civil Aviation*, 8th ed.; Vol. I: Aircraft Noise, Appendix 2; ICAO: Montreal, QC, Canada, 2017.
64. ICAO. Environmental Protection. In *Annex 16 to the Convention on International Civil Aviation*, 8th ed.; Vol. I: Aircraft Noise, Appendix 1; ICAO: Montreal, QC, Canada, 2017.
65. Torija, A.; Woodward, R.; Flindell, I.; McKenzie, A.; Self, R. On the assessment of subjective response to tonal content of contemporary aircraft noise. *Appl. Acoust.* **2018**, *146*, 190–203. [CrossRef]
66. ECAC. CEAC Doc 29, 4th ed. Report on Standard Method of Computing Noise Contours Around Civil Airports. In Proceedings of the European Civil Aviation Conference, European Civil Aviation Conference (ECAC) Document 29, Paris, France, 7 December 2016. Available online: https://www.ecac-ceac.org/ecac-docs (accessed on 30 September 2021).
67. Citarella, R.; Federico, L.; Cicatiello, A. Modal acoustic transfer vector approach in a FEM–BEM vibro-acoustic analysis. *Eng. Anal. Bound. Elem.* **2007**, *31*, 248–258. [CrossRef]
68. Kontos, K.B.; Janardan, B.A.; Gliebe, P.R. *Improved NASA-ANOPP Noise Prediction Computer Code for Advanced Subsonic Propulsion Systems-Vol. 1 ANOPP Evaluation and Fan Noise Model Imporvements*; NASA Contractor Report 195480; NASA: Cincinnati, OH, USA, 2013.
69. Krejsa, E.A.; Stone, J.R. *Enhanced Fan Noise Modeling for Turbofan Engines*; NASA Contractor Report 218421; NASA: Cincinnati, OH, USA, 2014.
70. ICAO. Environmental Protection. In *Annex 16 to the Convention on International Civil Aviation*, 8th ed.; Vol. I: Aircraft Noise, Chapter 14; ICAO: Montreal, QC, Canada, 2017.
71. Saiyed, N.; Mikkelsen, K.; Bridges, J. Acoustics and Thrust of Separate-Flow Exhaust Nozzles with Mixing Devices for High Bypass- Ratio Engines. *AIAA J.* **2003**, 41, 372–378. [CrossRef]
72. Callender, B.; Gutmark, E.; Martens, S. Far-Field Acoustic Investigation into Chevron Nozzle Mechanisms and Trends. *AIAA J.* **2005**, 43, 87–95. [CrossRef]

73. Rask, O.; Kastner, J.; Gutmark, E. Understanding How Chevrons Modify Noise in Supersonic Jet with Flight Effects. *AIAA J.* **2011**, 49, 1569–1576. [CrossRef]
74. Cuppoletti, D.R. Supersonic Jet Noise Reduction with Novel Fluidic Injection Techniques. Ph.D. Thesis, University of Cincinnati, Cincinnati, OH, USA, 2013.
75. Semlitsch, B.; Cuppoletti, D.R.; Gutmark, E.J.; Mihăescu, M. Transforming the Shock Pattern of Supersonic Jets Using Fluidic Injection. *AIAA J.* **2019**, 57, 1851–1861. [CrossRef]
76. Sahai, A.K. Consideration of Aircraft Noise Annoyance During Conceptual Aircraft Design. Ph.D. Thesis, Rheinisch-Westfälische Technische Hochschule Aachen, Aachen, Germany, 2016. Available online: urn:nbn:de:hbz:82-rwth-2016-071181 (accessed on 25 October 2021).

aerospace MDPI

Article

A Numerical Study on Aircraft Noise Mitigation Using Porous Stator Concepts

Christopher Teruna *,†, Leandro Rego †, Damiano Casalino, Daniele Ragni and Francesco Avallone

Department of Wind Energy, Delft University of Technology, 2628 CD Delft, The Netherlands; l.rego@tudelft.nl (L.R.); d.casalino@tudelft.nl (D.C.); d.ragni@tudelft.nl (D.R.); F.Avallone@tudelft.nl (F.A.)
* Correspondence: c.teruna@tudelft.nl
† These authors contributed equally to this work.

Abstract: This manuscript presents the application of a recently developed noise reduction technology, constituted by poro-serrated stator blades on a full-scale aircraft model, in order to reduce rotor-stator interaction noise in the fan stage. This study was carried out using the commercial lattice Boltzmann solver 3DS-SIMULIA PowerFLOW. The simulation combines the airframe of the NASA High-Lift Common Research Model with an upscaled fan stage of the source diagnostic test rig. The poro-serrations on the stator blades have been modeled based on a metal foam with two different porosity values. The results evidence that the poro-serrations induce flow separation on the stator blades, particularly near the fan-stage hub. Consequently, the thrust generated by the modified fan stage is lower and the broadband noise emission at low frequencies is enhanced. Nevertheless, the tonal noise components at the blade-passage frequency and its harmonics are mitigated by up to 9 dB. The poro-serrations with lower porosity achieve a better trade-off between noise emission and thrust penalty. An optimization attempt was carried out by limiting the application of porosity near the tip of the stator blades. The improved leading-edge treatment achieves a total of 1.5 dB in sound power level reduction while the thrust penalty is below 1.5 %. This demonstrates that the aerodynamic effects of a leading-edge treatment should be taken into account during the design phase to fully benefit from its noise reduction capability.

Keywords: aeroacoustics; rotor-stator interaction noise; porous material

Citation: Teruna, C.; Rego, L.; Casalino, D.; Ragni, D.; Avallone, F. A Numerical Study on Aircraft Noise Mitigation Using Porous Stator Concepts. *Aerospace* **2022**, *9*, 70. https://doi.org/10.3390/aerospace9020070

Academic Editors: Lothar Bertsch and Adrian Sescu

Received: 28 December 2021
Accepted: 24 January 2022
Published: 27 January 2022

1. Introduction

Aircraft noise emission is an important aspect in aircraft design as aviation regulations become more stringent. On a modern airliner, the propulsion system, such as a turbofan engine, contributes to a large portion of the total noise emission [1]. The periodic impingement of the fan wake on the outlet guide vane (OGV) is one of the major noise generation mechanisms in a high-bypass turbofan that is responsible for both tonal and broadband noise emissions [2]. Mitigating such aeroacoustic mechanism is even more important during the approach phase of an aircraft, wherein the jet noise contribution is smaller [3], due to the noise impact on areas surrounding airports. With progressively increasing bypass ratios to improve fuel efficiency, the fan wake–OGV interaction noise in a turbofan, especially the tonal component, is expected to become stronger as the axial distance between the fan and the OGV is reduced in order to limit the total engine weight [4]. This happens for multiple reasons [5]: (1) turbulence structures in the fan wake are more coherent closer to the fan blades, leading to more intense aerodynamic perturbations on the OGVs; (2) the unsteady loading on the fan blades due to flow distortion induced by the downstream OGVs (i.e., potential flow effect); (3) the fan blades scatter the sound waves emitted by the OGVs, generating additional cut-on acoustic modes. Existing noise reduction approaches, such as acoustic liners, might face additional challenges for this application due to the reduced amount of space available in the fan stage. Therefore, it is necessary to introduce appropriate solutions directly at the source level.

The fan wake–OGV interaction mechanism can be considered as a turbulence-impingement noise (TIN) [6] problem, for which several mitigation approaches have been suggested in literature, such as serrations [7–10] and permeable treatments [8,11–15]. The serrations are modifications of the leading-edge shape along the blade span, such as by implementing a sinusoidal planform instead of a regular straight one. The serrations allow for reducing the coherence of the noise scattering phenomena along the spanwise direction [16]. In addition, the source intensity at the tip and at the side of the serrations is weakened in comparison to that on a straight leading edge (LE) [17]. A sinusoidal serration can be characterized by its wavelength Λ (spanwise tip-to-tip distance) and amplitude H (tip-to-root distance). Recent studies have shown that both Λ and H can be optimized for a given turbulence length scale L in the flow field, such that $\Lambda/L_z \approx 4$ and $H/L_x \leq 2$ [18,19], where the subscripts x and z denote the streamwise and spanwise directions, respectively.

Permeable treatments, on the other hand, are modifications of the surface and of the internal volume of a body to allow for flow communication. The usage of permeable treatments is aimed at mitigating the distortion of impinging vortical structures on a leading edge by realizing a gradual dissipation of these structures as they penetrate into the porous medium [8]. As a consequence, the noise scattering on a permeable surface is milder than on a solid one. Recent investigations have shown that TIN reduction generally becomes higher as the permeability of the leading edge is increased [12,14], although this would also cause a more significant aerodynamic penalty. Unfortunately, to the authors' best knowledge, the optimization procedure for the design and implementation of a permeable LE is still inconclusive, unlike for LE serrations. In fact, novel permeable LE designs are still being actively investigated at the time of writing [14,15,20,21]. Regardless, these studies have shown that permeable LE applications are very promising in both efficacy and versatility aspects.

As a proof-of-concept, the authors have previously investigated the application of LE serrations and porous LE in a rod-airfoil configuration (RAC) [15] using the lattice Boltzmann method. The RAC was chosen as it could emulate the TIN mechanisms of the fan wake–OGV interaction in a turbofan. A thin and cambered airfoil was employed for resembling the typical features of a turbomachinery blade. The aforementioned study was intended for comparing the effects of LE serrations and porous LE on TIN emission and on the aerodynamic performance of the airfoil. It was found that the serrations suppressed the noise source intensity near the serration tip and induced a strong decorrelation of the noise sources that are distributed along the LE span. The porous LE, which was modeled after a metal foam, was also capable of mitigating the intensity of the noise sources at the porous medium surface, although the spanwise coherence distribution was only slightly affected. Moreover, additional noise scattering was observed near the solid–porous junction (i.e., the transition location between the porous LE and the solid afterbody). As a result, the usage of LE serrations achieved higher noise reduction than that of the porous LE, particularly for the tonal noise component. An attempt to improve the porous LE design was carried out by coupling it with a serration-like planform. With this combination, it was found out that a larger broadband noise reduction could be achieved while the tonal noise remained similar as that of the regular serrations.

While the usage of a permeable edge treatment has been considered promising in literature, most investigations were performed using simplified, laboratory-scale models, which cannot completely mimic the physical environment in real-life systems. This is partly due to the costs and the enormous complexity when dealing with manufacturing, instrumentation, and testing of a full-scale configuration. On the other hand, such investigation is crucial for assessing the impact of these noise mitigation solutions, for instance, on a realistic aircraft model, particularly during the initial design phase. Nevertheless, using the combination of a high-fidelity numerical method and parallel computing, aeroacoustic investigations of complete aircraft models have recently become attainable [22,23]. By employing a similar approach, this manuscript presents a numerical investigation on porous LE modifications applied to the stators in the fan stage of an aircraft model. The simulation considers the combined NASA High-Lift Common Research Model (HL-CRM) and source diagnostic test (SDT) [2] geometries, wherein the latter has been upscaled to match the dimensions that resemble its real-life counterpart. Hence, this investigation aims to assess the efficacy of the permeable LE modification on stators in a turbofan fan stage, while also providing deeper insights into the consequences of such noise mitigation approach from both acoustics and aerodynamics perspectives.

The remainder of this manuscript is structured in the following order. Section 2 provides an overview of the lattice Boltzmann solver that has been employed, followed by the description of the simulation setup. The simulation is later verified in Section 3. The outcome of the simulation is discussed in Section 4. The manuscript is concluded with Section 5.

2. Methods

2.1. Description of Numerical Method

This study was performed using the commercial solver 3DS-SIMULIA PowerFLOW 6-2019, based on a lattice Boltzmann method (LBM). It has been previously used for studying the aerodynamics of an aircraft fuselage [24] and the aeroacoustics of a turbofan model [4,5]. Further details on the methodology can be found in [25].

The LBM is derived from the Boltzmann's kinetic theory that describes a fluid as a collection of particles that continuously evolve towards a thermodynamic equilibrium state. The state of the particles is not tracked at an individual level, but rather by using a probability distribution function. This process can be described using the Boltzmann transport equation (BTE), which has the following expression (neglecting body forces):

$$\frac{\partial F}{\partial t} + \mathbf{V} \cdot \nabla F = \mathbf{C}, \tag{1}$$

where $F(\mathbf{x}, t)$ is the particle distribution function in spatial ($\mathbf{x}$) and temporal (t) dimensions, $\mathbf{V}$ is the particle velocity vector, and $\mathbf{C}$ is the collision operator.

The BTE is discretized onto a Cartesian grid (i.e., lattice) where the particles are confined to the nodes, and the particle velocity vectors are limited to a predetermined (discrete) number of directions. The discretized form of the BTE is given as

$$F_{\mathrm{n}}(\mathbf{x} + \mathbf{V}_{\mathrm{n}}\Delta t, t + \Delta t) - F_{\mathrm{n}}(\mathbf{x}, t) = \mathbf{C}_{\mathrm{n}}(\mathbf{x}, t), \tag{2}$$

where $\mathrm{n} = 1, 2, \ldots, \mathrm{Q}$ are the discrete directions. While PowerFLOW employs a D3Q19 (3 spatial dimensions, 19 discrete velocity vectors) solver for low Mach number flows, the present study employs a high-subsonic version which incorporates an entropy relationship similar to the formulation of Zhang et al. [26]. The left hand side of equation 2 refers to a time-explicit advection. The collision term $\mathbf{C}_{\mathrm{n}}$ is based on that of the Bhatnagar–Gross–Krook (BGK) model [27]:

$$\mathbf{C}_{\mathrm{n}} = -\frac{\Delta t}{\tau}[F_{\mathrm{n}}(\mathbf{x}, t) - F_{\mathrm{n}}^{\mathrm{eq}}(\mathbf{x}, t)]. \tag{3}$$

In essence, the BGK formulation states that the local distribution function F eventually returns to the equilibrium one F^{eq} within a time scale characterized by τ. This single relaxation time is also related to the dimensionless kinematic viscosity ν and temperature Θ as follows [28]:

$$\nu = \Theta\left(\tau - \frac{\Delta t}{2}\right). \tag{4}$$

The equilibrium distribution function F^{eq} can take the form of the regular Maxwell–Boltzmann distribution, but it has been approximated with a third-order expansion as [29]

$$F_{\mathrm{n}}^{\mathrm{eq}} \approx \rho \mathrm{w}_{\mathrm{n}} \left[1 + \frac{\mathbf{V}_{\mathrm{n}} \cdot \mathbf{u}}{\Theta} + \frac{(\mathbf{V}_{\mathrm{n}} \cdot \mathbf{u})^2}{2\Theta^2} - \frac{\mathbf{u}^2}{2\Theta} + \frac{(\mathbf{V}_{\mathrm{n}} \cdot \mathbf{u})^3}{6\Theta^3} - \frac{(\mathbf{V}_{\mathrm{n}} \cdot \mathbf{u})\mathbf{u}^2}{2\Theta^2} \right], \tag{5}$$

where w_{n} are the weighting functions depending on the number of discrete velocity vectors. Eventually, macroscopic flow quantities, such as density ρ and velocity $\mathbf{u}$, can be recovered after solving Equation (2) for F_{n}.

$$\rho(\mathbf{x}, t) = \sum_{\mathrm{n}} F_{\mathrm{n}}(\mathbf{x}, t), \tag{6}$$

$$\rho \mathbf{u}(\mathbf{x}, t) = \sum_{\mathrm{n}} \mathbf{V}_{\mathrm{n}} F_{\mathrm{n}}(\mathbf{x}, t). \tag{7}$$

PowerFLOW employs a very large eddy simulation (VLES) model, which can be classified as a hybrid approach (e.g., hybrid RANS/LES and detached eddy simulation—DES), but it is fundamentally different from its Navier–Stokes-based counterparts. The VLES model was derived [30] by drawing an analogy between the phenomenon of turbulence (eddy viscosity hypothesis) and the kinetic theory of gases. In this approach, a two-equation $k - \epsilon$ renormalization group (RNG) [31] was applied to locally adjust the turbulent relaxation time τ_{eff}:

$$\tau_{\mathrm{eff}} = \tau + C_{\mu} \frac{k^2/\epsilon}{(1 + \eta^2)^{0.5}}, \tag{8}$$

where $C_{\mu} = 0.09$ and η is based on a local strain parameter ($k|\mathbf{S}/\epsilon|$), a local vorticity parameter ($k|\omega/\epsilon|$), and local helicity parameters. Using Equation (8), it is implied that the effective relaxation time governs the relaxation properties of the BTE and, consequently, it also determines the characteristic time scales of turbulence in the flow field. Notice that the turbulence model has not been explicitly used to calculate the Reynolds stresses, unlike in RANS where the turbulence model is generally used to solve a closure problem. The VLES model allows for capturing the nonlinearity of Reynolds stresses, which enables the development of large eddies in the simulation domain [30,32].

The wall boundary condition is approximated by specular-reflection and bounce-back processes for slip and no-slip walls, respectively [28]. On a no-slip wall, a wall function is applied based on the generalized law-of-the-wall model [33], which has been extended to account for pressure gradient and surface roughness. The formulation is given by

$$u^+ = \frac{1}{k} \ln\left(\frac{y^+}{A}\right) + B, \tag{9}$$

where

$$A = 1 + f\left(\frac{dp}{dx}\right), \quad B = 5.0, \quad k = 0.41, \quad y^+ = \frac{u_{\tau} y}{\nu} \tag{10}$$

such that A is a function of the pressure gradient.

The BTE is solved on a lattice of cubic elements, which are referred to as voxels. Currently, a variable resolution scheme is implemented such that the simulation domain can be subdivided into regions with different voxel sizes as long as the voxel resolution varies by a factor of two between adjacent regions. When the voxels intersect with solid bodies, planar surfaces are created along the solid–fluid boundaries. The solution produced by the lattice Boltzmann scheme is inherently unsteady and compressible. Moreover, the scheme has low dispersion and dissipation properties, allowing for acoustic phenomena to be resolved in the simulation domain. This implies that the solver offers a direct noise computation capability with a cutoff frequency that corresponds to approximately 15 voxels per wavelength. Nevertheless, the computational cost associated with extending high-resolution voxel region into the acoustic far-field often remains expensive. Hence, an acoustic analogy based on Farrasat's formulation 1A [34] of the Ffowcs-Williams and Hawkings (FW-H) analogy [35] was employed and solved using the forward-time formulation [36].

2.2. Simulation Setup

The present study considers a full-scale aircraft model that combines two main elements, both of which were developed by NASA. The airframe (e.g., fuselage and lifting surfaces) is obtained from the common research model (CRM), which resembles a typical transonic wide-body airliner. The default CRM configuration is equipped with flow-through (empty) nacelles. Hence, the port nacelle is replaced with the fan stage model from the source diagnostic test (SDT) rig. Figure 1 illustrates the present CRM/SDT configuration. It is worth mentioning that the CRM geometry lacks the horizontal and vertical stabilizers. The high-lift devices (slats and flaps) are deployed as the main wings are set in an approach condition. This airframe configuration is also referred to as the high lift CRM (HL-CRM) [37]. In the present study, the SDT fan stage is upscaled by a factor of 5.5 to match the dimensions of typical modern turbofans. Upscaling is not performed for the CRM geometry as it was made available in full scale. The dimensions of the NASA HL-CRM and the upscaled NASA-SDT geometries are compared against their closest real-life counterparts in Table 1. Note that the fan in the current CRM/SDT setup operates at a lower rotational speed (i.e., 1420 RPM) than that in the original SDT study [2] in order to maintain the same fan tip speed $U_{\text{ref}} = 237.57\,\text{m/s}$.

This study is particularly interested in addressing the noise produced by the fan wake–OGV interaction mechanism, and for this purpose, the LE of the OGVs was modified using poro-serrations. A closer view of the poro-serrated OGV in the fan stage is provided in Figure 2. Such LE modification was chosen based on authors' past findings [15]; the serration planform is very effective at attenuating the tonal noise component at low frequencies, while the porosity of the LE reduces the broadband noise generation in the higher frequency range. The porous treatment is modeled after a Ni–Cr–Al metal foam [38] with two different mean pore diameters, i.e., 800 µm and 450 µm; the former was also employed in the authors' previous investigation [15]. The properties of the metal foam are reported in Table 2. In the subsequent sections, the aircrafts equipped with poro-serrated OGVs are referred to as the M800 and M450 configurations respectively, emphasizing the mean pore size. The serrations in these configurations have a sinusoidal planform that is characterized by a wavelength $\Lambda = 0.76c$ and amplitude $H = 0.38c$, where $c = 218.5\,\text{mm}$ is the average chord length of the OGV. The serration parameters are selected based on the turbulence length scales in the axial and radial directions upstream of the OGV. The serrations modify the planform of the OGV only at the first 20 % of the OGV chord. Downstream of the serration root, the porous medium region wraps around a streamlined solid–porous junction that extends in between 35 % to 60 % of the OGV chord.

Aside from the previous two, the M450E configuration was introduced to mitigate the adverse aerodynamic impact from the M800 and M450 designs. In this particular case, the porous medium region is applied only at the outer 85 % of the OGV radius, where the strength of the noise sources is expected to be the strongest [4]. The inner span of the OGV is equipped with solid serrations to prevent the cross-flow in the porous medium region due to the pressure difference between the suction and pressure sides of the blade [15], and therefore the blade loading characteristic is preserved. The serration planform remains uniform across the OGV blade, where the serration wavelength is half of that of the M450 configuration but the amplitude remains the same.

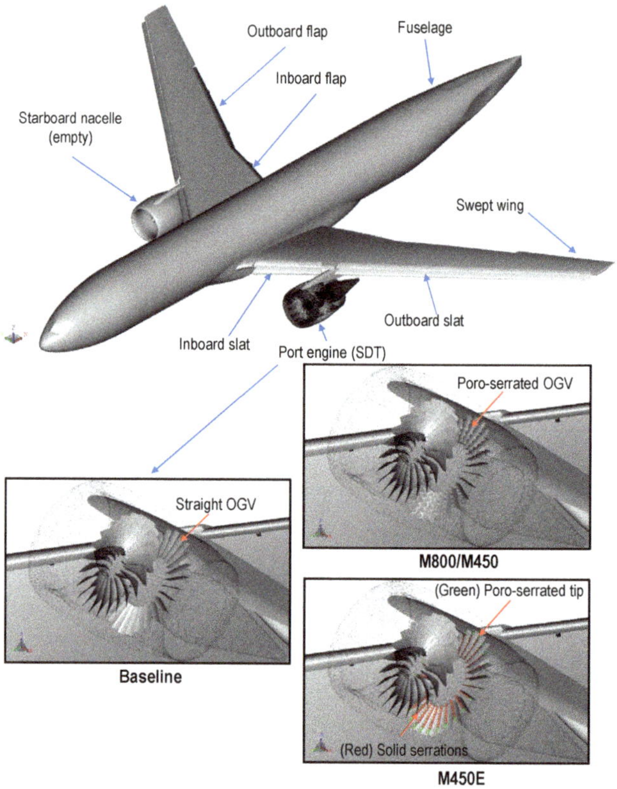

Figure 1. The NASA HL-CRM/SDT configuration in the present study. The lower part of the figure shows a comparison between the baseline configuration and others equipped with poro-serrated OGVs. LE serrations in the M450E inset are colored red and green to distinguish the solid region from the porous one.

Table 1. Comparisons between the NASA-CRM/SDT geometries against some real-life examples.

	NASA HL-CRM	Airbus A330-300
Fuselage length (m)	62.8	63.6
Wingspan (m)	58.7	60.3
	NASA SDT (5.5:1 scale)	**Rolls-Royce Trent 1000**
Fan diameter (m)	2.93	2.85
Bypass ratio	-	10.8–11.0
Fan blade count	22	20
OGV blade count	26	46

Table 2. The properties of porous materials employed for the poro-serrated OGVs.

Type	$d_p(\mu m)$	$\phi(\%)$	$K(m^2)$	$C\ (m^{-1})$
Metal foam (M800)	800	91.65	2.7×10^{-9}	2613
Metal foam (M450/M450E)	450	89.28	6.11×10^{-10}	9758

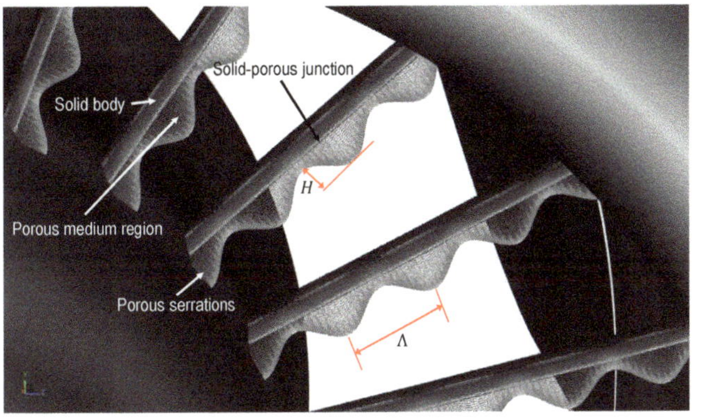

Figure 2. A closer view of the modified OGV (M800/M450 configurations) in the fan stage, highlighting the amplitude H and wavelength Λ of the serration planform.

The simulation domain is shown at the top of Figure 3. The CRM/SDT model is positioned at the center of a cubic domain whose sides are 1400 m long, or roughly 24 times the aircraft wingspan. All domain boundaries are specified as "inlets" with a freestream velocity of 68 m/s and a static pressure of 101, 325Pa. Despite the designation, PowerFLOW allows some inlet boundaries to behave as outlets depending on the numerical solution. The airframe is installed at an incidence of 7 deg to realize an angle of attack of 4 deg and a standard rate of descent of 3 deg [39]. For computing far-field noise using the acoustic analogy, a permeable surface is defined enclosing the exterior of the turbofan and an inboard segment of the port wing as shown in Figure 3. A total of six end caps are added at the downstream end of the permeable surface to filter pseudo-sound contamination from the jet plume. These surface samples' pressure and velocity fluctuate at 5.93 kHz, resulting in a Nyquist frequency that equals 5.7 times the expected blade-passage frequency ($BPF_1 \approx 520$ Hz). Data recording was performed for 1.32 s, which is roughly equivalent to 31 fan rotational cycles. This allows for capturing acoustic data at frequencies as low as 100 Hz ($\approx 0.2\ BPF_1$). An acoustic sponge region was specified starting from a radius of 258 m (i.e., four times the length of the CRM fuselage) from the aircraft's nose to prevent acoustic reflection at the domain boundaries.

The simulation domain was subdivided into 16 grid refinement regions, in which the finest voxels are located inside the fan stage and at regions surrounding the jet plume. A snapshot of the voxel distribution in the simulation domain is shown in Figure 4. The smallest voxel size is equal to 2.54 mm, or $0.0116c$, and thus there are approximately 33 voxels along the axial distance between the serration tip and root. The simulation is carried out for a total of 2.56 s with an initial transient of 1.24 s. The simulations were carried out using the Dutch National Supercomputer facility (Cartesius) with 720 cores of Intel-Haswell Xeon E5-2690 v3. For the finest grid configuration, the computational cost is approximately 95,000 CPU hours for the configuration with poro-serrated OGVs (M800/M450).

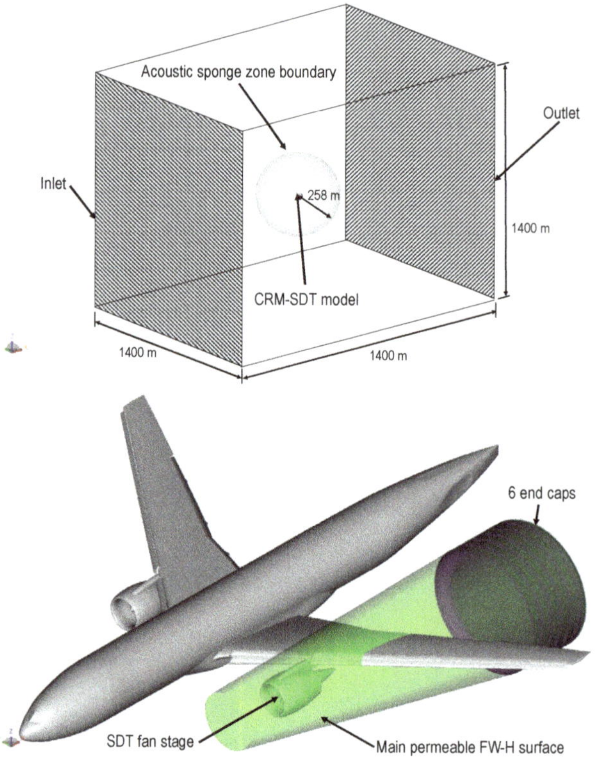

Figure 3. (**Top**) The simulation domain enclosing the CRM/SDT model. (**Bottom**) The configuration of permeable FW-H surface for noise analyses.

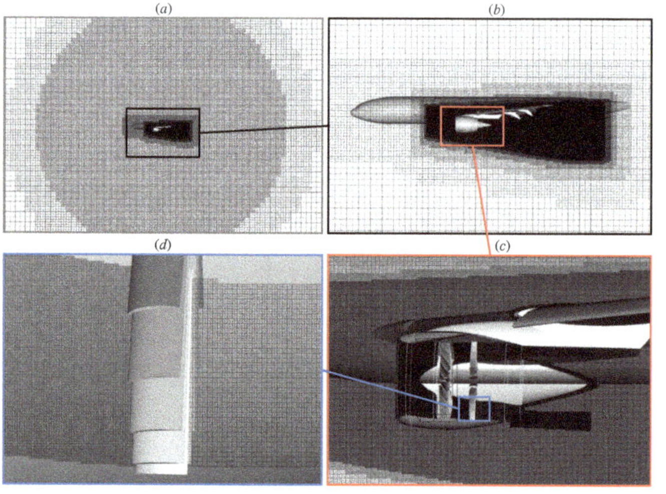

Figure 4. The voxel distribution in the simulation domain: (**a**) near the borders of the acoustic sponge boundary; (**b**) surrounding the airframe; (**c**) surrounding the SDT fan stage; (**d**) near the OGVs.

3. Simulation Verifications

3.1. Grid-Convergence Study

A grid-convergence study of the simulation results is carried out using three different mesh resolutions, namely *coarse*, *medium*, and *fine*, with a refinement ratio of $\sqrt{2}$. The simulation settings are reported in Table 3. Note that the grid resolution level is defined using the number of voxels assigned over the chord length (c) of the baseline SDT OGV.

Table 3. Simulation domain specifications for grid convergence study

Grid Type	Resolution (Voxels/c)	Voxel Count (Millions)	
		Baseline	M800
Coarse	41	104.83	104.89
Medium	61	279.28	279.44
Fine	86	744.05	744.43

The effect of varying the grid resolution level on the mean flow field is evaluated using the time-averaged thrust values. The fan-stage thrust was computed using the following equation.

$$\text{Thrust} = (\dot{m}_{\text{outlet}} U_{x,\text{outlet}} - \dot{m}_{\text{inlet}} U_{x,\text{inlet}}) + (p_{\text{outlet}} - p_{\text{inlet}}) A_{\text{outlet}} \quad (11)$$

The mass flow at the inlet $\dot{m}_{\text{inlet}}$ is equal to that at the outlet $\dot{m}_{\text{outlet}}$ as the SDT configuration lacks a core engine stage, and thus the thrust contribution from the first term of the Equation (11) is only due to the difference between the average axial velocity (U_x) at the fan-stage inlet and that at the outlet. The second term in Equation (11) refers to the pressure-area term, with A_{outlet} being the cross-section area of the fan-stage outlet. Figure 5 shows the variation of the thrust with the number of voxels in the simulations for both baseline and M800 configurations. The thrust value approaches convergence as the number of voxels is increased, such that the differences between the values for the fine and medium grid resolution levels are 0.87 % and 0.28 % for the baseline and M800 cases, respectively.

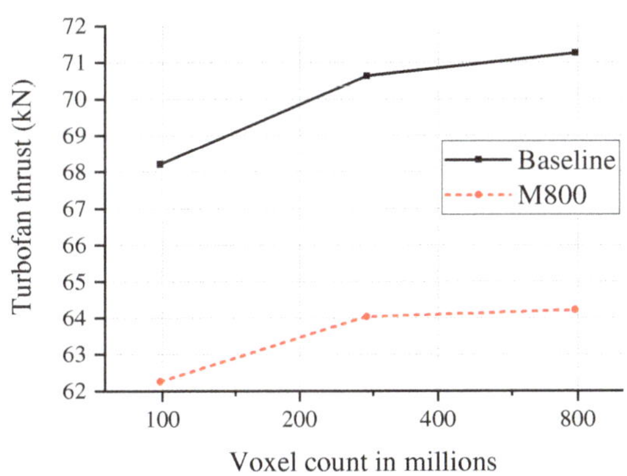

Figure 5. The variation of the thrust produced by the SDT fan stage against the grid resolution level.

The influence of the grid resolution on the sound emission characteristics of the CRM/SDT configuration is evaluated using the sound power level (PWL) spectra. The source power level (PWL) was estimated by integrating the far-field noise intensity over a fictitious spherical surface with a radius of 120 m (i.e., ≈ twice the length of the CRM fuselage) centered at the fan-stage inlet. A total of 648 measurement points are distributed on the surface with an increment of 10° in both azimuthal and polar directions. The variation of the PWL spectra with the grid resolution level is depicted in Figure 6, where converging trends can be observed for both the baseline and M800 cases. For the M800 configuration, the variations in the tonal noise component are more noticeable than the broadband ones. The spectra also show the presence of tones that are unrelated to the harmonics of BPF_1 only when the medium or fine grid setting is employed. These tones might be related to a physical phenomenon that cannot be resolved properly without a sufficiently fine grid. Nevertheless, they only have a minor contribution to the overall PWL (OPWL), such that the differences in OPWL between the medium and coarse grid settings is 0.3 dB, and 0.15 dB between the fine grid setting and the medium one.

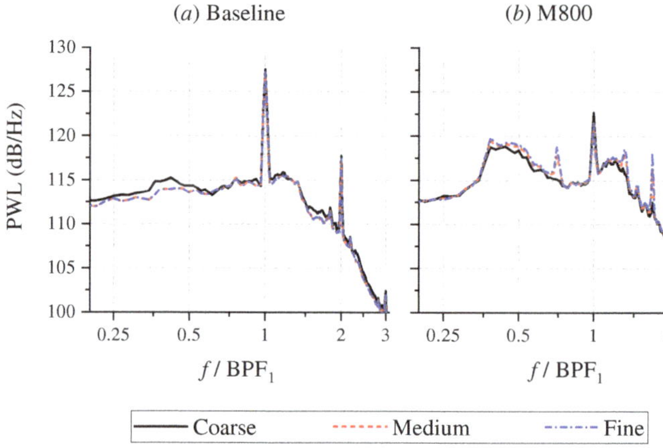

Figure 6. The variation of the sound power level (PWL) of the SDT fan stage against the grid resolution level.

3.2. Integral Length Scales and Serration Design Verification

Obtaining the integral length scales in the fan wake is essential for optimizing the serration design parameters. At present, the integral length scales are computed at axial locations halfway between the fan and OGV blades, which will be referred to as the interstage in subsequent sections. The procedure mimics that of Podboy et al. [40], where axial and radial velocity fluctuations are measured locally using a probe. The integral length scales are then calculated by invoking Taylor's frozen turbulence hypothesis as follows:

$$L_i = \langle \mathbf{U} \rangle \int_0^\infty \frac{\langle u_i(t)u_i(t+\Delta t)\rangle}{u_i^2} dt, \tag{12}$$

where $\mathbf{U}$ is the local velocity magnitude, u_i is the velocity fluctuation in the ith direction, and $\langle \cdot \rangle$ is the ensemble averaging operator.

The variation of axial and radial length scales along the fan-stage radius for the baseline configuration is shown in Figure 7. In the axial direction, the integral length scale is around $L_x = 0.15c$, although it can reach up to $0.19c$ near the hub. Regardless, the usage of a serration amplitude $H = 0.38c$ still satisfies the optimal ratio $H/L_x \leq 2$. The radial length scale (L_z) tends to decrease towards the outer fan-stage radius, varying in between $0.17c$ and $0.1c$. The serration wavelength in both M800 and M450 configurations is $\Lambda = 0.76c$, which satisfies the optimal ratio $\Lambda/L_z \approx 4$ near the hub, but overestimates near the tip. In order to maximize noise attenuation at the outer radius of the OGV, where local noise sources are expected to be the strongest, the serrations in the M450E are designed using a shorter wavelength of $\Lambda = 0.38c$.

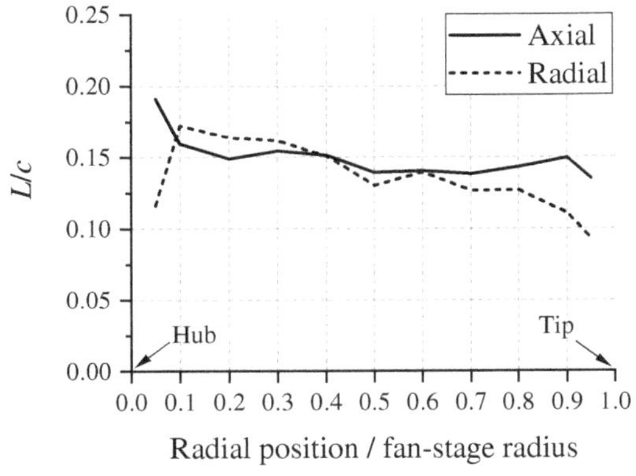

Figure 7. The axial and radial integral length scales of turbulence in the interstage for the baseline configuration.

4. Results

4.1. Aerodynamic Performance and Flow Statistics

This subsection discusses the impact of the poro-serrations on the aerodynamics of the fan stage, such as the mean flow conditions and turbulence behaviors. The overall aerodynamic performance of the fan stage is examined based on the amount of thrust generated. The thrust produced by the baseline configuration is 70.64 kN. It is worth mentioning that this thrust level corresponds to a fan rotational speed that is 61.7 % of the maximum value [2]. Although a simulation with the fan at maximum rotational speed has yet to be performed, the maximum thrust of the SDT fan stage was estimated to be 325 kN using a typical RPM-thrust curve for a turbofan engine [41]. This amount is comparable to the take-off thrust rating of a Trent 1000 turbofan [42]. In comparison, the M800 configuration achieves a lower thrust of 64.22 kN. Consequently, the poro-serrations in the M800 case led to a 9 % thrust penalty. In comparison, the thrust penalty for the lower-porosity configuration (M450) is 7.5 %. Assuming that the thrust penalty does not vary for a small change in the fan RPM, and the aforementioned RPM–thrust relationship is also valid for the fan stages with modified stators, it is estimated that the rotational speed of the fan in the M800 configuration should be increased to 1462 RPM (i.e., an increase of 1.8 %) to retain the same amount of thrust as the baseline. Correspondingly, the fan rotational speed should be increased by 1.5 % to 1454 RPM for the M450 configuration. Nevertheless, it is conjectured that the overall fan-stage noise would only increase by less than 0.5 dB in both cases, assuming that the dipoles at the fan and OGV surfaces are the main noise sources [43]. As expected, the thrust penalty for the M450E configuration is the smallest compared to others, which are 1.4 %. This also evidences the opposite trend between the extent of the porous treatment and aerodynamic performance [12].

Flow statistics are examined at different locations in the fan stage as provided in Figure 8. Figure 8a shows three axial slices corresponding to the fan-stage inlet, the interstage between the fan blades and the OGVs, and the fan-stage outlet. The interstage slice is equally distanced from both the trailing edge (TE) of the fan blades and the leading edge (LE) of the OGVs. Aside from the axial slices, radial slices are also made at 10 %, 45 %, and 80 % of the outer radius of the fan stage, as shown in Figure 8b.

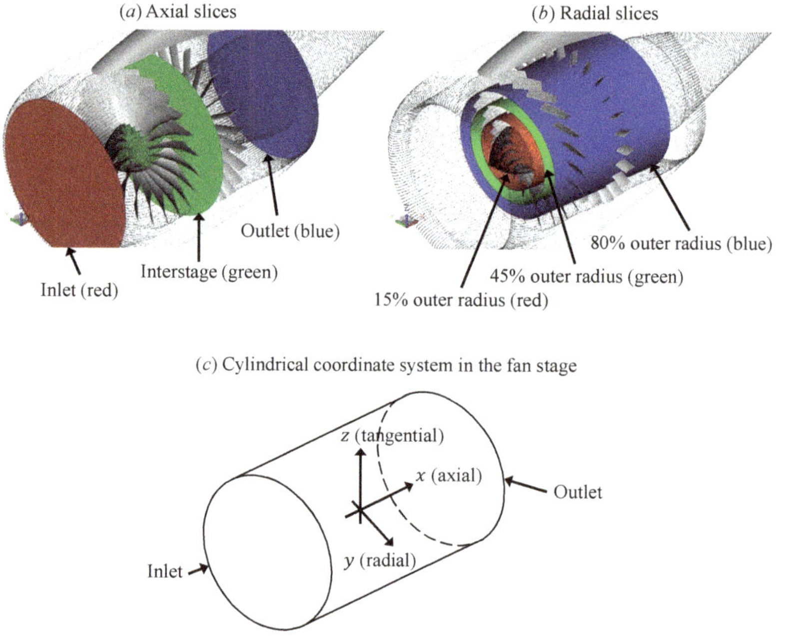

Figure 8. A sketch of the slice planes where contours are plotted in the subsequent figures: (**a**) axial slices and (**b**) radial slices. The cylindrical coordinate system in the fan stage is described in (**c**).

Firstly, the phase-locked mean velocity magnitude contours on the axial slices are provided in Figure 9. The usage of phase-locked statistics allows for emphasizing the periodic flow phenomena at each fan blade passage. Note that the phase-locked average velocity magnitude values $\langle||\mathbf{U}||\rangle$ are normalized against a reference velocity $U_{\mathrm{ref}} = 237.57\,\mathrm{m/s}$, which is the average fan tip speed. As there is an incidence angle between the airframe and the incoming flow, the velocity distribution at the fan-stage inlet is not uniform; the flow at the lower side of the fan stage has a higher velocity than that at the upper side. The average velocity magnitude for the M800 and M450 configurations is slightly lower than in the baseline case, indicating that the air intake at the inlet decreases due to the poro-serrated OGVs. The mean flow field at the interstage is characterized by a spiral pattern due to the swirl induced by the fan rotation. At this location, the mean velocity is, on average, lower for the M800 and M450 configurations, although the swirling patterns do not appear to be affected by the presence of the poro-serrated OGVs. A more noticeable discrepancy can be found at the outlet of the fan stage, where the velocity magnitude near the hub for the M800 case is around 10 % lower than that of its baseline counterpart. On the other hand, the reduction of the mean velocity is less severe for the M450E configuration due to the limited spanwise extent of porous LE treatment. Given that the fan RPM is identical for all cases, it is reasonable to deduce that the poro-serrations have introduced a stronger flow resistance, which results in the thrust reduction discussed earlier.

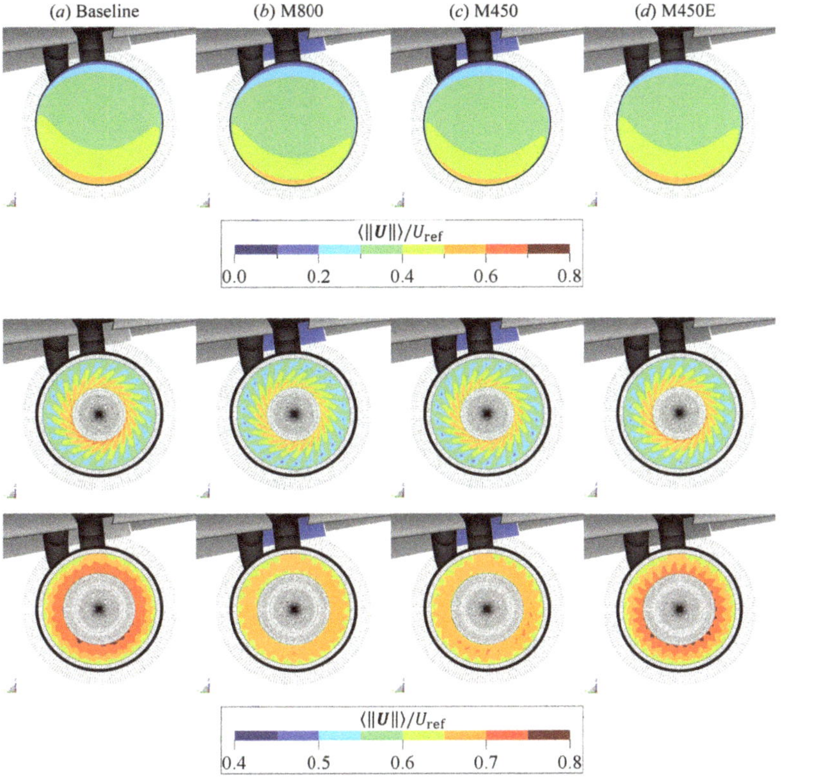

Figure 9. (Front view) Contours of phase-locked average of velocity magnitude ($||\mathbf{U}||$) at the SDT inlet (**top** row), interstage (**middle** row), and outlet (**bottom** row).

In Figure 10, the contours of phase-locked average of velocity magnitude are plotted on the different radial slices. Due to the variation in the tangential velocity along the radius of the fan, the fan wake approaches the OGV with an oblique angle, resulting in a sweeping motion from the bottom of each plot to the top.

The contours at the top row of the figure (the slice at 10 % of the outer radius) reveal that separated-flow regions at the suction sides of the poro-serrated OGVs (in M800 and M450 cases) result in significantly wider wake trails than those in the baseline configuration. As a consequence, the average mean velocity in the fan-stage cross section downstream of the poro-serrated OGVs becomes lower with respect to the baseline one, which is associated with the aforementioned thrust penalty for the configurations with modified stators. A comparable behavior was reported by the authors previously in a rod-airfoil configuration [15], where an isolated airfoil was used instead of a high-solidity configuration. The flow separation is induced by the cross-flow through the porous medium region, which itself is driven by the surface pressure imbalance between the suction and pressure sides of the poro-serrations. Consequently, the poro-serrated OGVs are also less effective at recovering the swirl in the fan wake. This is particularly noticeable in the middle row of the figure (i.e., the slice at 45 % of the outer radius), where the inclination of the wake trail behind the poro-serrated OGV relative to the axial direction is steeper than that of the unmodified OGV. On the other hand, the swirl recovery process in the M450E configuration appears to be comparable to the baseline one. Nonetheless, the solid serrations also cause wider wake trails compared to those of the baseline. This is associated with the enhanced adverse pressure gradient, particularly downstream of the serration root, as reported previously in literature [15,16,19].

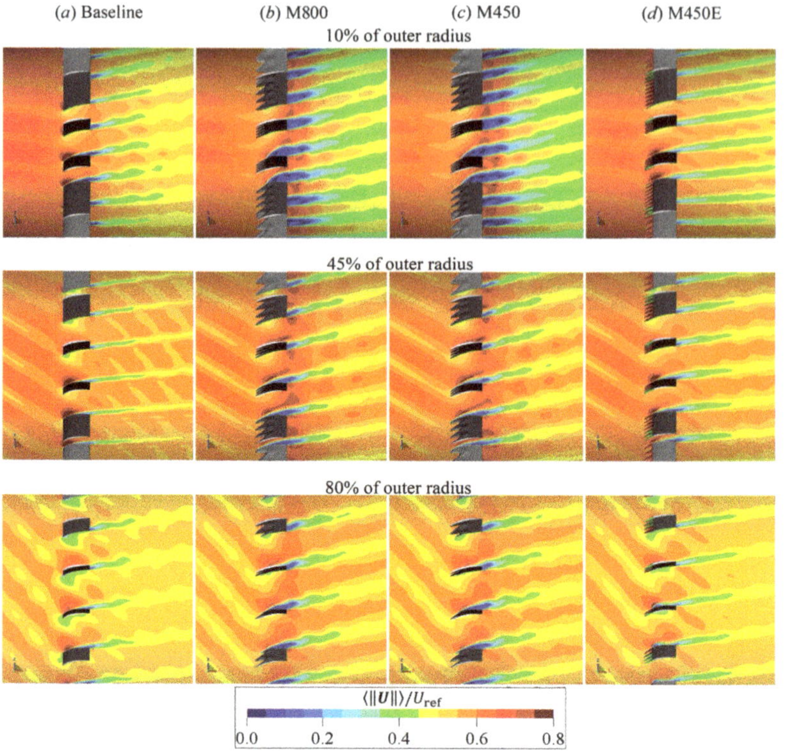

Figure 10. (Side view) Contours of phase-locked average of velocity magnitude $||\mathbf{U}||$ at the different radial sections (**top**—10 %, **middle**—45 %, and **bottom**—80 % of the outer radius) in the fan stage.

The effects of the poro-serrations on the unsteady velocity fluctuations in the flow field are examined using the contours of the RMS (i.e., standard deviation) of velocity magnitude $||\mathbf{U}_{\mathrm{RMS}}||$ in Figures 11 and 12. In the top row of Figure 11, the spiral patterns corresponding to the fan wake can be clearly identified. It is also evident that the level of velocity fluctuations tends to be higher towards the outer radius of the fan stage [4], and thus it is expected that the TIN sources near the outer span of the OGV blades are stronger than the inner ones. This is the reason behind the design of the M450E configuration, where the serrations are made porous only at the outer span of the stator blade.

The contours at the interstage of all four configurations are relatively similar, implying that the turbulence level in the fan wake is relatively unaffected by the LE modifications at the OGV blades. In the bottom row of Figure 11, the contours at the fan stage outlet clearly show noticeably higher velocity fluctuations for both M800 and M450 cases, which can be associated with the wake trails downstream of the separated-flow region. The same phenomena are depicted by the radial slices in Figure 12. The contours at 10 % of the outer radius (top row) clearly depict the separated-flow region at the suction sides of the OGVs in the M800 and M450 cases. Interestingly, the flow fluctuations downstream of the OGV in the M450E case appear to be weaker than in the baseline. This behavior might be associated with the serrations' capability in re-energizing the boundary layer at the suction side of the OGV due to the streamwise vortices generated from the serration tip [44,45]. At other radial positions, the influence of the separated-flow regions in the M800 and M450 configurations is still evident. Nevertheless, the enhancement of the velocity fluctuations in the separated-flow regions is slightly milder for the M450 case due to the lower porosity value.

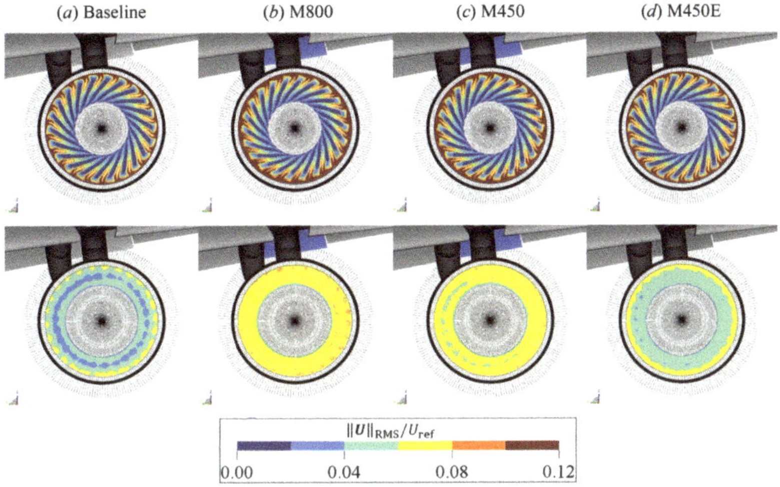

Figure 11. (Front view) Contours of phase-locked standard deviation of velocity magnitude ($||\mathbf{U}_{\mathrm{RMS}}||$) at the SDT interstage (**top** row) and outlet (**bottom** row).

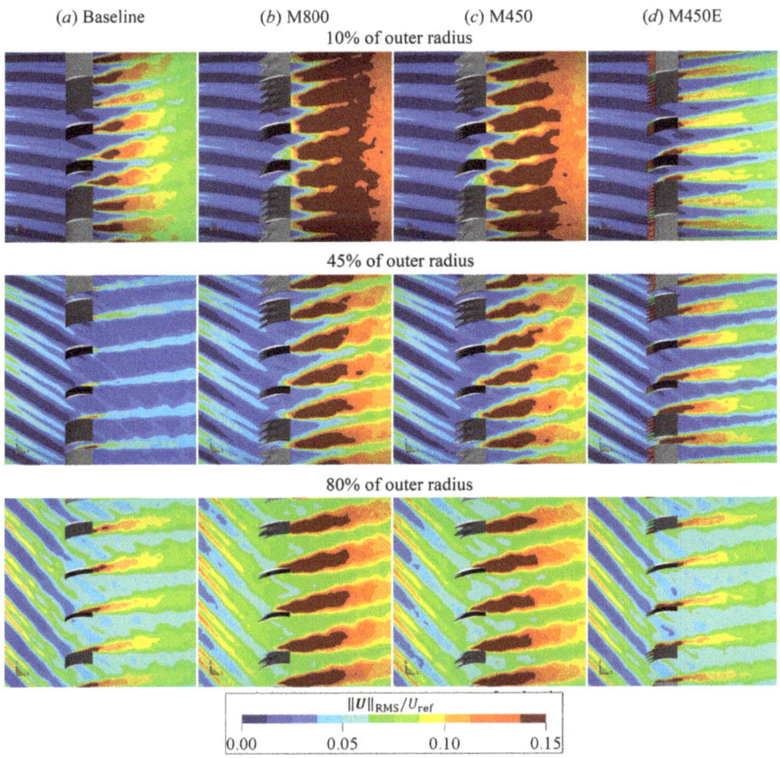

Figure 12. (Side view) Contours of phase-locked standard deviation of velocity magnitude ($||\mathbf{U}_{\mathrm{RMS}}||$) at the different radial sections (**top**—10 %, **middle**—45 %, and **bottom**—80 % of the outer radius) in the fan stage.

To better understand the flow characteristics in the separated-flow region above the suction side of the poro-serrated OGVs, the spectra of of the axial (S_{uu}) and tangential (S_{ww})

velocity fluctuations are examined in Figure 13. For this purpose, the velocity fluctuations are sampled at a distance of 50 mm above the OGV TE. The frequency axis is expressed in multiples of the blade-passage frequency $BPF_1 \approx 520.5$ Hz. The spectra for the M450E configuration is only shown under plot (c) for the sake of clarity of the other plots.

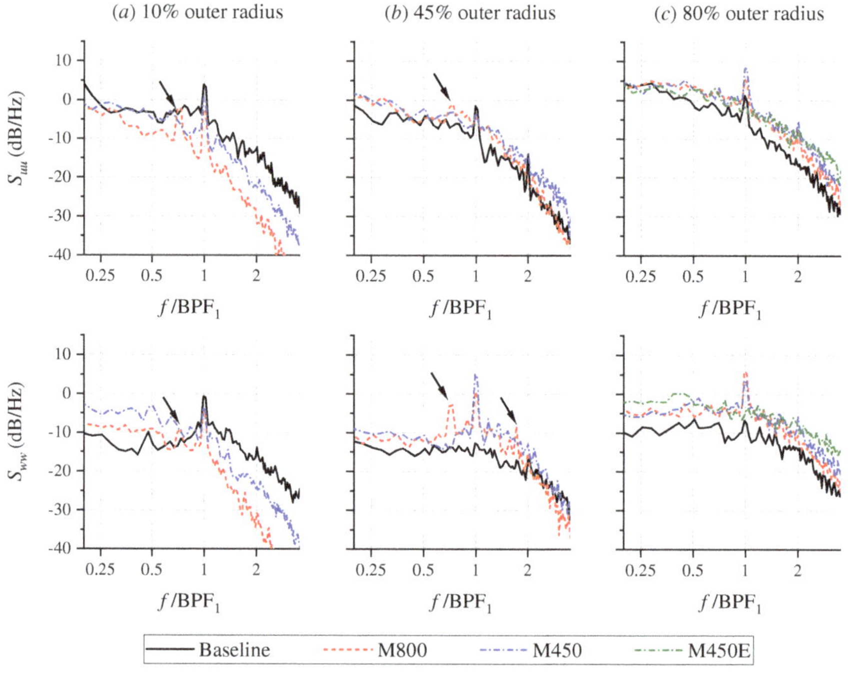

Figure 13. Plots of axial (S_{uu}) and tangential (S_{ww}) velocity fluctuations spectra at a height of 50 mm (0.23c) above the OGV TE. The power spectral density is normalized against a reference velocity of 1 m/s. The arrows in (**a**,**b**) indicate narrowband velocity fluctuations components that are unrelated to the BPFs.

At 10 % of the outer radius, it is apparent that the velocity fluctuations in the separated-flow region are dominated by the low-frequency components. This behavior is particularly noticeable in the S_{ww} spectra, which can be associated with the unsteady upwash due to the flow ejection at the suction side of the poro-serrations. The M450 configuration also exhibits higher axial and tangential velocity fluctuations compared to those in the M800 case at this radial position, reflecting the contours in Figure 12. Unlike at the previous location, the S_{uu} and S_{ww} levels of both M800 and M450 configurations at 45 % of the outer radius are, on average, stronger than those of the baseline in almost the entire frequency range of interest. However, the spectra of the M800 case show an additional tonal peak at $f/BPF_1 = 0.71$, which is especially prominent in the S_{ww} plot. Nevertheless, this tonal peak is completely absent in the M450 case, implying the influence of porosity. Interestingly, the frequency of the tonal peak corresponds to a Strouhal number of approximately 0.1 when it is normalized against the mean velocity magnitude upstream of the poro-serrated OGV and the maximum blade thickness. This behavior might be attributed to a quasi-periodic vortex shedding at a solid–porous junction (see Figure 2), as previously reported by Carpio, et al. [46]. At 80 % of the outer radius (plot c), all three modified configurations exhibit almost identical spectra, although their velocity fluctuation intensity is generally stronger compared to that of the baseline.

The aerodynamic loading characteristics on the OGVs are examined using the contours in Figure 14 and surface pressure distribution plots in Figure 15. The time-averaged surface pressure is expressed as $C_{p,\text{mean}} = (\langle p \rangle - p_\infty)/(0.5\rho_\infty U_{\text{ref}}^2)$, where $p_\infty = 101.325$ Pa. Meanwhile the RMS of surface pressure fluctuations is expressed as $C_{p,\text{RMS}} = p_{\text{RMS}}/(0.5\rho_\infty U_{\text{ref}}^2)$.

The mean surface pressure contour for the baseline configuration (Figure 14a) evidences the clearly-defined suction regions near the LE (i.e., area with low $C_{p,\text{mean}}$). In both M800 and M450 configurations, prominent suction peaks are no longer present on the poro-serrations. The tip of the solid serrations in the M450E configuration also exhibits the same behavior. However, a noticeable suction region can still be found at the serration root, which compensates for the loss of aerodynamic loading at the serration tip; this can be clearly seen in plot (**a**) of Figure 15. In the same plot, the separated-flow regions at the suction side of the OGVs for both M800 and M450 configurations are evidenced by the relatively flat $C_{p,\text{mean}}$ distributions downstream of the poro-serrations ($x/c > 0.2$). On the pressure side of the M800 and M450 OGVs, the $C_{p,\text{mean}}$ values decrease with a steeper slope compared to that of the baseline, indicating flow acceleration caused by the narrowing of the flow passage in the inter-OGV channel due to the separated-flow region (see the velocity contours in Figure 10). The pressure distribution on the porous segment of the serrations in the M450E configuration is shown in Figure 15b. This location also corresponds to the serrations root as in Figure 15a. Nevertheless, the permeability of this region causes the suction peak at the blade LE to vanish, similar to that in both M800 and M450 configurations.

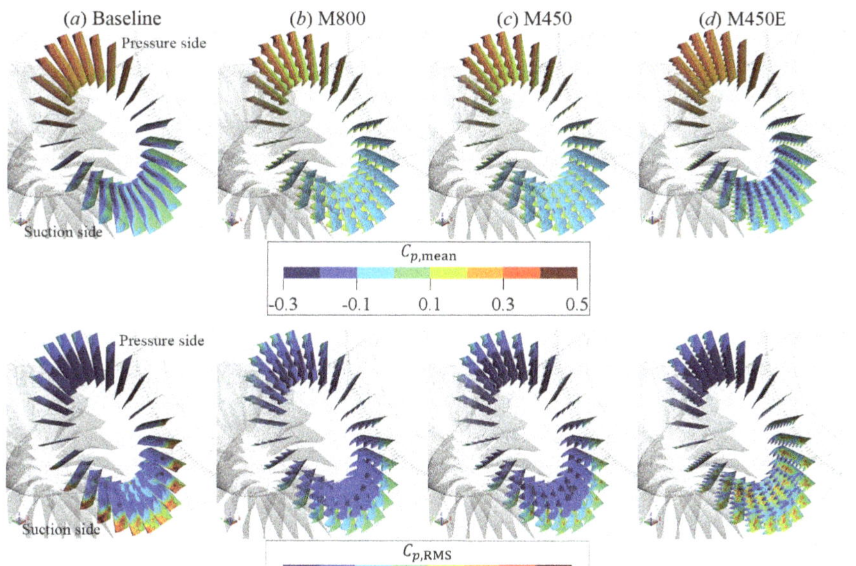

Figure 14. (Isometric view) Contours of time-averaged surface pressure coefficient $C_{p,\text{mean}}$ and the RMS of surface pressure fluctuations $C_{p,\text{RMS}}$ on the OGV blades.

The lower halves of Figures 14 and 15 illustrate the distribution of pressure fluctuations on the OGVs. The $C_{p,\text{RMS}}$ contour for the baseline case shows that the largest pressure fluctuations can be found near the outer span of the blades. When the baseline OGV is replaced with the poro-serrated one (M800 and M450 configurations), the surface pressure fluctuations near the LE tip become less intense. Towards the TE, however, the $C_{p,\text{RMS}}$ values on the poro-serrated OGVs tend to be higher than those on the baseline blades, especially at the suction side. Such behavior seems to be associated with the flow separation, which eventually induces a stall-noise mechanism [47] that is responsible for generating excess broadband noise in the low-frequency range. In Figure 15a, the root segment of the solid serrations in the M450E case exhibits a higher $C_{p,\mathrm{RMS}}$ peak than that of the baseline. This is caused by the reinforcement of upwash/downwash at the serration root due to the production of a secondary streamwise vortex system from the serration tip [15,17]. This mechanism is reflected in the $C_{p,\mathrm{RMS}}$ contour in Figure 14d, where regions with relatively high pressure fluctuations can be found immediately downstream of the serration root. As shown in Figure 15b, the usage of poro-serrations at the outer span of the OGV in the M450E configurations appears to be beneficial as the overall $C_{p,\mathrm{RMS}}$ levels along the chord are lower compared to the those in the baseline case.

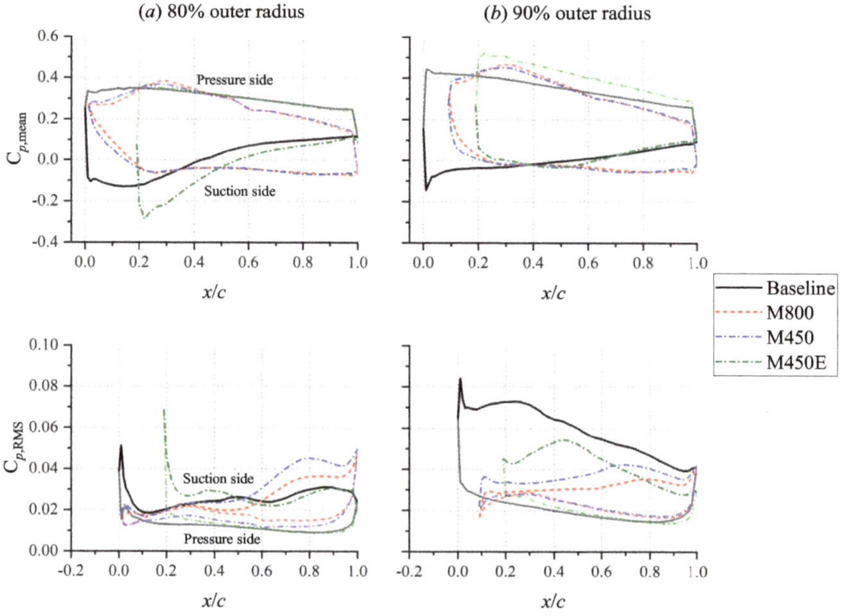

Figure 15. Plots of (**top** row) time-averaged surface pressure coefficient and (**bottom** row) RMS of surface pressure fluctuations at the outer span of the OGV. Pressure coefficient values shown are averaged over the 26 blades. The pressure distributions at the pressure side are plotted with lower opacity. For the M450E configuration, note that the LE porosity is only implemented at >85 % of the blade outer radius.

4.2. Far-Field Noise Characteristics

The effects of the porous treatments on the acoustic characteristics of the CRM/SDT configuration are discussed in this subsection. The sound generated by the CRM/SDT model can be visualized using the dilatation field contour, which is provided in Figure 16. The figure only shows the contour for the baseline configuration as it has been found that the contours for the other configurations are qualitatively similar. The contour clearly shows wavefronts originating from the inlet and the outlet of the nacelle, which correspond to the sound produced by the fan stage components. After leaving the fan stage, sound waves are also diffracted by the high-lift devices. Additional noise is also generated from within the jet plume downstream of the nacelle.

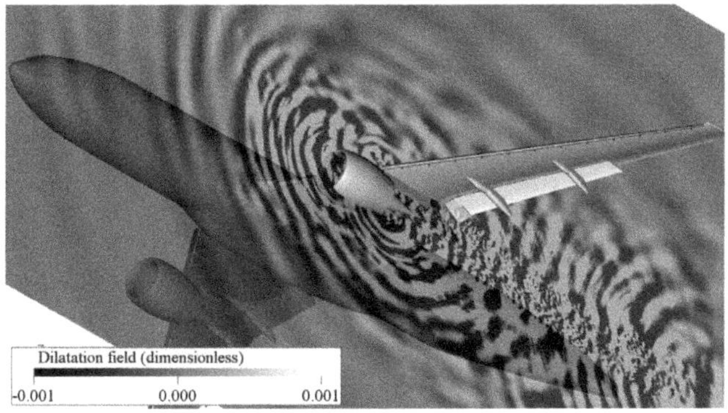

Figure 16. The contour of the dilatation field (time derivative of pressure) surrounding the nacelle containing the SDT fan stage (baseline configuration). Note that the dilatation field is expressed in nondimensional unit.

In Figure 17, the noise emitted by the aircraft is evaluated at two observers that are referred to as *approach* and *lateral* locations, respectively. The *approach* reference position is located at a distance of 120 m directly underneath the aircraft. On the other hand, the *lateral* position is located at a distance of 450 m alongside the aircraft. These locations are based on the reference measurement distances outlined in ICAO Annex 16 [48,49]. The sound pressure level (SPL) spectra in the figure show that the sound intensity at the *approach* location is 12 dB higher on average than at the *lateral* one, which is roughly equal to the difference assuming a spherical spreading by a point source (i.e., both measurement points are located in the acoustic far-field). The tones corresponding to the blade-passage frequency are less prominent at the *lateral* position, due to the noise directivity. By comparing the noise spectra between the different configurations (see Figure 17b), it can be deduced that the application of poro-serrations can substantially reduce the tonal noise component at BPF_1 and its harmonics. However, in the M800 and M450 cases, it is also evident that the broadband noise component in the low-frequency range has been enhanced, particularly at frequencies below BPF_2.

The acoustic energy emitted by the fan stage is characterized by the source power level (PWL) spectra in Figure 18a. The overall PWL (OPWL) values in the frequency range of $0.2 < f/BPF_1 < 3.5$ are also reported in the figure. It is evident that only the M450E configuration achieves a reduction in acoustic power emission, whereas the M800 and M450E ones produce comparable OPWL as the baseline case. As depicted in Figure 18b, this is due to the fact that the reduction of PWL at the BPFs (i.e., the main tonal noise component) has been counterbalanced by the enhanced broadband noise component. The plot also shows that changes in the ΔOPWL are proportional to the porosity of the poro-serrations. For instance, based on Figure 18a, the PWL values in the frequency range "A" ($0.3 < f/BPF_1 < 0.7$) are increased by an average of 2 dB and 1.5 dB for the M800 and M450 configurations, respectively. The broadband noise increase is absent in the M450E case because its solid serrations do not induce any severe flow separation. Above $2BPF_1$ (frequency range "D"), broadband noise attenuation can be observed in the spectra of all three modified configurations. It is worth mentioning that the spectra of the M800 configuration exhibit additional tones that are unrelated to the harmonics of BPF_1, which can be found at $f/BPF_1 = 0.71$, 1.33, 1.48, and 1.71. Following the velocity fluctuations spectra in Figure 13, the tones are likely associated with the vortex shedding process taking place at the solid–porous junction of the poro-serrated OGV [46]. Nevertheless, these additional tones are absent in the M450 configuration, which results in a lower OPWL value relative to its higher-porosity counterpart.

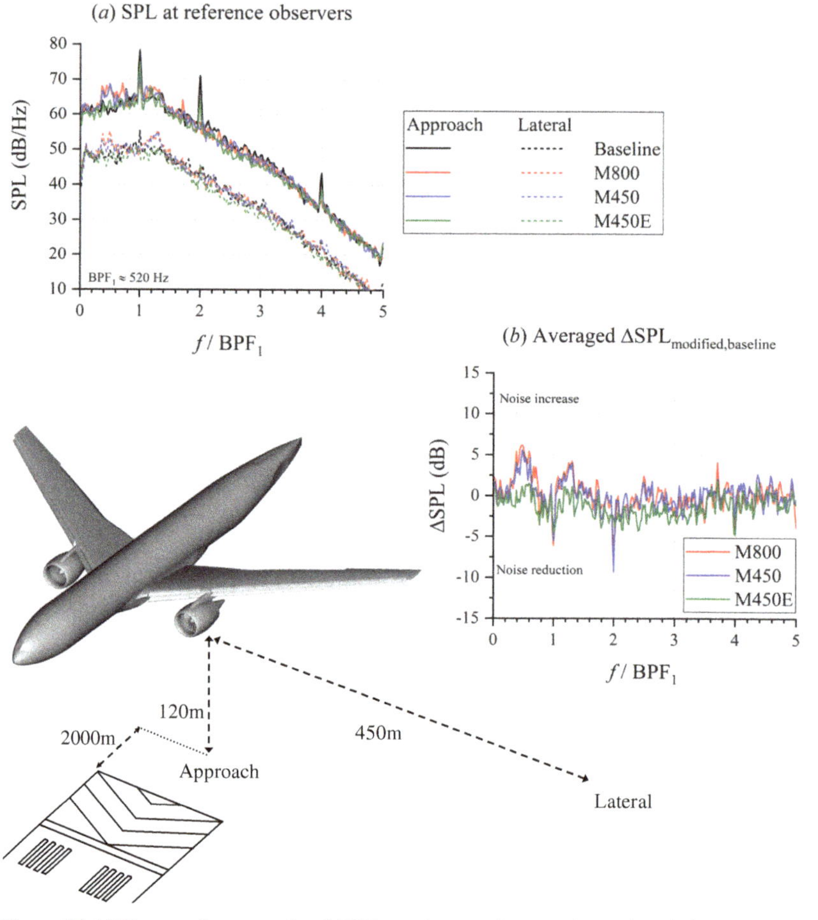

Figure 17. (**a**) The sound pressure level (SPL) at reference observers beneath and alongside the aircraft. (**b**) The SPL difference between the baseline and the two modified cases. Note that the dimensions in the illustration at the lower left of the figure are not to scale.

The noise directivity on the lateral plane of the aircraft is shown in Figure 19. The overall SPL (OSPL) values are reported at a distance of 120 m from the fan-stage inlet corresponding to the *approach* reference location. The nose of the aircraft is aligned with the 180° angle. The directivity plot implies that noise is radiated predominantly towards the upstream and downward directions, the latter of which is attributed to the shielding effect from the wing [50]. For most observer angles, both M800 and M450 configurations produce higher OSPL by 1 to 2 dB relative to the baseline configuration, whereas the M450E configuration produces a reduction of 1 dB. Noise attenuation in the M800 and M450 cases is present only in between 310° and 330°, i.e., in the downstream direction.

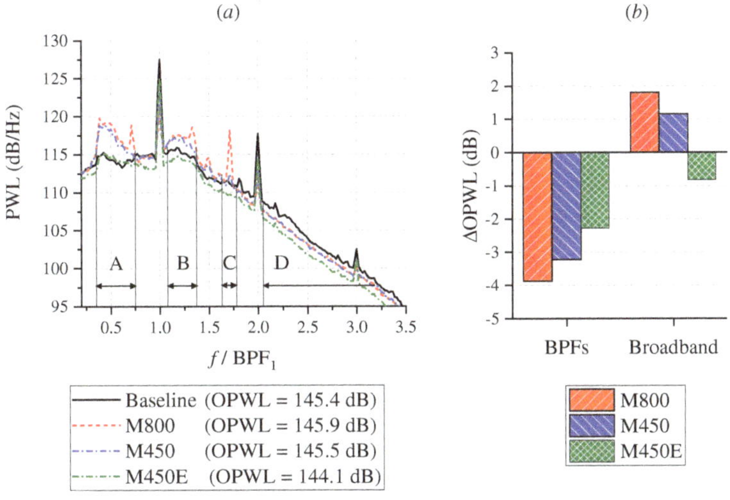

Figure 18. (**a**) Source power level (PWL) of the CRM/SDT configuration. The capital letters denote the frequency ranges where the directivity patterns are plotted in Figure 19. Plot (**b**) shows the OPWL difference between the modified configurations and the baseline, divided into two categories: BPFs (the tonal peaks BPF_1 to BPF_3) and broadband (the broadband noise component and other tones up to $f/BPF_1 = 3.5$).

The noise directivity analysis is expanded in the lower part of the figure, where the directivity patterns are plotted for the different frequency ranges that are previously highlighted in Figure 18a. Plot A, which corresponds to $0.35 < f/BPF_1 < 0.75$, evidences that the noise increase caused by the M800 and M450 OGVs can be observed in all directions. Similar trends are found in plot B ($1.08 < f/BPF_1 < 1.34$), except that the noise increase is only present in directions normal to the fan stage axis. In both plots, nonetheless, the excess noise produced by the M450 case is less severe than that by the M800 one. On the other hand, a slight noise reduction in the M450E case can be found in these frequency ranges. A more noticeable discrepancy between the M800 and M450 cases is shown in plot C, where the OSPL distribution of the latter is relatively similar to that of the baseline, unlike the former. This is related to the additional tone produced by the M800 configuration, whose intensity is comparable to the BPF_2 of the baseline configuration, which is not present with the lower-porosity case. Plot D shows the broadband noise reduction in the high-frequency range, which is mainly observed in the lower downstream direction for the M800 and M450 configurations. Conversely, the M450E also exhibits reduced noise level in the upstream direction.

Basic noise metrics, such as OSPL, do not fully describe the noise characteristics as perceived by a human listener. As a matter of fact, a person with normal hearing capability is most sensitive to sound at frequencies in between 2 and 4 kHz [48]. Furthermore, tonal noise component, in general, induces more annoyance than the broadband one. In the interest of evaluating the impact of the porous treatments on the perceived noise features of the CRM/SDT setup, the perceived noise level (PNL) metric is used, followed by a noise footprint analysis. This analysis was performed using the 3DS-SIMULIA *Opty∂B*-FOOTPRINT tool in conjunction with 3DS-SIMULIA PowerACOUSTICS. The tool has been used to perform digital aircraft noise pre-certification [51], as well as to investigate the impact of trajectory and flight conditions on the noise footprint generated by eVTOL vehicles [52].

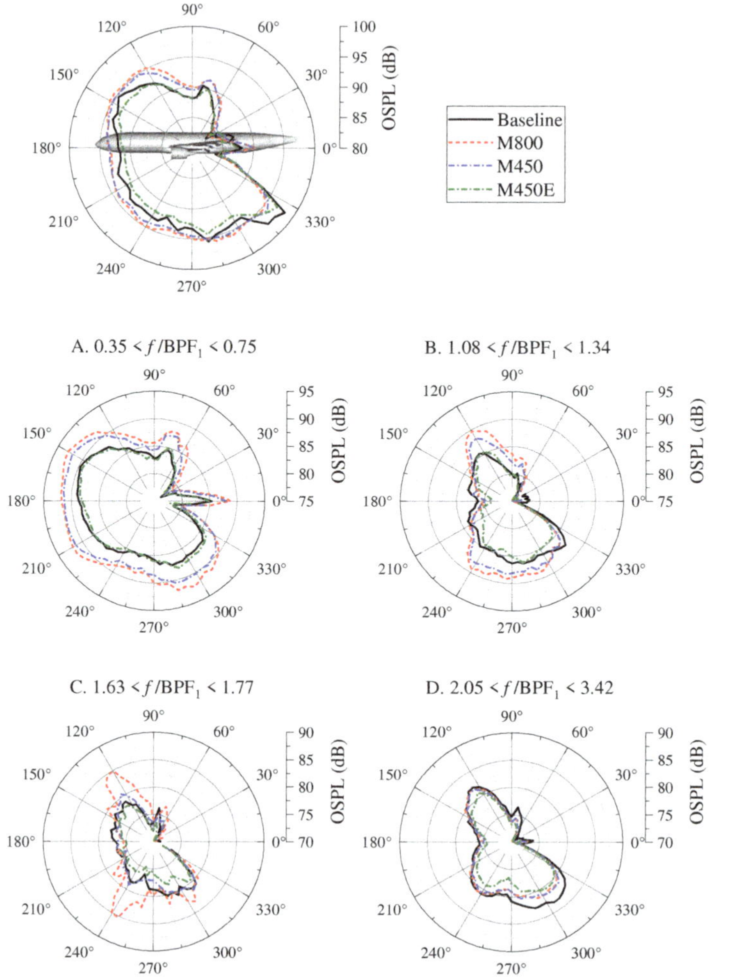

Figure 19. The noise directivity pattern along the lateral axis. Directivity plots A to D correspond to the frequency bands where notable discrepancies between the modified and baseline configurations are present in Figure 18.

Currently, the noise footprint analysis assumes that the aircraft travels along an *approach* trajectory, as illustrated in Figure 20a, which is a descending flight path with a constant slope of 3 degrees. While the present simulation only considers a single fan stage that is installed underneath the port wing of the aircraft, its far-field noise signal (i.e., computed using FW-H analogy) has been mirrored along the symmetry plane of the aircraft to emulate the noise contribution of a second (starboard) fan stage. As shown in Figure 17, the aircraft altitude is 120 m when it is directly above the reference *approach* microphone location, which is indicated as the zero downrange position in Figure 20.

In Figure 20b,c, the perceived noise level (PNL) and its tone-corrected level (PNLt) were used to evaluate the noise level measured at the reference microphone as the aircraft travels along its trajectory. The downrange coordinate is defined from −850 m to 850 m and the PNL values are computed in steps of 0.5 s. When the aircraft is approaching the microphone (i.e., downrange > 0), the M800 and M450 configurations tend to produce higher PNL values (e.g., by up to 2 dB) than the baseline, while reduced PNL can be observed in the M450E case. The difference between the M450E and the baseline configurations becomes more apparent in the PNLt plot due to the mitigated tonal noise emission in the former. However, the PNLt distributions of both M800 and M450 configurations remain comparable to the baseline one, as the tonal noise attenuation is offset by the broadband excess noise. The PNL values for all cases peak slightly ahead of the reference microphone location due to the stronger tendency of noise radiation in the lower upstream direction as previously indicated in Figure 19. As the aircraft leaves the reference microphone location, the PNL and PNLt values for all three modified configurations become smaller than those of the baseline, which is attributed to the noise reduction in the lower downstream direction previously shown in Figure 19.

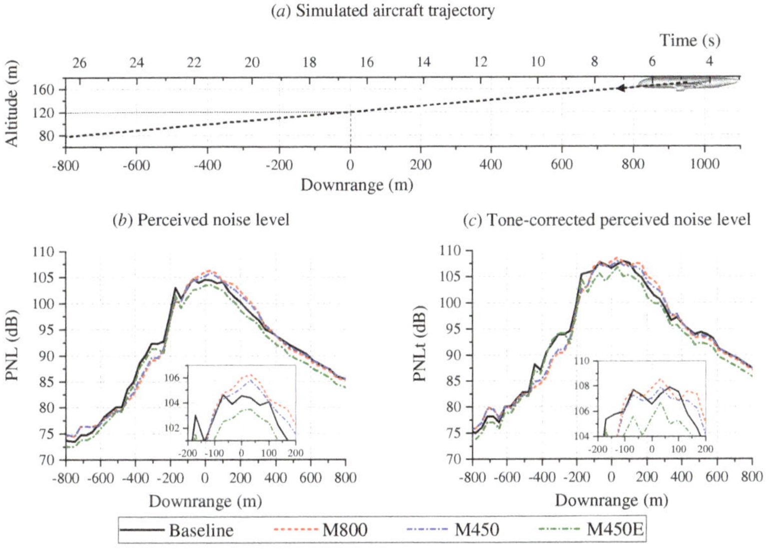

Figure 20. (**a**) The aircraft trajectory for noise footprint analysis, (**b**) the variation of perceived noise level (PNL), and (**c**) the tone-corrected PNL (PNLt) for a reference aircraft approach scenario. The reference microphone is located at the origin of the downrange axis, where the aircraft altitude is 120 m.

The noise footprint of the aircraft during the flyover is illustrated in Figure 21 in term of effective PNL (EPNL). The footprint was computed on a square grid that is 2.5 km by 2.5 km wide. For the M800 and M450 configurations, the contours also evidence the enhancement of noise radiation in front of the aircraft, whereas noise reduction is only present downstream of it. The figure shows that the noise increase is more severe in the M800 case than the M450 one, which reflects the source power spectra trends in Figure 18. On the other hand, the M450E configuration exhibits EPNL reduction in the entire map, although noise attenuation is also larger behind the aircraft. Based on the contours, the EPNL values at the origin (i.e., *approach* reference location) are 105.9 dB, 106.8 dB, 106.2 dB, and 104.4 dB for the baseline, M800, M450, and M450E configurations, respectively. For comparison, the *approach* EPNL for a Boeing 777-300ER is $\approx$ 101 dB [53]. While this is substantially lower than the EPNL value for the baseline CRM-SDT model, it is worth mentioning that the SDT fan stage might no longer be acoustically optimized due to the upscaling. Furthermore, the SDT fan stage also lacks acoustic liner in the nacelle, unlike in typical modern turbofans.

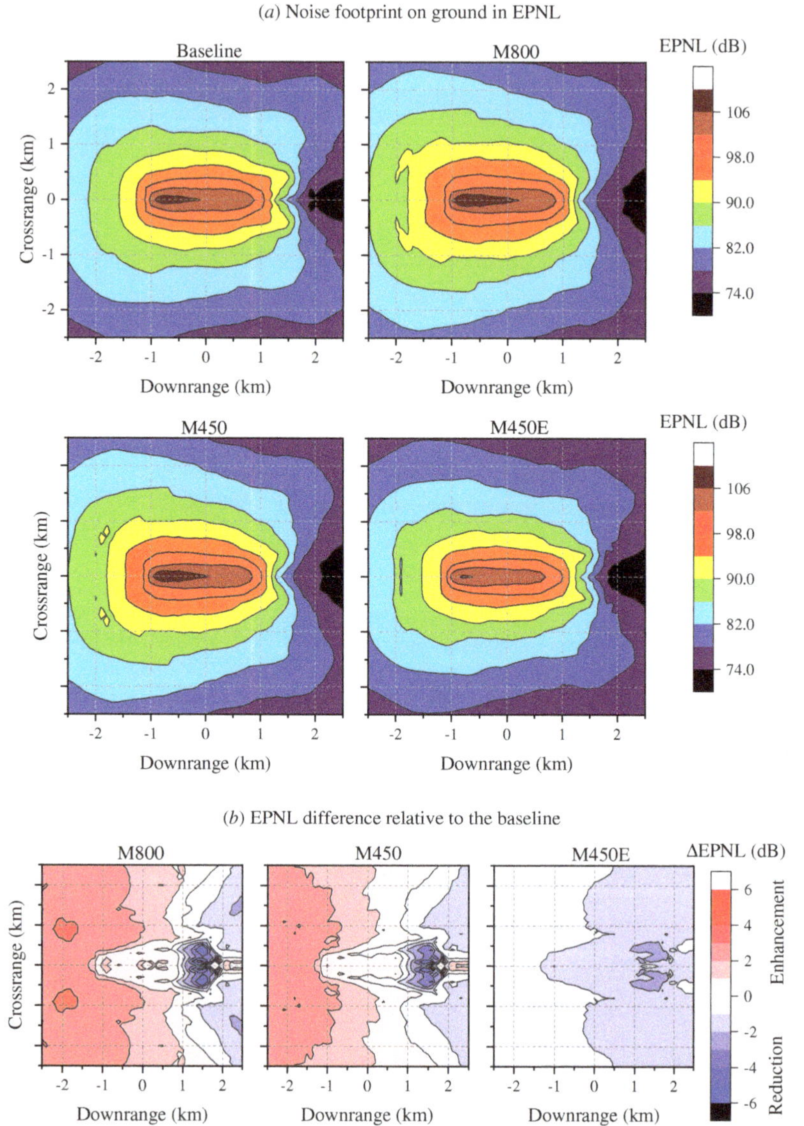

Figure 21. Noise footprint contours, in term of effective PNL (EPNL), corresponding to the aircraft trajectory shown in Figure 20.

5. Conclusions

The present study focuses on a numerical investigation of the application of a porous leading-edge (LE) treatment in a full-scale aircraft simulation. The model consists of the NASA HL-CRM airframe combined with the NASA-SDT fan stage. The outlet guide vanes (OGV) of the fan stage are modified to incorporate a poro-serration treatment to mitigate the fan wake-OGV interaction noise. The poro-serrations were modeled after a metal foam with two different porosity values (e.g., M800 and M450E configurations) to examine how porosity affects the efficacy of the noise reduction treatment. A third modified configuration (M450E) was proposed where the serrations are made permeable only at the outer 85 % of the OGV blade span, where the noise source intensity is expected to be the strongest.

The fan stages with the modified OGVs are found to produce a substantial tonal noise reduction at the blade-passage frequency (BPF) and its harmonics (e.g., by up to 7 dB at BPF_1 and 9 dB at BPF_2). However, the poro-serrations in the M800 and M450 configurations are also responsible for enhancing the broadband noise component in the low-frequency range. Moreover, additional tones that are not associated with the BPFs are also emitted in the higher porosity case (i.e., M800). As a consequence, the overall source power levels of both M800 and M450 cases are relatively unchanged with respect to the baseline one. Conversely, the M450E configuration achieves 1.5 dB reduction in term of overall source power level. The noise directivity analysis reveals that the modified fan stages exhibit noise reduction mainly in the lower downstream direction. This behavior is also reflected in the noise footprint contours where the noise reduction level is generally higher behind the aircraft.

By analyzing the flow field in the fan stage, it has been shown that the broadband noise enhancement in the M800 and M450 configurations is attributed to the flow separation above the suction side of the poro-serrated OGVs. The separated-flow region is induced by an unsteady cross-flow in the porous medium region, which is driven by a pressure balancing process. A vortex shedding phenomenon near the solid-porous junction was identified for the poro-serrations with higher porosity, which causes additional tones to be emitted. The flow separation on the poro-serrated OGVs eventually decreases the mean axial velocity and the mass flow rate in the fan stage, lowering the amount of thrust generated. The thrust penalty is found to be proportional with the porosity of the LE treatment, which is 9 % in the M800 case and 7.5 % in the M450 case. Thus, the poro-serration with the lower porosity value exhibits a more favorable trade-off between acoustic and aerodynamic characteristics compared to that with higher porosity.

Using the M450E configuration, it was demonstrated to be more beneficial to apply porous treatment only at the outer span of the OGV for two reasons. Firstly, the fan-wake impingement at these locations generates stronger surface pressure fluctuations on the OGV, which can be efficiently mitigated by the porous medium. Secondly, by limiting the spanwise extent where the leading edge is permeable, the adverse aerodynamic impact associated with the porous medium region, such as the mean cross-flow between the suction and pressure sides of the poro-serrations, can be mitigated.

Author Contributions: Conceptualization, D.C., C.T. and L.R.; methodology, D.C., C.T. and L.R.; software, D.C.; validation, C.T.; formal analysis, C.T. and L.R.; investigation, C.T. and L.R.; resources, D.C., D.R. and F.A.; data curation, C.T. and L.R.; writing—original draft preparation, C.T.; writing—review and editing, C.T., D.C., D.R. and F.A.; visualization, C.T. and D.C.; supervision, D.C., D.R. and F.A.; project administration, D.C. and D.R.; funding acquisition, D.R. and F.A. All authors have read and agreed to the published version of the manuscript.

Funding: This study is supported by the project SMARTANSWER (Smart Mitigation of flow-induced Acoustic Radiation and Transmission for reduced Aircraft, surface traNSport, Workplaces and wind enERgy noise) which has received funding from the European Union's Horizon 2020 research and innovation program under the Marie Skłodowska-Curie grant agreement No. 722401. More information can be found on https://www.h2020-smartanswer.eu/ (accessed on 27 December 2021).

Institutional Review Board Statement: Not applicable.

Informed Consent Statement: Not applicable.

Data Availability Statement: The data presented in this study are available on request from the corresponding author. The data are not publicly available due to the usage of commercial software packages and proprietary data formats.

Acknowledgments: The authors acknowledge the computing time on the Dutch National Computer Facilities, *Cartesius*, which was provided by the Dutch Research Council (NWO) through the project 2021/ENW/01050159.

Conflicts of Interest: The authors declare no conflict of interest.

Abbreviations

The following abbreviations are used in this manuscript:

BPF	Blade-passage frequency
CRM	Common research model airframe
EPNL	Effective perceived noise level
FW-H	Ffowcs-Williams and Hawkings analogy
LE	Leading edge
OGV	Outlet guide vane
OPWL	Overall source power level
OSPL	Overall sound pressure level
PNL	Perceived noise level
PNLt	Tone-corrected perceived noise level
PWL	Source power level
SDT	Source diagnostic test rig
SPL	Sound pressure level
TE	Trailing edge
TIN	Turbulence-impingement noise

References

1. Casalino, D.; Diozzi, F.; Sannino, R.; Paonessa, A. Aircraft noise reduction technologies: A bibliographic review. *Aerosp. Sci. Technol.* **2008**, *12*, 1–17. [CrossRef]
2. Woodward, R.; Hughes, C.; Jeracki, R.; Miller, C. Fan Noise Source Diagnostic Test–Far-field Acoustic Results. In Proceedings of the 8th AIAA/CEAS Aeroacoustics Conference & Exhibit, Breckenridge, CO, USA, 17–19 June 2002; p. 2427.
3. Zaporozhets, O.; Tokarev, V.; Attenborough, K. *Aircraft Noise: Assessment, Prediction and Control*; CRC Press: Boca Raton, FL, USA, 2011.
4. Casalino, D.; Hazir, A.; Mann, A. Turbofan broadband noise prediction using the Lattice Boltzmann Method. *AIAA J.* **2018**, *56*, 609–628. [CrossRef]
5. Casalino, D.; Avallone, F.; Gonzalez-Martino, I.; Ragni, D. Aeroacoustic study of a wavy stator leading edge in a realistic fan/OGV stage. *J. Sound Vib.* **2019**, *442*, 138–154. [CrossRef]
6. Paterson, R.W.; Amiet, R.K. Noise and surface pressure response of an airfoil to incident turbulence. *J. Aircr.* **1977**, *14*, 729–736. [CrossRef]
7. Hersh, A.S.; Soderman, P.T.; Hayden, R.E. Investigation of acoustic effects of leading-edge serrations on airfoils. *J. Aircr.* **1974**, *11*, 197–202. [CrossRef]
8. Roger, M.; Schram, C.; De Santana, L. Reduction of airfoil turbulence-impingement noise by means of leading-edge serrations and/or porous material. In Proceedings of the 19th AIAA/CEAS Aeroacoustics Conference, Berlin, Germany, 27–29 May 2013; p. 2108.
9. Narayanan, S.; Chaitanya, P.; Haeri, S.; Joseph, P.; Kim, J.; Polacsek, C. Airfoil noise reductions through leading edge serrations. *Phys. Fluids* **2015**, *27*, 025109. [CrossRef]
10. Bampanis, G.; Roger, M.; Ragni, D.; Avallone, F.; Teruna, C. Airfoil-turbulence interaction noise source identification and its reduction by means of leading edge serrations. In Proceedings of the 25th AIAA/CEAS Aeroacoustics Conference, Delft, The Netherlands, 20–23 May 2019; p. 2741.
11. Lee, S. Reduction of blade-vortex interaction noise through porous leading edge. *AIAA J.* **1994**, *32*, 480–488. [CrossRef]
12. Sarradj, E.; Geyer, T. Noise generation by porous airfoils. In Proceedings of the 13th AIAA/CEAS Aeroacoustics Conference (28th AIAA Aeroacoustics Conference), Rome, Italy, 21–23 May 2007; p. 3719.
13. Geyer, T.; Sarradj, E.; Giesler, J.; Hobracht, M. Experimental assessment of the noise generated at the leading edge of porous airfoils using microphone array techniques. In Proceedings of the 17th AIAA/CEAS Aeroacoustics Conference (32nd AIAA Aeroacoustics Conference), Portland, OR, USA, 5–8 June 2011; p. 2713.
14. Geyer, T.F.; Lucius, A.; Schrödter, M.; Schneider, M.; Sarradj, E. Reduction of turbulence interaction noise through airfoils with perforated leading edges. *Acta Acust. United Acust.* **2019**, *105*, 109–122. [CrossRef]
15. Teruna, C.; Avallone, F.; Casalino, D.; Ragni, D. Numerical investigation of leading edge noise reduction on a rod-airfoil configuration using porous materials and serrations. *J. Sound Vib.* **2021**, *494*, 115880. [CrossRef]
16. Chen, W.; Qiao, W.; Tong, F.; Wang, L.; Wang, X. Numerical investigation of wavy leading edges on rod–airfoil interaction noise. *AIAA J.* **2018**, *56*, 2553–2567. [CrossRef]
17. Turner, J.M.; Kim, J.W. Aeroacoustic source mechanisms of a wavy leading edge undergoing vortical disturbances. *J. Fluid Mech.* **2017**, *811*, 582–611. [CrossRef]
18. Gea Aguilera, F.; Gill, J.R.; Angland, D.; Zhang, X. Wavy leading edge airfoils interacting with anisotropic turbulence. In Proceedings of the 23rd AIAA/CEAS Aeroacoustics Conference, Denver, CO, USA, 5–9 June 2017; p. 3370.

19. Chaitanya, P.; Joseph, P.; Narayanan, S.; Vanderwel, C.; Turner, J.; Kim, J.W.; Ganapathisubramani, B. Performance and mechanism of sinusoidal leading edge serrations for the reduction of turbulence–aerofoil interaction noise. *J. Fluid Mech.* **2017**, *818*, 435–464. [CrossRef]
20. Palleja-Cabre, S.; Paruchuri, C.C.; Joseph, P.; Priddin, M.J.; Ayton, L.J. Downstream Perforations for the Reduction of Turbulence-Aerofoil Interaction Noise: Part I-Experimental Investigation. In Proceedings of the AIAA AVIATION 2021 FORUM, Washington, PA, USA, 2–6 August 2021; p. 2149.
21. Ayton, L.J.; Colbrook, M.J.; Geyer, T.F.; Chaitanya, P.; Sarradj, E. Reducing aerofoil–turbulence interaction noise through chordwise-varying porosity. *J. Fluid Mech.* **2021**, *906*. [CrossRef]
22. Lockard, D.P.; Choudhari, M.M.; O'Connell, M.D.; Duda, B.M.; Fares, E. Noise simulations of the high-lift common research model. In Proceedings of the 23rd AIAA/CEAS Aeroacoustics Conference, Denver, CO, USA, 5–9 June 2017; p. 3362.
23. Ribeiro, A.F.; Ferris, R.; Khorrami, M.R. Aeroacoustic Computations of a Generic Low Boom Concept in Landing Configuration: Part 2-Airframe Noise Simulations. In Proceedings of the AIAA AVIATION 2021 FORUM, Washington, PA, USA, 2–6 August 2021; p. 2196.
24. König, B.; Fares, E. Exa powerflow simulations for the sixth AIAA drag prediction workshop. *J. Aircr.* **2018**, *55*, 1482–1490. [CrossRef]
25. Succi, S. *The Lattice Boltzmann Equation: For Fluid Dynamics and Beyond*; Oxford University Press: Oxford, UK, 2001.
26. Zhang, R.; Fan, H.; Chen, H. A lattice Boltzmann approach for solving scalar transport equations. *Philos. Trans. R. Soc. A* **2011**, *369*, 2264–2273. [CrossRef] [PubMed]
27. Bhatnagar, P.L.; Gross, E.P.; Krook, M. A model for collision processes in gases. I. Small amplitude processes in charged and neutral one-component systems. *Phys. Rev.* **1954**, *94*, 511. [CrossRef]
28. Chen, H.; Teixeira, C.; Molvig, K. Realization of fluid boundary conditions via discrete Boltzmann dynamics. *Int. J. Mod. Phys. C* **1998**, *9*, 1281–1292. [CrossRef]
29. Chen, H.; Chen, S.; Matthaeus, W.H. Recovery of the Navier-Stokes equations using a lattice-gas Boltzmann method. *Phys. Rev. A* **1992**, *45*, R5339. [CrossRef]
30. Chen, H.; Orszag, S.A.; Staroselsky, I.; Succi, S. Expanded analogy between Boltzmann kinetic theory of fluids and turbulence. *J. Fluid Mech.* **2004**, *519*, 301–314. [CrossRef]
31. Yakhot, V.; Orszag, S.A. Renormalization group analysis of turbulence. I. Basic theory. *J. Sci. Comput.* **1986**, *1*, 3–51. [CrossRef]
32. Teixeira, C.M. Incorporating turbulence models into the lattice-Boltzmann method. *Int. J. Mod. Phys. C* **1998**, *9*, 1159–1175. [CrossRef]
33. Launder, B.E.; Spalding, D.B. The numerical computation of turbulent flows. In *Numerical Prediction of Flow, Heat Transfer, Turbulence and Combustion*; Elsevier: Amsterdam, The Netherlands, 1983; pp. 96–116.
34. Farassat, F.; Succi, G.P. A review of propeller discrete frequency noise prediction technology with emphasis on two current methods for time domain calculations. *J. Sound Vib.* **1980**, *71*, 399–419. [CrossRef]
35. Ffowcs-Williams, J.; Hawkings, D.L. Sound generation by turbulence and surfaces in arbitrary motion. *Phil. Trans. R. Soc. Lond. A* **1969**, *264*, 321–342.
36. Casalino, D. An advanced time approach for acoustic analogy predictions. *J. Sound Vib.* **2003**, *261*, 583–612. [CrossRef]
37. Lacy, D.S.; Sclafani, A.J. Development of the high lift common research model (hl-crm): A representative high lift configuration for transonic transports. In Proceedings of the 54th AIAA Aerospace Sciences Meeting, San Diego, CA, USA, 4–8 January 2016; p. 0308.
38. Rubio Carpio, A.; Merino Martinez, R.; Avallone, F.; Ragni, D.; Snellen, M.; van der Zwaag, S. Broadband Trailing-Edge Noise Reduction Using Permeable Metal Foams. In Proceedings of the INTER-NOISE and NOISE-CON Congress and Conference, Hong Kong, China, 27–30 August 2017; Volume 255, pp. 2755–2765.
39. Regulations, F.A. *Noise Standards: Aircraft Type and Airworthiness Certification*; Technical Report, Technical Report 14 CFR Part 36; United States Federal Aviation Administration: Washington, DC, USA, 2009.
40. Podboy, G.; Krupar, M.; Helland, S.; Hughes, C. Steady and unsteady flow field measurements within a NASA 22 inch fan model. In Proceedings of the 40th AIAA Aerospace Sciences Meeting & Exhibit, Hampton, VA, USA, 14–17 January 2002; p. 1033.
41. FAA. *Airplane Flying Handbook (FAA-H-8083-3A)*; Skyhorse Publishing Inc.: New York, NY, USA, 2011.
42. EASA. *Type-Certificate Data Sheet for Trent 1000 series engines (EASA.E.036)*; European Union Aviation Safety Agency: Cologne, Germany, 2019.
43. Curle, N. The influence of solid boundaries upon aerodynamic sound. *Proc. R. Soc. London. Ser. A* **1955**, *231*, 505–514.
44. Hansen, K.L.; Kelso, R.M.; Dally, B.B. Performance variations of leading-edge tubercles for distinct airfoil profiles. *AIAA J.* **2011**, *49*, 185–194. [CrossRef]
45. Kim, J.W.; Haeri, S.; Joseph, P.F. On the reduction of aerofoil–turbulence interaction noise associated with wavy leading edges. *J. Fluid Mech.* **2016**, *792*, 526–552. [CrossRef]
46. Carpio, A.R.; Avallone, F.; Ragni, D.; Snellen, M.; van der Zwaag, S. Quantitative criteria to design optimal permeable trailing edges for noise abatement. *J. Sound Vib.* **2020**, *485*, 115596. [CrossRef]
47. Brooks, T.F.; Pope, D.S.; Marcolini, M.A. *Airfoil Self-Noise and Prediction*; NASA Reference Publication, NASA-RP-1218; NASA: Washington, DC, USA, 1989.

48. Boettcher, J. Noise Certification Workshop. In *Session 2: Aircraft Noise Certification*; International Civil Aviation Organization: Montreal, QC, Canada, 2006.
49. Nöding, M.; Bertsch, L. Application of Noise Certification Regulations within Conceptual Aircraft Design. *Aerospace* **2021**, *8*, 210. [CrossRef]
50. Bertsch, L.; Heinze, W.; Guérin, S.; Lummer, M.; Delfs, J.W. 10 Years of Joint Research at DLR and TU Braunschweig toward Low-Noise Aircraft Design-What Did we Achieve? *Aeronaut. Aerosp. Open Access J.* **2019**, *3*, 89–105. [CrossRef]
51. De La Puente, F.; Guerra Crespo, J.J.; Rodríguez Ahlquist, J.; García Merino, I.; Mancini, S.; Kolb, A. On C295 FWSAR military transport aircraft acoustic certification by CAA. In Proceedings of the AIAA AVIATION 2021 FORUM, Washington, PA, USA, 2–6 August 2021; p. 2202.
52. Casalino, D.; van der Velden, W.C.; Romani, G. Community noise of urban air transportation vehicles. In Proceedings of the AIAA Scitech 2019 Forum, San Diego, CA, USA, 7–11 January 2019; p. 1834.
53. Michel, U. Correlation of aircraft certification noise levels EPNL with controlling physical parameters. In Proceedings of the 19th AIAA/CEAS Aeroacoustics Conference, Reno, NV, USA, 27–29 May 2013; p. 2014.

aerospace

Article

A Coupled LES-Synthetic Turbulence Method for Jet Noise Prediction

Joshua D. Blake [1], Adrian Sescu [1,*], David Thompson [1] and Yuji Hattori [2]

[1] Department of Aerospace Engineering, Mississippi State University, Starkville, MS 39762, USA; jdb621@gmail.com (J.D.B.); dst10@msstate.edu (D.T.)
[2] Institude of Fluid Science, Tohoku University, Katahira 2-1-1, Aoba, Sendai 980-8577, Japan; yuji.hattori@tohoku.ac.jp
* Correspondence: sescu@ae.msstate.edu

Abstract: Large-eddy simulation (LES)-based jet noise predictions do not resolve the entire broadband noise spectra, often under-predicting high frequencies that correspond to un-resolved small-scale turbulence. The coupled LES-synthetic turbulence (CLST) model is presented which aims to model the missing high frequencies. The CLST method resolves large-scale turbulent fluctuations from coarse-grid large-eddy simulations (CLES) and models small-scale fluctuations generated by a synthetic eddy method (SEM). Noise is predicted using a formulation of the linearized Euler equations (LEE), where the acoustic waves are generated by source terms from the combined fluctuations of the CLES and the stochastic fields. Sweeping and straining of the synthetic eddies are accounted for by convecting eddies with the large turbulent scales from the CLES flow field. The near-field noise of a Mach 0.9 jet at a Reynolds number of 100,000 is predicted with LES. A high-order numerical algorithm, involving a dispersion relation preserving scheme for spatial discretization and an Adams–Bashforth scheme for time marching, is used for both LES and LEE solvers. Near-field noise spectra from the LES solver are compared to published results. Filtering is applied to the LES flow field to produce an under-resolved CLES flow field, and a comparison to the un-filtered LES spectra reveals the missing noise for this case. The CLST method recovers the filtered high-frequency content, agreeing well with the spectra from LES and showing promise at modeling the high-frequency range in the acoustic noise spectrum at a reasonable expense.

Keywords: jet noise; turbulence modeling; synthetic turbulence; CFD

Citation: Blake, J.D.; Sescu, A.; Thompson, D.; Hattori, Y. A Coupled LES-Synthetic Turbulence Method for Jet Noise Prediction. *Aerospace* **2022**, 9, 171. https://doi.org/10.3390/aerospace9030171

Academic Editor: Lawrence S. Ukeiley

Received: 8 February 2022
Accepted: 17 March 2022
Published: 21 March 2022

1. Introduction

Numerical prediction of the broadband noise from a subsonic turbulent jet remains a challenge for Computational Aeroacoustics (CAA) due to the need to resolve a wide range of disparate spatial and temporal scales. Furthermore, the scale disparity between the fluid dynamic and acoustic disturbances requires greater numerical accuracy to prevent high frequencies from being damped by dissipation inherent in the numerical schemes [1]. With computational fluid dynamics (CFD), resolving the smallest turbulence scales needed for acoustic predictions drastically increases the computational cost. In fact, resolution of all the turbulent scales via direct numerical simulation (DNS) for a moderately high Reynolds number jet is not currently feasible due to the extreme cost.

The current standard for CFD-based jet simulations is large-eddy simulation (LES). Recent advances in simulation methodology, such as wall modeling for the nozzle interior boundary layer [2], have led to more reliable LES predictions (often at flight Reynolds numbers), and application to more complicated problems [3–6]. Despite these improvements, the problem still remains that LES is computationally expensive yet still cannot resolve high-frequency acoustic disturbances [3,4,7–9]. Since LES only resolves the large turbulent scales (which generate low-frequency noise [10]), modeling is required for the smallest turbulent scales that cannot be resolved by the computational grid. Subgrid-scale (SGS) models for LES do not generate acoustics that correspond to these modeled small scales, which leads to the problem of missing high-frequency content. Increasing the grid resolution to resolve small scales drives up the expense exponentially, making LES impractical for industrial use [11]. Present-day LES simulations can accurately resolve frequencies up to Strouhal numbers of 3 to 5 (St = fD_j/V_j) [4], although St = 7 has been reported recently [6]. This range does not even account for all of the high-frequency content in the jet noise spectra, which can extend up to St = 30 [9,12,13].

For subsonic jet noise, it is important to note that the frequency content varies with the observer angle to the jet [10]. This spectral directivity arises in part due to two separate source mechanisms for subsonic jet noise [10,14–18]. Large coherent eddies and turbulent structures near the end of the potential core generate low-frequency noise, which radiates primarily in the downstream direction. This low-frequency noise dominates the jet noise spectra, comprising the majority of observed sound from a subsonic jet.

A second, less intense sound source arises from small-scale turbulent eddies mixing randomly in the shear layer. Fine-scale mixing generates broadband noise (including high frequencies [10,19,20]) that radiates nearly uniformly. The strength of low-frequency noise at downstream angles renders high-frequency noise insignificant at those observation locations. However, high-frequency noise becomes more influential at sideline angles (90°), where low frequencies do not dominate the spectra [15,20,21]. A more thorough review of possible noise mechanisms for these two sources and their locations within the jet can be found in Blake [22].

For full-scale jets, the high frequencies produced by small-scale turbulence can correspond to the frequency range most annoying to humans and cannot fully be ignored [9,15,23]. From the previous discussion, jet noise predictions based on LES simulations cannot resolve high frequencies and, therefore, under-predict a less intense yet important portion of the jet noise spectra. The inability of LES to resolve a full noise spectrum in a cost-effective manner suggests the need for an alternative method for LES-based noise prediction, such as a model for the missing high-frequency noise.

Several authors have investigated different approaches for modeling the missing high-frequency noise content. Seror et al. [24] apply both a priori and a posteriori methods to filtered DNS results of isotropic turbulence in order to evaluate a hybrid LES/Lighthill noise prediction approach. They observed that the noise from unresolved turbulent scales (or residual SGS stresses) is essential to predict the full acoustic spectra. Batten et al. [25,26] combine resolved fluctuations from LES and subgrid fluctuations from a statistical, Fourier-mode model to generate noise sources. Bodony and Lele [7] develop equations to model the velocity and pressure arising from the missing subgrid scales, using an acoustic analogy approach. Their method is based on the idea that interaction between resolved and missing scales is important to account from the noise arising from the missing scales. Bodony and Lele [11] further developed this method into a statistical noise model for source terms corresponding to the missing noise, relating the space-time correlations of the source terms to the far-field power spectral density of the fluctuating density, obtaining the far-field noise by an adjoint formulation of Goldstein's generalized acoustic analogy. Independently, Bodony [9] uses a Gabor transform to derive equations that approximate the behavior of the subgrid turbulent scales above a cut-off frequency wave number. Additionally, a Lagrangian particle approach estimates subgrid fluctuations and reconstructs the Reynolds stress contributions from the modeled subgrid scales. The subgrid stresses, once obtained, can then be used with an acoustic analogy to form noise source terms for the subgrid turbulence. Unfortunately, the work only presented the method and did not show results.

The most relevant literature on this topic is a paper by Yao and He [27], who propose the use of synthetic fluctuations to account for the missing LES noise content. They apply the kinematic simulation (KS) method of Fung et al. [28] to model subgrid-scale fluctuations that are not resolved by LES, taking care to match both time and space statistics in order to account for the random sweeping hypothesis [29,30]. A hybrid CAA method is used to compute noise by Lighthill's acoustic analogy. Comparisons with DNS for isotropic turbulence show that a combination of LES and synthetic fluctuations can approximately recover the high-frequency content missing from the noise spectra. However, this method was not applied to more complex flows and cannot be used for turbulent shear flows (i.e., jets) because their implementation of the sweeping hypothesis does not account for the influence of the unsteady resolved fluctuations on the unresolved, modeled fluctuations.

The work of Yao and He [27] shows the possibility for synthetic turbulence methods to model the missing high frequencies from LES-predicted noise spectra in general. While several different approaches have been taken to solve the missing noise problem, none have shown results for jet noise prediction. With regards to synthetic turbulence methods, several authors have investigated the use of SEM to model jet noise [31,32].

Fukushima [31] modify the Stochastic Noise Generation and Radiation (SNGR) method (a synthetic turbulence method built on the summation of Fourier modes [33]) to use the synthetic eddy method (SEM) to model turbulent fluctuations in the jet shear layer via a superposition of eddies instead of a superposition of Fourier modes. The SEM method was proposed by Jarrin et al. [34] as a simple and computationally inexpensive method to generate more-realistic turbulence inflow conditions for homogeneous isotropic LES simulations [34–36]. Synthetic eddy methods allow locally-specified length scales and turbulent properties and therefore respond better to local stretching and deforming, which is ideal for the sweeping and straining of turbulence present in jet shear layers. A Reynolds-averaged Navier–Stokes (RANS)-generated background shear flow is used by Fukushima [31] and the Linearized Euler equations (LEE) with source terms are used to propagate the synthetically-generated noise sources.

The method of Hirai et al. [32] shows promise for the use of SEM methods in jet noise prediction, building on the work of Fukushima [31] by investigating the statistical properties of the synthesized turbulent velocity field. Hirai et al. introduce a turbulence dissipation mechanism (a time-decorrelation process) that modulates the intensity of each eddy in time to better match the temporal decorrelation observed in the experimental data of Fleury et al. [37]. This is needed because the original SEM method (by Jarrin et al. [34]) does not correctly model the dissipation of turbulent structures in time. In the method by Hirai et al. [32], each eddy is convected by a constant mean flow velocity. The size and strength of each eddy are recalculated in time based on convection and the background RANS field TKE and ϵ. The authors reported significantly over-predicting the jet noise when the model was first tuned to match the spatial and temporal statistics of the turbulent flowfield. The noise spectra was only matched when the turbulent properties of the synthetic field were adjusted. However, their results show that SEM-based methods can be used for jet noise predictions.

Given the highlighted issues with current LES approaches and the potential of SEM-based methods to model jet noise, the coupled LES-synthetic turbulence (CLST) method is presented to supplement the missing high-frequency noise for LES-based jet noise predictions. The CLST method is based on the assumption that the unresolved small-scale turbulence are isotropic and can therefore be modeled sufficiently by randomized synthetic turbulence rather than developing a model from first principles. In contrast to the methods of Fukushima [31] and Hirai et al. [32], the CLST method resolves large turbulent fluctuations (corresponding to low-frequency acoustic waves) with LES or CLES (coarse-grid large-eddy simulations) for the mean flow and generates small turbulent fluctuations (corresponding to high-frequency acoustic waves) with synthetic turbulence modeling. A representation of the process is shown in Figure 1. The resulting resolved and modeled fluctuations are combined and are used to generate noise sources, which are then propagated with a linearized Euler equation (LEE) solver in a hybrid CAA method. By design, the CLST method convects small-scale synthetic eddies by the resolved, large-scale velocities from LES or CLES, which accounts for sweeping and distortion of the small turbulent scales by the large turbulent scales [38–42].

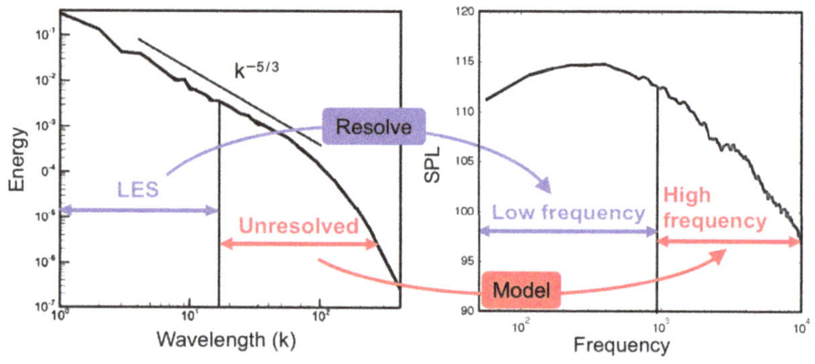

Figure 1. Representation of Resolved and Modeled Turbulent Scales in the CLST Method.

The primary goal of this work is to present a model for the high-frequency spectral content that is missing from LES far-field jet noise predictions and to demonstrate that the CLST method can correctly model near-field jet noise spectra. Such an approach has already shown promise with an SNGR-based turbulence model [43]. A secondary goal of the CLST method is to reduce overall LES simulation costs while still maintaining a high level of fidelity. A cost reduction is possible when using a less-dense spatial grid resolution (CLES) to model the jet. Further cost reduction is possible when using the LEE to propagate noise instead of the Navier–Stokes (NS) equations because the LEE are a simplified equation set and are cheaper to solve on a given grid. Additionally, with an LEE solver, a coarser grid resolution can be used outside the jet source region while still resolving the necessary acoustic spectrum. LEE can also be extended to the far-field for a cheaper cost than the NS equations. Of course, alternative methods for acoustic propagation, such as a permeable Ffowcs–Williams and Hakwings formulation, Kirchoff's method, or Lighthill's acoustic analogy could be used. However, LEE is chosen for the CLST method because the acoustic source terms come directly from velocity fluctuations, making it simple to add the contributions of velocity fluctuations generated by synthetic eddies.

Predicting a more-complete noise spectrum and reducing simulation costs should lead to equally reliable results in a shorter amount of time, potentially allowing the use of LES jet noise predictions more readily in research and industry. This might enable quicker design cycles or the inclusion of wing and pylon installation effects in LES jet noise simulations, as is suggested might soon be possible for business jets [13]. Overall, the CLST method will provide a new framework for LES predictions of turbulence-related noise mechanisms.

This work introduces the CLST method and details the implementation of CLST with SEM-generated fluctuations. The CLST method is applied to a moderately high Reynolds number, Mach 0.9 jet to investigate the effectiveness supplementing the missing high-frequency portion of the jet noise spectra.

2. Problem Formulation and Numerical Framework

2.1. Overview

The main idea of CLST (as shown in Figure 2) is that resolved and modeled velocity fluctuations are coupled to generate a more complete noise spectrum. The CLST method combines coarse-grid large-eddy simulations (CLES) and a modified synthetic eddy method (SEM). Using a hybrid CAA method, the jet flow is simulated with the Navier–Stokes (NS) equations in a finite difference multi-block, structured LES solver. Velocity fluctuations produced by the NS LES/CLES solver are passed to the linearized Euler equations (LEE) to generate acoustic sources. The LEE solver then simulates the resulting noise field. The use of LES or CLES turbulence modeling ensures the resolution of large-scale turbulent fluctuations that correspond to low-frequency noise in the far field. Velocities generated by SEM are added to the source terms in the LEE to generate sources corresponding to the missing high-frequency noise. Figure 2 shows how the proposed CLST method modifies the CLES/LEE-based noise prediction scheme by adding the generation of synthetic fluctuations to model high-frequency content.

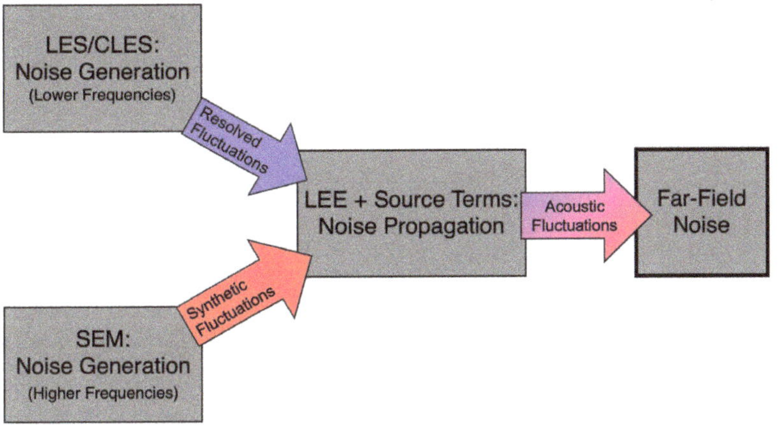

Figure 2. Diagram of the CLST Method.

The proposed CLST modeling approach involves the following general steps:

- Compute an axisymmetric RANS solution for a jet flow using a $k-\epsilon$ or $k-\omega$ turbulence model to generate a more realistic initial flowfield for the LES simulation.
- With the RANS solution as initialization, run the CLES solver until statistically-stationary results are obtained.
- Initialize synthetic turbulence from mean RANS flowfield and turbulence information.
- Continue the CLES simulation and interpolate CLES fluctuations onto the LEE grid.
- Convect synthetic turbulence with CLES velocities, accounting for sweeping and straining of the synthetic velocities.
- Compute velocity field produced by the summation of each synthetic eddy.
- Combine CLES fluctuations with synthetic fluctuations computed at each time step.
- Generate noise sources from LEE source terms and propagate noise with LEE to obtain far-field acoustic pressure data.

As shown in Figure 2, either LES or CLES (coarse-grid large-eddy simulations) can be used with the CLST method depending on the desired computational cost. For instance, given the high computational cost of LES, it is unlikely to see industrial application soon. However, if the CLST method were used with CLES or even Unsteady RANS (URANS) instead of LES, then it might be possible to save significant computational cost due to the lower-fidelity nature of CLES while achieving LES-level frequency resolution. Alternatively, applying the CLST method to LES, while still expensive, could possibly provide DNS-level frequency resolution at a fraction of the cost of full-resolution DNS.

Two assumptions underpin the CLST method and aid in limiting the complexity of this problem. The first assumption is that the two-source theory of jet noise holds and that high-frequency noise content is generated by fine-scale turbulence in the shear layer [10,15,17]. In other words, large scales of turbulence correspond to low-frequency noise, and small turbulent scales correspond to high-frequency noise. The second assumption is that a wide range of turbulent scales exists in the jet shear layer, and, consequently, there is a separation between large and small turbulent scales. Furthermore, it is assumed that the small scales are isotropic (per Kolmogorov's hypothesis) and statistically similar [44] and can therefore be modeled by a synthetic turbulence method.

Additionally, since the hybrid CAA formulation of the CLST method only allows coupling from the LES to LEE, the small scales can not influence the large scales and it is assumed that backscatter [45] should not be explicitly accounted for. However, the influence of the large scales on the small scales, known as sweeping and straining, is assumed to be significant for the radiated noise [40,41]. Propagation of acoustic waves is assumed to be a linear, inviscid phenomena even if generated by nonlinear phenomena. Random sweeping occurs when the passing of large energy-containing eddies leads to convection of small inertial-scale eddies [38–41]. Local straining is the idea that eddies are stretched and distorted by local fluctuations and accelerations [42].

It should be noted that the parameters of the CLST method presented here were developed after initial tests for the specific problem under investigation and that the parameters would likely change for different problems. The evolution of these parameters was not considered a topic of interest for this paper. A future sensitivity study could be used to develop more universal parameters. However, the parameters provided here serve as an initial starting point to demonstrate the potential effectiveness of the CLST method.

2.2. Synthetic Eddy Generation

In CLST, small-scale turbulent velocities are synthesized by a superposition of randomized Gaussian eddies, essentially a synthetic eddy method (SEM). An initial RANS simulation with the same flow field parameters as the LES simulation provides turbulent quantities to calculate the eddy amplitude, while the eddy size, orientation, initial position, and "lifetime" are calculated from random values. These randomized values are set on eddy initialization and do not change until an eddy is re-initialized after getting recycled. It should be noted that the values of CLST parameters in this work were chosen in part from preliminary analysis on this specific problem and further study is needed to generalize this method for different analysis conditions.

For eddy placement, a source region (the solid lines in Figure 3) is formed around the jet shear layer, centered on $D_j/2$, where D_j is the jet diameter. The source region is comprised of a cone, with $\alpha_s = 10.0°$ and $\beta_s = 6.3°$, $x_{min} = 1.5$ and $x_{max} = 12.5$ for this work. The parameters of the cone shape were chosen to mimic the initial shear layer and TKE contours from the LES-only flow field. Eddies are randomly placed in the three-dimensional source region by first choosing a random x_0 value from x_{min} to x_{max}. Given an x_0, y_0 and z_0 are chosen from inside the source region.

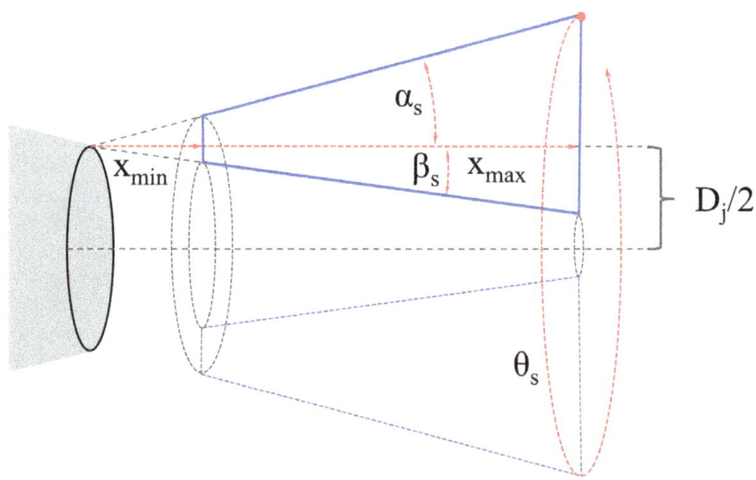

Figure 3. Source Region for Synthetic Eddies.

Synthetic eddies are generated using the following Gaussian shape function:

$$G(\mathbf{x},t) = exp\left[-\frac{x_p + y_p + z_p}{\sigma^2} - \frac{(t - t_0 - 0.5t_L)^2}{\sigma_t^2}\right] \quad (1)$$

$$x_p = (x - x_0 - u_c t)^2 \quad (2)$$

$$y_p = (y - y_0 - v_c t)^2 \quad (3)$$

$$z_p = (z - z_0 - w_c t)^2, \quad (4)$$

where the initial eddy center is located at $\mathbf{x}_0 = (x_0, y_0, z_0)$ at the time t_0, the eddy size is given by σ, the eddy life time is given by t_L, the time ramp width is given by σ_t, the eddy convection velocity is given by $\mathbf{u}_c = (u_c, v_c, w_c)$, and the current eddy position and time are given by $\mathbf{x}$ and t.

For N eddies, the following equations randomize the nth eddy's size and orientation for the synthetic velocity fluctuations $\mathbf{u}^n = (u^n, v^n, w^n)$:

$$u^n(\mathbf{x},t) = A(cz_p - by_p) \cdot G(\mathbf{x},t) \quad (5)$$

$$v^n(\mathbf{x},t) = A(bx_p - az_p) \cdot G(\mathbf{x},t) \quad (6)$$

$$w^n(\mathbf{x},t) = A(ay_p - cx_p) \cdot G(\mathbf{x},t), \quad (7)$$

where a, b, and c are randomized values between -2 and 2 to satisfy the divergence-free condition for a constant convection velocity. For reference, Hirai et al. used both N = 10,000 and N = 50,000 eddies, finding that more eddies improved the noise predictions [32].

The fluctuations generated by all eddies are summed in a point-wise manner to form the synthetic velocity field. At a point $\mathbf{x}$ and time t, the synthetic velocity fluctuations, $\mathbf{u}_s$, generated by all N eddies is given by

$$\mathbf{u}_s(\mathbf{x},t) = \sum_{n=1}^{N} \mathbf{g}^n(\mathbf{x},t), \quad (8)$$

where

$$\mathbf{g}^n(\mathbf{x},t) = \begin{cases} \mathbf{u}^n(\mathbf{x},t) & \text{if } (\mathbf{x}^n - \mathbf{x}_0^n - \mathbf{u}^n t) \leq \beta_c \sigma^n. \\ 0 & \text{otherwise.} \end{cases} \quad (9)$$

To minimize the computational cost incurred by performing calculating for thousands of eddies across the entire computational domain, the eddy calculation is omitted for any point further than a given distance from the eddy (Equation (9)), forming a calculation "window". The convection velocity is included in the distance calculation so that the calculation "window" follows the eddy downstream. Jarrin et al. [35] employ a similar cut-off in their SEM formulation. A value of β_c = 2.0 was observed to give good performance while not cutting off eddies stretched by sweeping.

2.3. Eddy Size Distribution

In realistic turbulent jet flows, small eddies are more numerous than large eddies. To mimic this behavior, the synthetic eddies are assigned to a "generation" or bin based on size. Three bins are chosen and distributed in the percentages shown in Table 1. Due to the Gaussian eddy shape, the actual eddy diameter is approxmiately 2σ, where σ is the eddy size. For the rest of this work, eddy size will refer to σ, not the actual eddy diameter.

Table 1. Eddy Bin Size Distribution.

Eddy Size (σ^n)	% of N Eddies	Minimum Size	Maximum Size
Small	90	σ_{min}	$\sigma_{min}(\delta_\sigma)$
Medium	9	$\sigma_{min}(\delta_\sigma)$	$\sigma_{min}(\delta_\sigma)^2$
Large	1	$\sigma_{min}(\delta_\sigma)^2$	σ_{max}

Given that the grid spacing is fine near the jet inlet and coarser downstream, an average minimum grid spacing, Δx_{min}, is calculated across the grid blocks in the eddy source region. The smallest eddy size is taken from the minimum average global grid spacing, $\sigma_{min} = 3.2\Delta x_{min}$, which is assumed to be the minimum eddy size that the grid can support practically. The largest eddy size, σ_{max}, is taken from the wavelength corresponding to the filter cut-off, $\lambda_c = \frac{2\pi\Delta x_{min} PPW}{\xi_c}$ (as is discussed in following sections, filtering is used to produce a CLES field for CLST). Wavelengths smaller than the filter cut-off wavelength are attenuated. The factor $PPW = 5/2$ is a ratio of the minimum number of points required to resolve the smallest wavelength by the currently-employed DRP scheme to the points per wavelength required in Kennedy and Carpenter [46]. To further control the eddy sizes, the maximum eddy size was adjusted for this work such that $\sigma_{max} = 0.45\lambda_c$. These parameters ensure that the smallest eddies are discretized by approximately four grid points on average.

The eddy bins are divided into equal ratios for each bin, with $\delta_\sigma = (\sigma_{max}/\sigma_{min})^{(1/3)}$. The bin ratio, δ_σ, is the size increase from one bin to the next. For this work, δ_σ is 1.145 based on the grid spacing and filter size. Table 1 shows the limits for each bin. The von-Karman turbulent kinetic energy spectra is plotted in Figure 4 for a representative point in the shear layer of a Mach 0.9 turbulent jet. The eddy size related to the fourth-order filter cut-off corresponds approximately to the most energetic large eddies, which is ideal for approximating CLES in the CLST framework. The distribution of eddies ensures that they reside within the range for isentropic turbulence and do not model the most energetic large eddies. The size of the synthetic eddies does not approach the dissipation range, but smaller eddies could be resolved by using a finer-spaced computational grid.

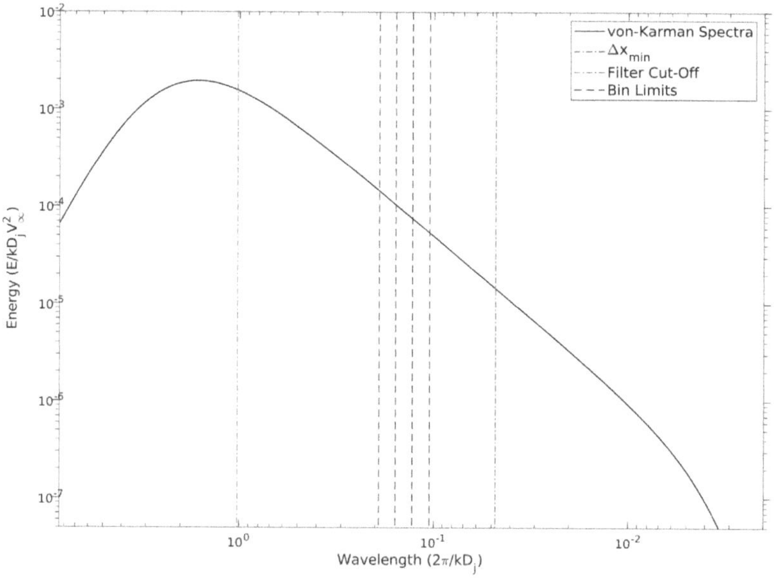

Figure 4. Eddy Size Distribution Shown for a Representative Turbulent Spectrum.

The eddy size is randomly chosen uniformly over the limits of the bin. Upon re-initialization, the eddies stay in their assigned bin, but the size can change within the limits of the bin. This ensures that the desired size distribution is maintained but allows for randomness in eddy sizes.

2.4. Amplitude Calculation

The amplitude for the nth eddy is calculated from the von Kármán–Pao energy spectrum for isotropic turbulence following the SNGR method of Bailly and Juvé [33]. For a given eddy size, σ^n, the wavenumber associated with the nth eddy is from $k^n = \frac{2\pi}{2\sigma^n}$. The energy associated with that wavenumber is given by the von Kármán–Pao energy spectrum as:

$$E(k^n) = \alpha \frac{u'^2}{k_e} \frac{(k^n/k_e)^4}{[1 + (k^n/k_e)^2]^{(17/6)}} \exp\left[-2\left(\frac{k^n}{k_\eta}\right)^2 \right], \tag{10}$$

where

$$\alpha \simeq 1.453, \quad u' = \sqrt{\frac{2}{3} k_{RANS}}, \quad k_e = 0.747/L_e. \quad \epsilon = \frac{u'^3}{L_e}, \quad k_\eta = \frac{\epsilon^{1/4}}{\nu^{3/4}}. \tag{11}$$

In these equations, $\nu = 1.4146 \times 10^{-5}\ \frac{m^2}{s}$ was taken as a constant and ϵ was calculated from the RANS-generated TKE (k_{RANS}) so that the amplitude of the synthetic eddies is only dependent upon the TKE. The wavelength of maximum energy, $L_e \approx 0.3 \times D_j$, is estimated from turbulent jet data in Pokora and McGuirk [47].

Finally, for the nth eddy,

$$A^n = \beta_{amp} \sqrt{E(k^n)/N}, \tag{12}$$

where β_{amp} is an amplification factor adjusted to produce fluctuations at the desired magnitude. The factor of $1/\sqrt{N}$ is taken from Jarrin et al. [34] and adjusts the amplitude based on the number of eddies. A value of $\beta_{amp} = 8.0 \times 10^6$ was used for the present simulations, although the value of this constant must be investigated in future work.

2.5. Eddy Convection

In the simplest SEM approach, eddy properties such as amplitude and size are fixed at initialization of the eddy, and each eddy is convected with a constant velocity or with the mean RANS flow velocity at the eddy center. These methods of convection do not allow for any distortion to the eddy shape that would occur via sweeping and straining.

Explicitly tracking and convecting an eddy with a distorted shape is not a straightforward task in a parallelized, multi-block grid approach if the convection velocity is non-constant due to sweeping by large eddies. To solve this difficulty, the eddy convection velocities in CLST are taken from each discrete point across the eddy. Figure 5 shows the pointwise velocity distribution for an eddy distorted by the presence of a non-uniform background flow. Convecting eddies in this pointwise manner saves complexity and computational time by avoiding the point searching and interpolation that would be needed to explicitly track an eddy, and, most importantly, accounts for the influence of sweeping and straining on the synthetic eddies.

To this point, an important feature of the CLST method is that sweeping and straining of the synthetic eddies are modeled. Sweeping occurs when the small-scale synthetic eddies are influenced by the movement and rotation of large-scale eddies. Straining is produced by non-uniformities in the local flow field due to large eddies. To accomplish this in the CLST method, eddies are convected via the instantaneous CLES velocities rather than the mean flow velocities. In other words, the synthetic eddies are convected by the mean jet flow plus any fluctuations resulting from large-scale eddies.

A representative diagram of sweeping due to pointwise convection and is given in Figure 6. Figure 6a shows the initial position of a small representative eddy in a flow field influenced by the rotation of a larger eddy. After applying pointwise convection to the eddy, displacement and distortion of the eddy shape due to sweeping and straining can be observed in Figure 6b.

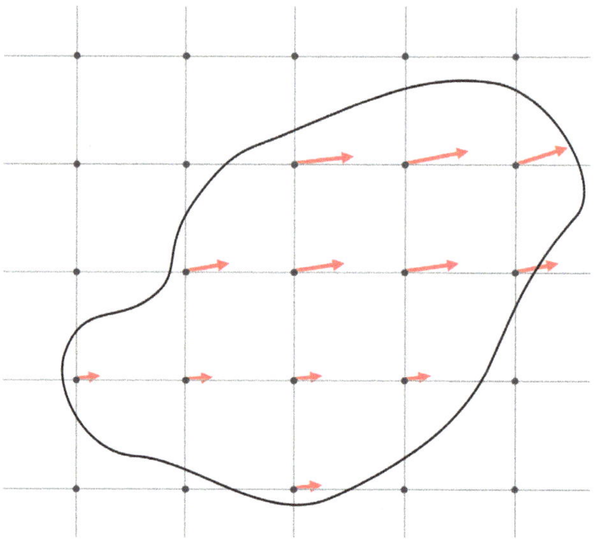

Figure 5. Point-wise velocity for a distorted synthetic eddy.

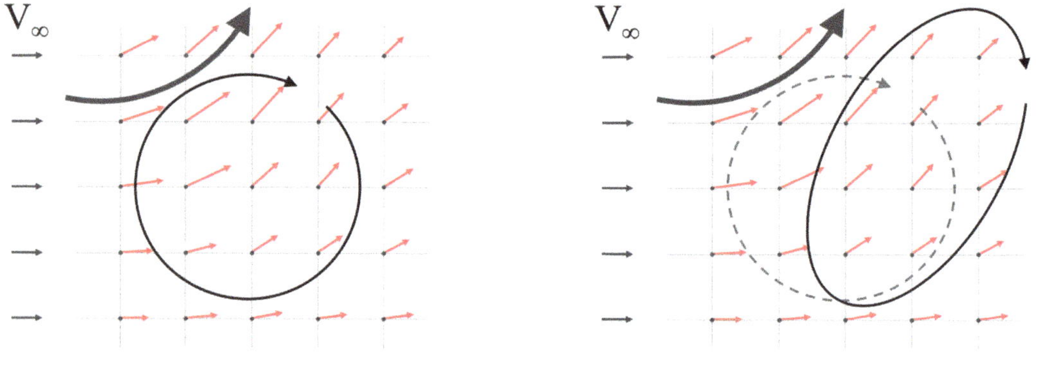

(**a**) (**b**)

Figure 6. Point-wise velocity for small eddy in a uniform flow with influence from a large neighboring eddy: (**a**) Initial Eddy Position. (**b**) Convected Eddy Position.

Pointwise convection leads to changes in certain eddy properties. As a given eddy is convected downstream, the eddy size remains constant, but the eddy amplitude and convection velocity can vary with spatial position. The initial Gaussian eddy shape can also change due to shear, sweeping, or straining effects. Each synthetic eddy is small in size compared to spatial changes in the mean flow field. The lifetime of each eddy is also small compared to the overall temporal fluctuations of the jet. Therefore, it is assumed that changes in the eddy properties over the eddy shape are small, as well as changes in eddy properties over the eddy lifetime since and eddy does not convect very far downstream. Any errors due to violating the divergent-free condition are expected to be minimal to the overall noise produced by the jet.

2.6. Recycling Eddies

At the beginning of the simulation, N eddies are created and inserted into the flow. They are initialized with an eddy lifetime based on the eddy size, such that small eddies exist for a short time. As the simulation progresses, the amplitude of each eddy is modulated in time with a Gaussian function (see Equation (1)) so that the eddy fades in on initialization and fades out as it reaches the end of its lifetime. At the end of its life, the eddy is recycled, meaning that the eddy is re-initialized with a new random size, orientation, lifetime, and position in the jet shear layer. The equations for eddy lifetime, $t_L^n = 10.5\sigma^n$, and the time ramp width, $\sigma_t^n = t_L^n/4$, were chosen to match the visually-observed time-development behavior of large-eddy structures in LES. Initially, eddies are inserted over a time range of 10-fold the lifetime of the maximum eddy size. Staggering the eddy insertion in this manner prevents similarly-sized eddies from syncing in time, which eliminates global pulsing behavior that can occur if thousands of eddies are recycled simultaneously.

2.7. Coupling of LES and SEM Velocity Fluctuations

In the calculation of the synthetic eddy field, at each point, the contribution from each synthetic eddy is summed at a given point. Once the total contribution of each synthetic eddy is computed, the synthetic velocity fluctuations are added to resolved LES fluctuations and passed to the LEE source terms to generate acoustic sources.

2.8. Numerical Methods

The Navier–Stokes (NS) equations are used for LES and the linearized Euler equations (LEE) are used for acoustic propagation. The three-dimensional Navier–Stokes equations are solved in a conservative form cast in curvilinear coordinates with a multi-block solver. The governing equations and flow variables are non-dimensionalized by the mean flow density $\bar{\rho}$, speed of sound a_∞, and reference length D_j. Both NS and LEE solvers use the same spatial and temporal discretization algorithms.

2.9. Source Terms for LEE

The LEE equations in primitive variable form are taken from Hirai et al. [32]. In this formulation, the mean shear terms and mean flow gradients are set to zero to damp instabilities resulting from unbounded homogeneous solutions to the original LEE [48]. The source terms are the same as Bogey et al. [48]:

$$\mathbf{S} = \beta \begin{bmatrix} 0 \\ S_1 - \overline{S_1} \\ S_2 - \overline{S_2} \\ S_3 - \overline{S_3} \\ 0 \end{bmatrix}, \tag{13}$$

where

$$S_i = -\frac{\partial \bar{\rho} u_i' u_j'}{\partial x_j}, \quad \overline{S_i} = -\frac{\partial \overline{\bar{\rho} u_i' u_j'}}{\partial x_j}. \tag{14}$$

The fluctuating velocities u_i' and u_j' in Equation (14) come from the combined LES and synthetic turbulent fluctuations from CLST, $u' = u'_{CLES} + u'_s$. The calibration parameter β is used to adjust the intensity of the sources [49]. From initial testing, a value of $\beta = 1.3125$ was chosen (it is taken as 1.0 by Lafitte et al. [49]) so that the amplitude of the LEE velocity fluctuations would match the amplitude of the LES velocity fluctuations. The mean density ($\bar{\rho}$) is used rather than the instantaneous density ($\rho = \bar{\rho} + \rho'$) in Equation (14) because the multiplication of the three fluctuating quantities $\rho' u' v'$ is negligible. The bar signifies a time-averaged quantity. The time average of the source terms is not zero, necessitating the subtraction of the mean of the source term [48].

2.10. Discretization Schemes

The spatial derivatives are discretized using the seven-point, fourth-order dispersion-relation-preserving schemes of Tam and Webb [50], which are designed to ensure that the schemes calculate the correct wave speeds and propagation characteristics (nondispersive, nondissipative, isotropic) for the main wave modes (acoustic, vorticity, and entropy) [50], trying to achieve the same dispersion relations as the original partial differential equations. The DRP schemes are simply optimized, explicit finite-difference schemes and are commonly used for LES-based jet noise simulations [51].

High-order, low-pass implicit spatial filters from Kennedy and Carpenter are used to damp spurious, non-physical waves present in the LES solver [46]. Explicit filters require less computational effort and are conceptually simpler than implicit filters. The high-order spatial filtering serves a second purpose as an implicit subgrid-scale (SGS) turbulence model for the LES code. Such implicit LES approaches are common in LES-based jet noise predictions [19,51]. The tenth-order filter applied to the Navier–Stokes variables produces the best LES results in the present code by suppressing spurious content and providing dissipation for subgrid stresses. Alternatively, using the fourth-order filter effectively produces a CLES flowfield by damping additional small-scale fluctuations.

The time integration is performed using an explicit second-order Adams–Bashforth method (Butcher [52]).

2.11. Filtering for CLES

In order to evaluate the effectiveness of synthetic velocities supplementing the jet noise spectra via the CLST method, the "missing" LES frequency content must first be measured or estimated. The simplest approach is to artificially generate an under-resolved CLES flowfield by removing resolved fluctuations from a LES simulation in a controlled manner. Applying a low-pass filter to the NS flowfield produces an approximate CLES flowfield by damping the turbulent structures smaller than a cut-off wavelength specified by the filter. Since the effects of the filter are known, the removed fluctuations are quantifiable and can be considered the "missing" noise produced by CLES on an effectively "coarse" grid resolution. A comparison of the acoustic spectra generated by LES (unfiltered) and CLES (filtered) results reveals the missing noise. Filtering thus provides a method to measure the "unresolved" noise and directly investigate the effects of adding synthetic velocities.

As stated previously, the fourth-order filter from Kennedy and Carpenter [46] can be used instead of the tenth-order filter to produce a CLES flowfield. The version of the fourth-order filter used in this investigation widens the stencil by a point in each direction to increase the filter's effectiveness. The filter cut-off frequency is estimated using $\xi_c = k_{max} \cdot \Delta x_{min}$, where ξ_c is the cut-off wavelength assuming the cut-off occurs at 80% of the maximum wave amplitude ($\xi_c \approx 1.47$ from Figure 6 in Kennedy and Carpenter [46]) and k_{max} is the maximum resolvable wave number of 2π. The exact manner in which the filtering is applied to produce CLES results in this investigation is noted in Section 3.3.

3. Results

Given that the goal of the CLST method is to account for the effects of fine-scale turbulence, the method was tested at a moderately high Reynolds number with a wide range of turbulent scales present in the jet. A round, isothermal jet was simulated at M = 0.9 and Re = 100,000. These parameters were chosen to match an LES study from Bogey and Marsden [53] that presented well-resolved results of both the jet flow physics and near-field jet noise spectra (including high-frequency content up to St = 5). Comparisons are made with results from Bogey and Marsden [53] and Bogey [19] to investigate the mean jet flow properties and the associated near-field radiated noise predicted by the current LES-LEE method. Finally, the CLST method is applied to the same jet case.

3.1. Simulation Setup

For purely comparative purposes, the CLES, SEM, and LEE flowfields are calculated on the same computational grid. This also eliminates errors or increased costs due to interpolation between the solutions. Validation for the LEE solver was previously performed both in 2D and 3D, with results (not included here for brevity) showing that the LEE solver predicts the correct acoustic wave propagation behavior [22].

The computational grid for the jet simulations has 148 blocks and 4.88 million grid points. The domain extends from $x = -15D_j$ to $x = 60D_j$ and out to $40D_j$ in the y- and z-directions at the widest point. Rather than resolve the interior flow of a nozzle, the jet is modeled as a hyperbolic tangent velocity profile ($0.05D_j$ thick) where the "exit" of the jet nozzle corresponds to the grid's inlet plane. It is acknowledged that neglecting a nozzle geometry will have an influence on the development of the shear layer, jet flow field, and thus on the radiated jet noise [2,20,54]. However, in the context of developing the CLST method, the use of an inflow boundary condition is seen as an initial step. Divergence-free fluctuations are forced at the inlet plane to promote the natural transition of the shear layer from quasi-laminar to turbulent [55]. The number of azimuthal vortex ring modes imposed in the present study is 15 (as suggested [55]) with an amplitude of 0.06. All other boundaries in the simulation domain employ far-field boundary conditions that use extrapolation. A combination of grid stretching and sponge layers are used at the far-field boundaries to dissipate and filter out fluctuations before they generate spurious acoustic reflections at the boundaries.

Pressure probes were placed in the near-field at the points listed in Table 2. The probes are divided into two categories. The first set is represented by four probes (P1–P4) equally spaced along the line $y = 5D_j$, which is close to the jet but far enough from the shear layer to avoid contamination of the acoustic data with hydrodynamic fluctuations. These points should give a good survey of observer locations where high-frequency noise contributes to the spectra. The second set is represented by two probes (P5 and P6) along the line $y = 7.5D_j$, which are use in a following section to validate the simulation with noise data from Bogey and Marsden [53]. Frequencies up to St = 5 were observed to be the highest frequencies resolvable on this computational grid at probe locations P1 to P4 [22] and St = 2 were the resolution limit for probes P5 and P6.

Table 2. Point Locations of Pressure Data Probes for CLST Investigation.

Point	P1	P2	P3	P4	P5	P6
x/D_j	1.0	2.0	3.0	4.0	0.0	10.0
y/D_j	5.0	5.0	5.0	5.0	7.5	7.5
z/D_j	0.0	0.0	0.0	0.0	0.0	0.0

3.2. Les Comparison

Data from Bogey and Marsden [53] and Bogey [19] are used to evaluate the accuracy of the present LES simulations for the Re = 100,000, Mach 0.9 jet. Bogey and Marsden [53] investigate the effect of different grid resolution parameters on the jet flow and noise spectra. Their results are quite detailed, using computational grids with up to one billion grid points and modeling the jet noise out to $75D_j$ from the nozzle. Additionally, Bogey and Marsden simulate the jet flow with a pipe nozzle geometry. Bogey [19] presents additional results from identical simulations.

In comparison, the grid of the present simulation does not extend to the far field with adequate resolution to resolve pressure fluctuations past $15D_j$ and has at least 50-fold fewer grid points (4.8 million vs. 250 million). The present simulation method employs an inlet plane for the jet rather than a physical nozzle geometry. Disturbances are imposed at the inlet to excite the jet turbulence. Only the near-field region of the computational grid for has sufficient resolution to capture noise data. Given the high level of resolution in Bogey and Marsden's results and the computational limitations of the present simulations, identical replication of Bogey and Marsden's results is not expected. However, if the same mean jet flow and turbulence properties are captured by the present simulation, the noise spectra should agree well for low and mid frequencies.

It is expected that Bogey and Marsden's use of a nozzle geometry leads to differences when comparing the present simulations. In the present LES, disturbances are imposed at the inlet to stimulate fluctuations at the inlet and produce a turbulent jet. The data from the present simulation are most similar to results from Bogey and Marsden for an initially laminar, transitioning jet. Given that a nozzle geometry is not included in the present simulation and that some small region of transition is expected when imposing a jet flow at an inlet plane, a comparison to an initially laminar jet is acceptable.

Combined contours of instantaneous vorticity and acoustic pressure provide an initial qualitative investigation of the results from the present simulation. Figure 7 is plotted in a style similar to results from Bogey and Marsden [53], with vorticity contours in the center (over the range y = ±2.5D$_j$) surrounded by the near-field pressure fluctuations. Vorticity is scaled by u_j/r_j and pressure is presented in Pascals. The potential core ends around 5D$_j$. Both large- and small-scale turbulent motions are observed prior to the merging of the shear layers. As the grid for the present simulation coarsens near x = 15D$_j$, the turbulence dies off sooner than in Bogey and Marsden's results.

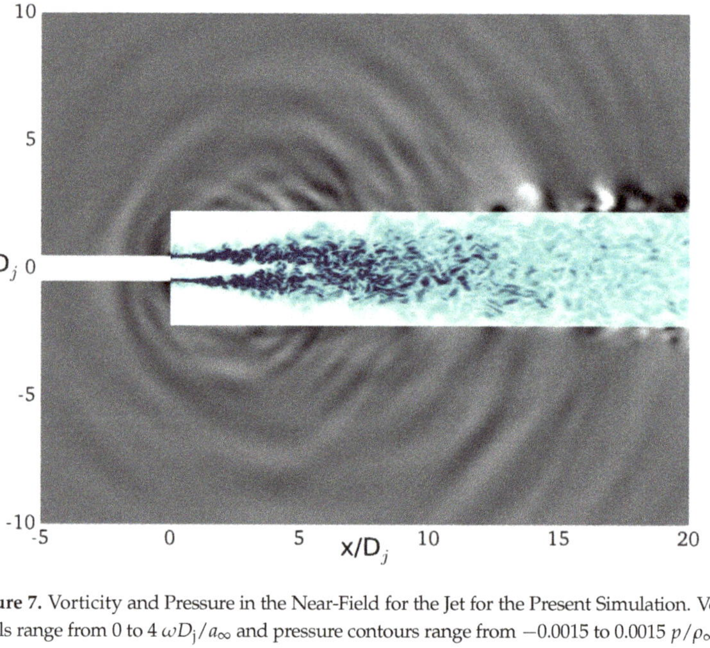

Figure 7. Vorticity and Pressure in the Near-Field for the Jet for the Present Simulation. Vorticy contour levels range from 0 to 4 $\omega D_j/a_\infty$ and pressure contours range from −0.0015 to 0.0015 $p/\rho_\infty a_\infty^2$.

In Figure 7, pressure waves with strong amplitudes are observed radiating downstream at approximately a 30° angle measured from the jet axis. These waves are related to the noise from the large-scale coherent structures [53]. Additionally, small-scale pressure waves are seen radiating from the shear layer before x = 5D$_j$. These two phenomena are the expected essential components of jet noise. The fine-resolution results of Bogey and Marsden show many small-scale pressure fluctuations. Although small waves are observed in the present simulation, they are not well resolved by the present computational grid.

For a more quantitative comparison, centerline flow properties are compared to replicated data (data reproduced with permission from author) from Bogey [19]. Mean axial velocity is plotted in Figure 8a. The velocity decay matches Bogey's data well, and the decrease in velocity around x = 4.5D$_j$ −5D$_j$ indicates the end of the potential core. More rapid turbulent development in this jet leads to a shorter potential core than is typically expected (around x = 7D$_j$ [53]), but the present simulation shows a slightly longer potential core than Bogey's data. The intensity of roll-up and pairing of vortical structures in the transitioning shear layer causes the jet to develop more rapidly, shortening the potential core.

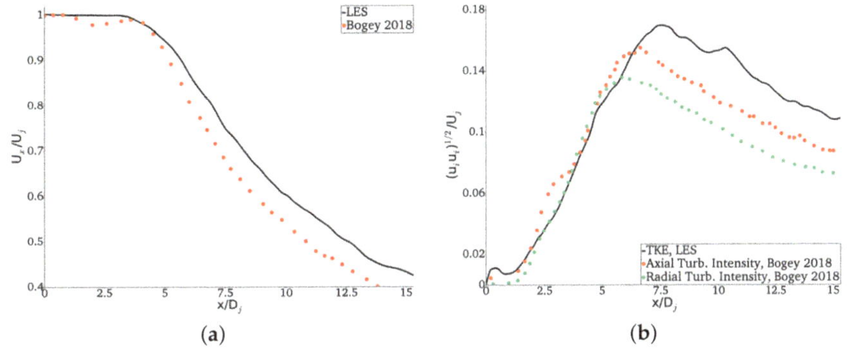

Figure 8. Centerline Jet Properties: (**a**) Mean Velocity. (**b**) Turbulence Intensity.

Root mean square values of axial and radial turbulence intensity from Bogey are plotted in Figure 8b. The square root of time-averaged TKE from the present simulation is plotted for comparison. Although the quantities are not identical, they are similarly derived from multiplications of the turbulent velocity fluctuations and demonstrate that the present simulation predicts a similar turbulence decay rate along the centerline. The rise in turbulence indicates the merging of the shear layers at the end of the potential core. The present simulation shows a slightly elongated potential core compared to Bogey's data. Both centerline velocity and turbulence intensity results demonstrate that the mean flow properties in the present simulation agree well with LES results from Bogey.

Two probe locations in the near-field of the jet along the line y = 7.5D$_j$ are used for comparison of the noise spectra from a time-history of pressure: point P5 (x = 0D$_j$) and Point P6 (x = 10D$_j$), shown in Table 2. Figure 9 shows SPL spectra for the two points from the present simulation and also from Bogey and Marsden [53]. The pressure from the present simulation is taken from the Navier–Stokes solver, since the probe locations are still in the near-field of the jet where the grid resolution is fine enough to capture noise data. For both probe locations, the frequency spectra from St = 0.2 to St = 2 are matched. Higher-frequency roll-off is observed above this frequency, and it is more pronounced in Probe P5, with an under-prediction of 20 dB at St = 5. The drastic roll-off in high-frequency content for probe P5 is due to coarser grid resolution near this probe location. The frequency decay is less obvious for probe P6, where grid resolution is finer. The noise at probe P6 under-predicts Bogey and Marsden's data by approximately 5 dB at St = 5. The high-frequency roll-off is expected because fewer points were used in this computational grid compared to Bogey and Marsden's grid.

With this favorable spectral comparison, it is expected that this grid and case setup can reliably resolve the jet spectra up to at least St = 2 at these probe locations, or up to St = 5 in locations where the grid spacing is finer. A perfect comparison with Bogey and Marsden's data is not expected, given the differences in grid resolution and simulation technique. However, these results demonstrate enough quantitative and qualitative similarities to assume that the correct jet physics and noise mechanisms are captured by this computational setup.

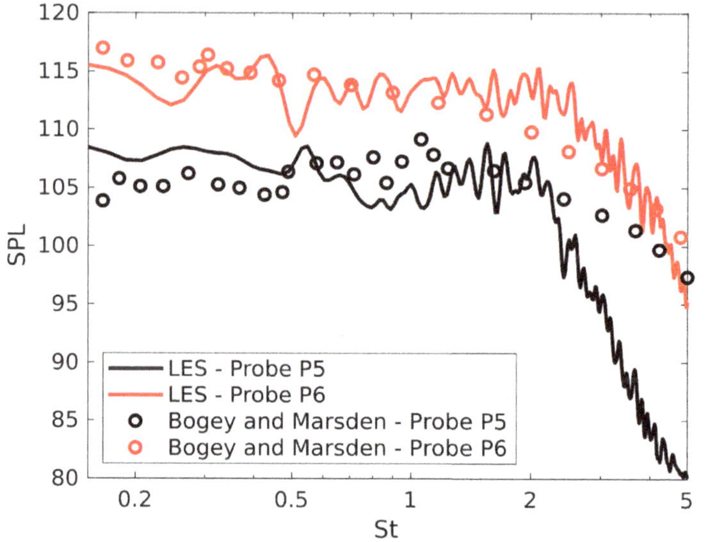

Figure 9. Comparing Near-Field LES Noise Data

3.3. CLST *Noise Spectra*

The primary goal of CLST is to supplement the high-frequency content of LES. First, the noise spectra is analyzed, followed by an investigation of the turbulent properties of CLST in a following section. The same M = 0.9 jet was used to evaluate the CLST method.

Conceptually, CLST works by adding synthesized small-scale fluctuations to under-resolved LES (referred to as CLES here). To properly evaluate the noise generated the the CLST method, three results are needed: LES, CLES, and CLST. The LES-generated noise spectra provides the known, expected target solution for the CLST-based noise results. The CLES-generated noise spectra represents the results of an under-resolved LES simulation, an example of when the CLST method would be useful in predicting a full noise spectra. In a typical application of CLST, the unresolved noise spectra would not be known. However, in this instance, the under-resolved portion of the noise spectra on this grid can be obtained by comparing to the LES-generated noise spectra. The CLST-generated noise spectra represents using the CLST method to model the high frequencies that are not resolved by the CLES noise prediction. The CLST-generated spectra can be compared to the LES-generated spectra to evaluate the characteristics of the high-frequency noise added by the CLST method.

In the following analysis, the data were obtained from one simulation run which gave LES, CLES, and CLST flow-fields simultaneously. Pressure histories for LES were taken directly from the NS solver because the grid has sufficient resolution in the near-field for noise. Although not shown here for brevity, near-field noise spectra obtained in this manner from the LES solver demonstrated good agreement with the noise obtained from the LEE solver [22].

Filtering was applied to the LES velocity field to obtain CLES pressure histories and velocities, which can be seen as under-resolved LES on the current computational grid. The jet noise produced by the LES and CLES flowfields was computed using the near-field pressure probes. The spectra from CLES can be considered the noise radiated from an under-resolved LES simulation. Synthetic turbulence was combined with resolved large-scale velocity fluctuations (filtered CLES velocities) into the LEE source terms, giving the noise produced by the CLST method. Pressure histories for CLST are taken from the

LEE solver. The frequency content added by the synthetic turbulence was revealed by comparing noise spectra from LES, CLES, and CLST.

This approach is representative of adding synthetic turbulence to an under-resolved LES simulation and recovering the "missing noise", but in this case, the added noise is quantifiable and comparable to the "missing noise" because the "missing noise" has been removed by a filter with known frequency properties. The excellent agreement that was previously demonstrated [22] between the NS and LEE pressure makes this is a valid comparison for all three methods.

For the computation, a non-dimensional time-step size of 8.01×10^{-4} was used, which equates to a CFL of 0.13. This was required for stability of the Adams–Bashforth time-marching algorithm. Obtaining all three results from one simulation is only performed for comparison purposes in this investigation.

In this evaluation of the CLST method, 8000 eddies were inserted into the shear layer. Preliminary results revealed that adding 8000 eddies formed a reasonable balance between increased computational cost and additional high-frequency noise in this simulation. A fourth-order filter was applied to the LES field twice for every time step to obtain the CLES field. The average minimum grid spacing in the source region was Δx_{min} = 5.546 $\times 10^{-2}$. All other CLST parameters can be found in the methods section.

A qualitative comparison of the jet turbulence and near-field pressure spectra are shown in Figure 10 at the same instance in the simulation. Contours of vorticity (shown as non-dimensional quantities scaled by D_j/U_j) are shown for a range of y = $\pm 2D_j$. Pressure field contours are shown in units of non-dimensional pressure (scaled by $\frac{1}{\rho_\infty a_\infty{}^2}$). LES results in Figure 10a reveal the developing turbulent jet, the merging of the potential core around x = $5D_j$, and the slow dissipation of turbulence downstream past $15D_j$.

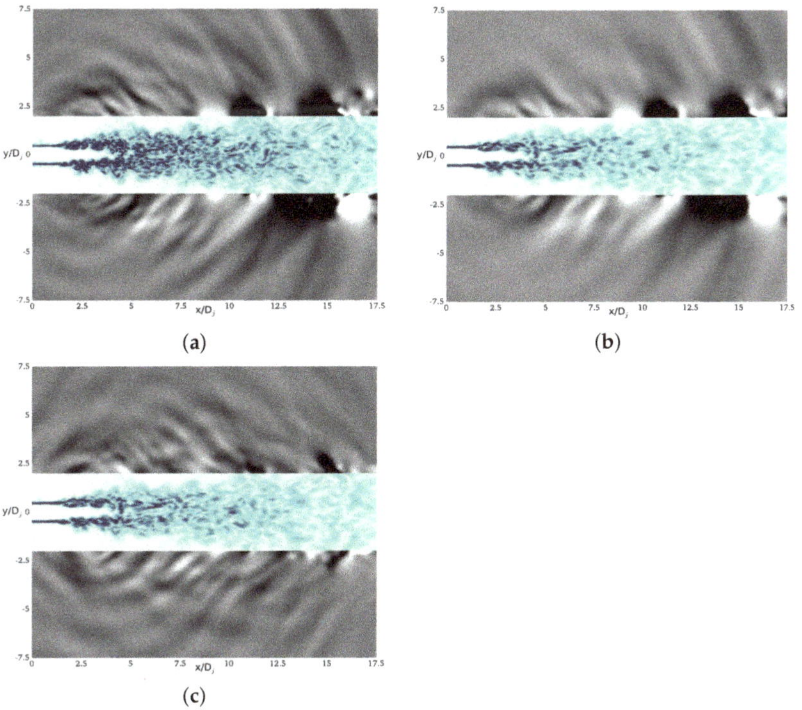

Figure 10. Contours of Vorticity in the Shear Layer and Pressure in Near-Field for LES, CLES, and CLST. Vorticy contours levels range from 0 to 4 $\omega D_j/a_\infty$ and pressure contours range from -0.0015 to 0.0015 $p/\rho_\infty a_\infty^2$: (**a**) LES. (**b**) CLES. (**c**) CLST.

The pressure field displays several expected noise phenomena: large-scale waves radiating at 30°, large-scale pressure fluctuations, and smaller, spherical waves radiating from the early shear layer. The CLES results in Figure 10b reveal that filtering removes the fine-scale turbulence while leaving large-scale turbulent structures untouched. The large-scale pressure fluctuations are still evident in the CLES pressure field, but damping is observed and small-amplitude pressure fluctuations are not observed, especially upstream of the potential core. In Figure 10c, with the synthetic turbulence added to the CLES flow field, there is a nearly-negligible increase in the intensity of turbulent velocities within the first 5–7 nozzle diameters. However, additional small-scale pressure waves are discernible in Figure 10c, indicating that the synthetic eddies from CLST are generating additional noise. Note the angling of the large-scale pressure waves toward the downstream, indicating the expected directivity of the most significant noise. Slight visualization artifacts are present in the flow field in Figures 10a,c, but these do not contaminate the pressure spectra.

Pressure data was recorded from the CLST simulation every 15 time-steps over a total of 140,000 iterations. The probe locations are given in Table 2. The noise spectra from the pressure histories were obtained by a Fast Fourier Transform (FFT). The spectra for probe locations P1 through P4 in Table 2 are shown in Figure 11. Spectra from LES, CLES, and CLST are compared. Frequencies from St = 0.2 to St = 5 are present in the LES noise spectra. The noise produced by CLES filtering shows strong high frequency at roll-off starting at frequencies of St = 1 to St = 2. At St = 2, CLES is around 10 dB below LES for all four points. The CLES filtering also reduces portions of the low-frequency noise spectrum by up to 5 dB in places, potentially indicating that the filter influences a wider range of frequencies than intended. For all four probes, the addition of synthetic turbulence leads to the recovery of the filtered frequencies in the range St = 1 to St = 5. Several small differences (less than 5 dB) are observed between the LES and CLST spectra at various points. Notably, at point P1 (Figure 11a), there is an over-prediction around St = 2. Overall, at all four locations along the line y = 5D_j, the shape and level of the LES spectra is modeled well with CLST.

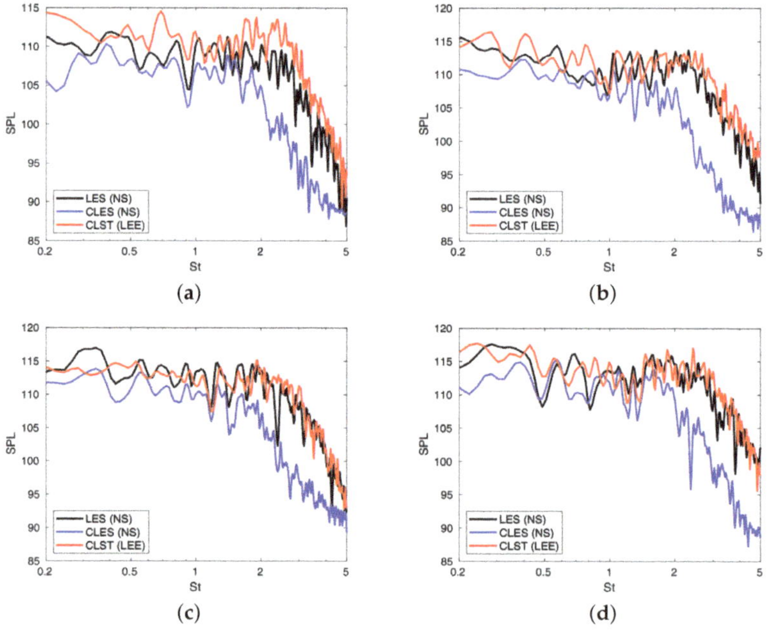

Figure 11. SPL Spectra for CLST along the Line y = 5D_j: (**a**) Point P1 (x = 1D_j). (**b**) Point P2 (x = 2D_j). (**c**) Point P3 (x = 3D_j). (**d**) Point P4 (x = 4D_j).

The spectra from points P5 and P6 are shown in Figure 12. Again, filtering removes frequencies above St = 1. Some evidence of roll-off from grid resolution can be observed above St = 4 for both probe locations. Again, the noise from CLST recovers the missing spectra and matches LES spectra. Point P5 shows slight over-prediction around St = 2.

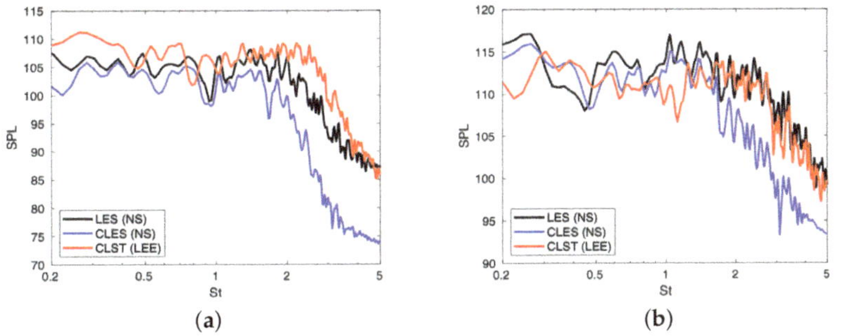

Figure 12. SPL Spectra for CLST along the Line y = 7.5D_j: (**a**) Point P5 (x = 0D_j). (**b**) Point P6 (x = 10D_j).

In two of the probe locations directly to the sideline of the nozzle, Point P1 (Figure 11a) and Point P5 (Figure 12a), an over-prediction of 3–5 dB was observed around St = 2–3. It is possible that excess noise is generated from the synthetic eddies at these frequencies, but the exact cause of these slight differences is unclear.

The intensity of the synthetic eddies was tuned to match the amplitude of the pressure waves, not the LES velocity fluctuations. The synthetic velocities were combined with the CLES velocity fluctuations in post processing, but these results are not presented here because the the synthetic velocities did not make a visibly discernible difference in the CLES velocity field. This is an indication that the synthetic turbulence field generates more noise than expected, which could be the result of inserting unrealistic eddies into the flow. Alternatively, the distortion of eddies due to sweeping and straining could violate the divergence-free condition and generate additional noise.

Ideally, matching the turbulent flow field first would provide an anchor point for further applications of the CLST method, eliminating the need to re-tune the amplitudes of the synthetic eddies to match the pressure waves for each simulation. It may be difficult to match both turbulent flow properties and the noise spectra. Hirai et al. [32], who also employed a synthetic turbulence model for jet noise predictions, had a similar issue with over-predicting the noise levels when first tuning their model to match turbulent statistics. However, given that the noise results match well in the previous section and meet the primary goal of the CLST method, the differences in the turbulent flow field are acceptable for the current stage of CLST development.

4. Discussion

In the CLST framework, synthetic turbulence is used to supplement an under-resolved jet noise spectra. Even with a moderate number of synthetic eddies, the present simulation results demonstrate the effectiveness of this approach for a high subsonic jet. For all recorded probe locations in the Mach 0.9 jet simulation, the CLST method recovered the high-frequency noise removed by filtering. Despite small differences in the noise spectra, it is clear that CLST can supplement the missing noise spectra from under-resolved turbulence and capture the correct noise spectra curves for near-field observer locations.

Previous applications of synthetic turbulence to jet noise predictions (Fukushima [31] and Hirai et al. [32]) demonstrated the effectiveness of synthetic turbulence models (including synthetic eddies) for jet noise predictions. However, both of these approaches use steady RANS simulations to solve for the mean background flow, which means that the influence of convecting, large-scale anisotropic turbulent structures cannot be modeled properly. Hirai et al. [32] partially account for this with a time decorrelation process that models the influence of the shear from the mean background flow on the synthetic turbulence. This approach is still limited since the background flow is obtained from steady RANS. A more physically-accurate model would include localized, time-varying influences on the synthetic turbulence from the background flow, such as the effect of large-scale convected eddies. Recall that large-scale turbulent structures are significant to jet noise production [10,14,15,17,18], and therefore, their influence on the flowfield must be properly accounted for.

The CLST method advances the synthetic turbulence modeling approach by directly simulating the large anisotropic turbulence scales in the mean flow and using their motions to account for sweeping of the synthetic turbulence, including additional key physics in the jet noise modeling. Overall, this introduces a new framework for modeling the missing high-frequency content in jet noise simulations.

The primary drawback of the CLST method is that the eddies appear to generate more noise than anticipated when the intensity of the synthetic turbulent field is matched to the LES turbulence intensity. This led to reducing the amplitude of velocity fluctuations to the point that the combined resolved and synthetic velocity fluctuations from CLST did not approach the levels of resolved velocity fluctuations observed in the LES solution. A major focus of future investigations will therefore be to match the turbulent flow quantities of LES without over-predicting the noise, as well as fine-tuning and generalizing the synthetic eddy parameters. The extra synthetic eddy noise could result from inserting unrealistic synthetic eddies into the flow. Additionally, distortion of the eddies due to sweeping and stretching may violate the divergence-free condition, leading to extra noise.

A second drawback of the CLST method was that the cost was higher than expected due to a large memory overhead. The extra numerical cost incurred by calculating CLST with 8000 eddies was approximately 8-fold that of the LES + LEE solvers. With optimization, it is estimated that CLST would only cost 1.5 or 2-fold more than a baseline LES + LEE solution. It would be desirable to use more eddies (upwards of 50,000 were used by Hirai et al. [32]), so future work will also focus on a more efficient code optimization to reduce the high-memory cost of the current implementation of CLST.

It is recognized that significant cost savings could be realized by using a coarser LEE grid since the resolution requirements are lower for LEE than for a scale-resolving NS solution, although for this study all grids were kept the same for simplicity. Even so, the computational cost increase for CLST vs. CLES was moderate in the Mach 0.9 jet simulation. In contrast, halving the grid spacing for a LES + LEE simulation may lead to a 16-fold increase (8x for LES, 8x for LEE) in computing time if the same grid is used for LEE as LES. In this light, the added cost of CLST is still cheaper than simply increasing the grid resolution of the LES solver to achieve a similar noise spectra with high-frequency spectral content.

5. Conclusions

The main idea of CLST is to resolve large-scale turbulent fluctuations with LES (or CLES) and to model small-scale turbulent fluctuations via a synthetic turbulence method (SNGR or SEM). Since fine-scale turbulence gives rise to the highest frequencies of the jet noise spectra, the synthetic turbulence serves as a model for the high-frequency noise that is often under-predicted by LES. Given that synthetic turbulence methods are computationally inexpensive, the CLST method models the high-frequency noise spectra at a reduced computational cost compared to resolving the same frequencies with LES. In CLST, the noise field is predicted using a formulation of the linearized Euler equations (LEE), where the acoustic waves are generated by source terms from combining the fluctuations from CLES and the stochastic fields.

The CLST method was applied to a Mach 0.9 jet, and the LEE solver was validated against the LES noise field with excellent agreement. For the moderately high Reynolds number jet (Re = 100,000), the computational setup was shown to match LES results by

Bogey and Marsden [53]. The CLST method was evaluated by using a filter to produce an under-resolved CLES flow field. Synthetic fluctuations generated by CLST were added to the CLES flowfield. A comparison with LES revealed that the CLST method works well in supplementing the high-frequency noise spectra for under-resolved LES jet simulations. A slight over-prediction of high frequencies was observed at some sideline probe locations for the SEM-based method.

Although additional testing and development are needed, this initial application of the SEM-based CLST method shows promise for a novel approach to modeling a more-complete jet noise spectra with only a moderate increase in cost.

Author Contributions: Conceptualization, J.D.B., A.S. and Y.H.; methodology, A.S. and J.D.B.; software, A.S. and J.D.B.; validation, A.S. and J.D.B.; investigation, A.S. and J.D.B.; resources, A.S.; writing—original draft preparation, J.D.B.; writing—review and editing, J.D.B., A.S., D.T. and Y.H.; visualization, J.D.B.; supervision, A.S. and D.T. All authors have read and agreed to the published version of the manuscript.

Funding: Part of the work was carried out under the Collaborative Research Project between the Institute of Fluid Science, Tohoku University, and the Center for Advanced Vehicular Systems, Mississippi State University. Additional support was provided by the Mississippi Space Grant. We thank these institutions for their support.

Institutional Review Board Statement: Not applicable.

Informed Consent Statement: Not applicable.

Data Availability Statement: The data presented in this study are available on request from the corresponding author. The data are not publicly available due to data size and storage limitations.

Conflicts of Interest: The authors declare no conflict of interest.

Nomenclature

A	eddy amplitude
a, b, c	random values for eddy orientation
a_∞	speed of sound, m/s
D_j	jet diameter, m
E	von Kármán–Pao energy spectrum
f	frequency, Hz
G	Gaussian shape function
$\mathbf{g}^n$	cut-off function for nth eddy
k^n	nth wave number, 1/m
k_e	wave number for maximum energy, 1/m
k_η	wave number for Kolmogorov scale, 1/m
k_{RANS}	RANS turbulent kinetic energy
L_e	wavelength of maximum energy, m
N	total number of eddies
p	pressure, N/m^2
r_j	jet radius, m
S_i	ith-component of LEE source term
St	Strouhal number, fD_j/V_j
t_0	initial eddy insert time, s
t_L	eddy life time, s
$u_i{}'$, $u_j{}'$	fluctuating velocities, m/s
u^n, v^n, w^n	synthetic velocity components for the nth eddy, m/s
u_c, v_c, w_c	components of the eddy convection velocity, m/s
u'_s	synthetic velocity field, m/s
V_j	jet core exit velocity, m/s
x_{min}, x_{max}	minimum/maximum x for source region
x_0, y_0, z_0	initial location of eddy center

x_p, y_p, z_p	directional components of the Gaussian function
α_s, β_s	inner and outer source cone angles, ^{circ}
β	scaling parameter for LEE source terms
β_{amp}	SEM amplification factor
β_c	SEM cut-off parameter
Δx_{min}	minimum average grid spacing, m
δ_σ	eddy bin ratio
ϵ	turbulent dissipation rate, m^2/s^3
θ_s	azimuthal angle around jet, ^{circ}
λ_c	filter cut-off wavelength, m
ν	kinematic viscosity, m^2/s
ρ	density, kg/m^3
σ^n	nth eddy size, m
$\sigma_{min}, \sigma_{max}$	minimum/maximum eddy size, m
σ_t^n	nth eddy time ramp width
ξ_c	filter cut-off wavelength

References

1. Wang, M.; Freund, J.B.; Lele, S.K. Computational Prediction of Flow-Generated Sound. *Annu. Rev. Fluid Mech.* **2006**, *38*, 483–512. [CrossRef]
2. Brès, G.A.; Jordan, P.; Jaunet, V.; Le Rallic, M.; Cavalieri, A.V.G.; Towne, A.; Lele, S.K.; Colonius, T.; Schmidt, O.T. Importance of the Nozzle-Exit Boundary-Layer State in Subsonic Turbulent Jets. *J. Fluid Mech.* **2018**, *851*, 83–124. [CrossRef]
3. Brès, G.A.; Lele, S.K. Modelling of Jet Noise: A Perspective from Large-Eddy Simulations. *Philos. Trans. R. Soc. A Math. Phys. Eng. Sci.* **2019**, *377*, 20190081. [CrossRef]
4. Lyrintzis, A.S.; Coderoni, M. Overview of the Use of Large-Eddy Simulations in Jet Aeroacoustics. *AIAA J.* **2020**, *58*, 1620–1638. [CrossRef]
5. Bogey, C. Acoustic Tones in the Near-Nozzle Region of Jets: Characteristics and Variations Between Mach numbers 0.5 and 2. *J. Fluid Mech.* **2021**, *921*, A3. [CrossRef]
6. Stich, G.D.; Ghate, A.S.; Housman, J.A.; Kiris, C.C. Wall Modeled Large Eddy Simulations for NASA's Jet Noise Consensus Database of Single-Stream, Round, Convergent Jets. In Proceedings of the AIAA SCITECH 2022 Forum, San Diego, CA, USA, 3–7 January 2022.
7. Bodony, D.; Lele, S. Spatial Scale Decomposition of Shear Layer Turbulence and the Sound Sources Associated with the Missing Scales in a Large-Eddy Simulation. AIAA 2002-2454. In Proceedings of the 8th AIAA/CEAS Aeroacoustics Conference & Exhibit, Breckenridge, CO, USA, 17–19 June 2002.
8. Bodony, D.J.; Lele, S.K. On Using Large-Eddy Simulation for the Prediction of Noise from Cold and Heated Turbulent Jets. *Phys. Fluids* **2005**, *17*, 085103. [CrossRef]
9. Bodony, D.J. Developing a Subgrid Scale Noise Model for Use with Large-Eddy Simulation. In *Annual Research Briefs*; Center for Turbulence Research, Stanford University: Stanford, CA, USA, 2005.
10. Tam, C.; Viswanathan, K.; Ahuja, K.; Panda, J. The Sources of Jet Noise: Experimental Evidence. *J. Fluid Mech.* **2008**, *615*, 253–992. [CrossRef]
11. Bodony, D.; Lele, S. A Statistical Subgrid Scale Noise Model: Formulation. AIAA 2003-3252. In Proceedings of the 9th AIAA/CEAS Aeroacoustics Conference and Exhibit, Hilton Head, SC, USA, 12–14 May 2003.
12. Bogey, C.; Bailly, C. An Analysis of the Correlations Between the Turbulent Flow and the Sound Pressure Fields of Subsonic Jets. *J. Fluid Mech.* **2007**, *583*, 71–97. [CrossRef]
13. Tyacke, J.; Naqavi, I.; Wang, Z.N.; Tucker, P.; Boehning, P. Predictive Large Eddy Simulation for Jet Aeroacoustics–Current Approach and Industrial Application. *J. Turbomach.* **2017**, *139*, 081003. [CrossRef]
14. Mollo-Christensen, E.; Kolpin, M.A.; Martuccelli, J.R. Experiments on Jet Flows and Jet Noise Far-Field Spectra and Directivity Patterns. *J. Fluid Mech.* **1964**, *18*, 285–301. [CrossRef]
15. Narayanan, S.; Barber, T.J.; Polak, D.R. High Subsonic Jet Experiments: Turbulence and Noise Generation Studies. *AIAA J.* **2002**, *40*, 430–437. [CrossRef]
16. Panda, J.; Seasholtz, R.; Elam, K.A. Investigation of Noise Sources in High-Speed Jets Via Correlation Measurements. *J. Fluid Mech.* **2005**, *537*, 349–385. [CrossRef]
17. Bailly, C.; Bogey, C. Current Understanding of Jet Noise-Generation Mechanisms from Compressible Large-Eddy-Simulations. In *Direct and Large-Eddy Simulation VI*; Lamballais, E., Friedrich, R., Geurts, B., Métais, O., Eds.; Springer: Dordrecht, The Netherlands, 2006.
18. Viswanathan, K. Mechanisms of Jet Noise Generation: Classical Theories and Recent Developments. *Int. J. Aeroacoustics* **2009**, *8*, 355–408. [CrossRef]

19. Bogey, C. Grid Sensitivity of Flow Field and Noise of High-Reynolds-Number Jets Computed by Large-Eddy Simulation. *Int. J. Aeroacoustics* **2018**, *17*, 399–424. [CrossRef]
20. Bogey, C.; Sabatini, R. Effects of Nozzle-Exit Boundary-Layer Profile on the Initial Shear-Layer Instability, Flow Field and Noise of Subsonic Jets. *J. Fluid Mech.* **2019**, *876*, 288–325. [CrossRef]
21. Tam, C.K.; Auriault, L. Jet Mixing Noise from Fine-Scale Turbulence. *AIAA J.* **1999**, *37*, 145 –153. [CrossRef]
22. Blake, J. A Coupled Large Eddy Simulation-Synthetic Turbulence Method for Predicting Jet Noise. Ph.D. Thesis, Mississippi State University, Starkville, MS, USA, 2020.
23. Colonius, T.; Lele, S.K. Computational Aaeroacoustics: Progress on Nonlinear Problems of Sound Generation. *Prog. Aerosp. Sci.* **2004**, *40*, 345–416. [CrossRef]
24. Seror, C.; Sagaut, P.; Bailly, C.; Juvé, D. On the Radiated Noise Computed by Large-Eddy Simulation. *Phys. Fluids* **2001**, *13*, 476–487. [CrossRef]
25. Batten, P.; Goldberg, U.; Chakravarthy, S. Reconstructed Sub-Grid Methods for Acoustics Predictions at all Reynolds Numbers. AIAA 2002-2511. In Proceedings of the 8th AIAA/CEAS Aeroacoustics Conference & Exhibit, Breckenridge, CO, USA, 17–19 June 2002.
26. Batten, P.; Goldberg, U.; Chakravarthy, S. Interfacing Statistical Turbulence Closures with Large-Eddy Simulation. *AIAA J.* **2004**, *42*, 485–492. [CrossRef]
27. Yao, H.D.; He, G.W. A Kinematic Subgrid Scale Model for Large-Eddy Simulation of Turbulence-Generated Sound. *J. Turbul.* **2009**, *10*, 1–14. [CrossRef]
28. Fung, J.C.H.; Hunt, J.C.; Malik, N.; Perkins, R. Kinematic Simulation of Homogeneous Turbulence by Unsteady Random Fourier Modes. *J. Fluid Mech.* **1992**, *236*, 281–318. [CrossRef]
29. Kraichnan, R.H. Kolmogorov's Hypotheses and Eulerian Turbulence Theory. *Phys. Fluids* **1964**, *7*, 1723–1734. [CrossRef]
30. He, G.W.; Wang, M.; Lele, S.K. On the Computation of Space-Time Correlations by Large-Eddy Simulation. *Phys. Fluids* **2004**, *16*, 3859–3867. [CrossRef]
31. Fukushima, Y. Efficient Jet Noise Prediction Using Synthetic Eddy Method and Block-Structured Cartesian Mesh. In Proceedings of the 29th Congress of the International Council of the Aeronautical Sciences, St. Petersburg, Russia, 7–12 September 2014.
32. Hirai, S.; Fukushima, Y.; Obayashi, S.; Misaka, T.; Sasaki, D.; Ohmichi, Y.; Kanamori, M.; Takahashi, T. Influence of Turbulence Statistics on Stochastic Jet-Noise Prediction with Synthetic Eddy Method. *J. Aircr.* **2019**, *56*, 2342–2356. [CrossRef]
33. Bailly, C.; Juvé, D. A Stochastic Approach to Compute Subsonic Noise using Linearized Euler's Equations. AIAA 99-1872. In Proceedings of the 5th AIAA/CEAS Aeroacoustics Conference and Exhibit, Bellevue, WA, USA, 10–12 May 1999.
34. Jarrin, N.; Benhamadouche, S.; Laurence, D.; Prosser, R. A Synthetic-Eddy-Method for Generating Inflow Conditions for Large-Eddy Simulations. *Int. J. Heat Fluid Flow* **2006**, *27*, 585–593. [CrossRef]
35. Jarrin, N.; Prosser, R.; Uribe, J.C.; Benhamadouche, S.; Laurence, D. Reconstruction of Turbulent Fluctuations for Hybrid RANS/LES Simulations Using a Synthetic-Eddy Method. *Int. J. Heat Fluid Flow* **2009**, *30*, 435–442. [CrossRef]
36. Haywood, J.S.; Sescu, A.; Foster, J.; Farthing, M. Triple Hill's Vortex Synthetic Eddy Method for Turbulent Inflow Conditions. AIAA 2017-1225. In Proceedings of the 55th AIAA Aerospace Sciences Meeting, Grapevine, TX, USA, 9–13 January 2017.
37. Fleury, V.; Bailly, C.; Jondeau, E.; Michard, M.; Juvé, D. Space-Time Correlations in Two Subsonic Jets Using Dual Particle Image Velocimetry Measurments. *AIAA J.* **2008**, *46*, 281–292. [CrossRef]
38. Zhou, Y.; Rubinstein, R. Sweeping and Straining Effects in Sound Generation by High Reynolds Number Isotropic Turbulence. *Phys. Fluids* **1996**, *8*, 647–649. [CrossRef]
39. Chen, S.; Kraichnan, R.H. Sweeping Decorrelation in Isotropic Turbulence. *Phys. Fluids A Fluid Dyn.* **1989**, *1*, 2019–2024. [CrossRef]
40. Siefert, M.; Ewert, R. Sweeping Sound Generation in Jets Realized with a Random Particle-Mesh Method. AIAA 2009-3369. In Proceedings of the 15th AIAA/CEAS Aeroacoustics Conference, Miami, FL, USA, 11–13 May 2009.
41. Lafitte, A.; Garrec, T.L.; Bailly, C.; Laurendeau, E. Turbulence Generation from a Sweeping-Based Stochastic Model. *AIAA J.* **2014**, *52*, 281–292. [CrossRef]
42. He, G.; Jin, G.; Yang, Y. Space-Time Correlations and Dynamic Coupling in Turbulent Flows. *Annu. Rev. Fluid Mech.* **2017**, *49*, 51–70. [CrossRef]
43. Blake, J.; Sescu, A.; Thompson, D.S.; Hattori, Y. A Coupled LES/Stochastic Modeling Approach to Jet Noise Prediction. AIAA 2018-1248. In Proceedings of the 2018 AIAA Aerospace Sciences Meeting, Kissimmee, FL, USA, 8–12 January 2018.
44. Pope, S.B. *Turbulent Flows*; Cambridge University Press: Cambridge, UK, 2000.
45. Schumann, U.; Launder, B.E. Stochastic Backscatter of Turbulence Energy and Scalar Variance by Random Subgrid-Scale Fluxes. *Proc. R. Soc. A Math. Phys. Sci.* **1995**, *451*, 293–318.
46. Kennedy, C.A.; Carpenter, M.H. *Comparison of Several Numerical Methods for Simulation of Compressible Shear Layers*; Technical Report NASA-97-TP3484; NASA, Langley Research Center: Hampton, VA, USA, 1997.
47. Pokora, C.D.; McGuirk, J.J. Stereo-PIV Measurements of Spatio-Temporal Turbulence Correlations in an Axisymmetric Jet. *J. Fluid Mech.* **2015**, *778*, 216–252. [CrossRef]
48. Bogey, C.; Bailly, C.; Juvé, D. Computation of Flow Noise Using Source Terms in Linearized Euler's Equations. *AIAA J.* **2002**, *40*, 235–243. [CrossRef]

49. Lafitte, A.; Le Garrec, T.; Bailly, C.; Laurendeau, E. Prediction of Subsonic Jet Noise Relying on a Sweeping Based Turbulence Generation Process. AIAA 2012-2149. In Proceedings of the 18th AIAA/CEAS Aeroacoustics Conference, Colorado Springs, CO, USA, 4–6 June 2012.
50. Tam, C.K.; Webb, J.C. Dispersion-Relation-Preserving Finite Difference Schemes for Computational Acoustics. *J. Comput. Phys.* **1993**, *107*, 262–281. [CrossRef]
51. Bodony, D.J.; Lele, S.K. Current Status of Jet Noise Predictions Using Large-Eddy Simulation. *AIAA J.* **2008**, *46*, 364–380. [CrossRef]
52. Butcher, J. *Numerical Methods for Ordinary Differential Equations*; John Wiley and Sons, Ltd.: Hoboken, NJ, USA, 2003.
53. Bogey, C.; Marsden, O. A Study of the Grid Dependence of the Flow Field and Noise of Subsonic Jets. AIAA 2016-0261. In Proceedings of the 54th AIAA Aerospace Sciences Meeting, San Diego, CA, USA, 4–8 January 2016; pp. 1–19.
54. Viswanathan, K. Nozzle Shaping for Reduction of Jet Noise from Single Jets. *AIAA J.* **2005**, *43*, 1008–1022. [CrossRef]
55. Bogey, C.; Bailly, C. Effects of Inflow Conditions and Forcing on Subsonic Jet Flows and Noise. *AIAA J.* **2005**, *43*, 1000–1007. [CrossRef]

aerospace

MDPI

Article

Numerical Investigation of a Rectangular Jet Exhausting over a Flat Plate with Periodic Surface Deformations at the Trailing Edge

Colby Niles Horner [1], Adrian Sescu [1,*], Mohammed Afsar [2] and Eric Collins [3]

[1] Department of Aerospace Engineering, Mississippi State University, Starkville, MS 39759, USA; colbyhorner@hotmail.com

[2] Department of Mechanical & Aerospace Engineering, Strathclyde University, Glasgow G1 1XQ, UK; mohammed.afsar@strath.ac.uk

[3] Center for Advanced Vehicular Systems, Starkville, MS 39759, USA; emc@cavs.msstate.edu

* Correspondence: sescu@ae.msstate.edu

Abstract: Multiple competing factors are forcing aircraft designers to reconsider the underwing engine pod configuration typically seen on most modern commercial aircraft. One notable concern is increasing environmental regulations on noise emitted by aircraft. In an attempt to satisfy these constraints while maintaining or improving vehicle performance, engineers have been experimenting with some innovative aircraft designs which place the engines above the wings or embedded in the fuselage. In one configuration, a blended wing concept vehicle utilizes rectangular jet exhaust ports exiting from above the wing ahead of the trailing edge. While intuitively one would think that this design would reduce the noise levels transmitted to the ground due to the shielding provided by the wing, experimental studies have shown that this design can actually increase noise levels due to interactions of the jet exhaust with the aft wing surface and flat trailing edge. In this work, we take another look at this rectangular exhaust port configuration with some notional modifications to the geometry of the trailing edge to determine if the emitted noise levels due to jet interactions can be reduced with respect to a baseline configuration. We consider various horizontal and vertical offsets of the jet exit with respect to a flat plate standing in for the aft wing surface. We then introduce a series of sinusoidal deformations to the trailing edge of the plate of varying amplitude and wave number. Our results show that the emitted sound levels due to the jet–surface interactions can be significantly altered by the proposed geometry modifications. While sound levels remained fairly consistent over many configurations, there were some that showed both increased and decreased sound levels in specific directions. We present results here for the simulated configurations which showed the greatest decrease in overall sound levels with respect to the baseline. These results provide strong indications that such geometry modifications can potentially be tailored to optimize for further reductions in sound levels.

Keywords: large eddy simulations; jet noise; jet–surface interaction

Citation: Horner, C.N.; Sescu, A.; Afsar, M.; Collins, E. Numerical Investigation of a Rectangular Jet Exhausting over a Flat Plate with Periodic Surface Deformations at the Trailing Edge. *Aerospace* **2021**, *8*, 314. https://doi.org/10.3390/aerospace8110314

Academic Editor: Paul Bruce

Received: 25 August 2021
Accepted: 14 October 2021
Published: 24 October 2021

1. Introduction

Methods to reduce the noise generated by aircraft propulsion systems have long been a topic of research within the aerospace community. There have been many challenges faced by designers attempting to solve these types of problems due to the complexity of the flow phenomena as well as the costs involved with testing potential solutions. In some cases, noise reductions have only come at the expense of aircraft performance in other metrics (e.g., weight, cost).

Jet–surface interaction noise is usually associated with an increase in low frequency noise as a result of the interaction between a turbulent jet and a flat surface that is parallel to the jet axis. Jet noise is generally considered to be the result of two sources: scrubbing noise and trailing edge scattering noise (Brown and Wernet [1]). Scrubbing is generated by pressure fluctuations in the turbulent boundary layer impinging on a flat surface. Trailing edge scattering, which is the dominant noise source at low frequency (Brown [2], Podboy [3]), is produced when the hydrodynamic component of the upstream turbulence is scattered into acoustic waves after interacting with the trailing edge. The total installation noise can propagate upstream or downstream of the trailing edge.

High-speed jet exhausts interacting with nearby surfaces can generate flow distortions that significantly increase the overall noise emitted. In one such example, an aircraft engine placed on top of the wing in an attempt to shield the jet noise that would propagate to the ground while also improving aircraft performance actually resulted in an overall increase in sound levels. Another example of note is the tremendous noise generated by the jet exhausts of military aircraft interacting with the various surfaces of an aircraft carrier deck during take-off and landing operations.

Considerable research has been conducted on axisymmetric jets (Bridges [4], Brown and Bridges [5]), but, since the 1980s, a number of studies have shifted their focus to non-circular jet engine exit designs as potential passive flow control devices (Gutmark and Grinstein [6]). Geometric corrugations applied at the nozzle exit plane (e.g., tabbed studied by Gao et al. [7], cruciform by El Hassan et al. [8], and chevrons by Violato et al. [9]) have been observed to break down the large-scale structures present in the jet exhaust flow. This subsequently reduces the mixing between the jet and ambient-fluid. It was noted that these effects were most pronounced for larger wetted perimeters of the jet exit (Shakouchi and Iryama [10]) and an increased number of corners/sides (Gutmark and Grinstein [6]). The effect of external expansion ramps on supersonic flow emitted from high-aspect-ratio rectangular nozzles was analyzed by Malla et al. [11] to determine if the ramps could be used to reduce noise. Research conducted by Mancinelli et al. [12] and Proenca et al. [13] show further results regarding the interaction of rectangular jets with flat plates. Other aircraft propulsion concepts have utilized acoustically treated engine parts (inlet ducts, exhaust ducts, and inner walls), but this type of treatment tends to have a negative overall impact due to additional cost and weight.

While static chevron configurations at the jet exit have been shown to provide a significant reduction in noise without noticeable loss in thrust, some researchers have been looking into potential acoustic benefits that could be gained by exciting the chevrons with piezoelectric actuators. Mechanical perturbations of the flow by the actuated chevrons are believed to increase the production of small-scale disturbances while diminishing the large-scale turbulent structures thought to be responsible for the dominant portion of jet mixing noise. Mohan et al. [14] introduced piezoelectric actuators to chevrons in an attempt to modify the growth rate of the mixing layer. Butler and Calkins [15] examined four different types of nozzles that exist: round, faceted, faceted with static chevrons, and faceted with active chevrons. Their results demonstrated a 2 to 4 dB reduction in noise with the static chevrons, and an additional 2 dB reduction with the actuated chevron configurations.

Zaman [16] studied the integration of microjets near the engine exhaust port and found that a clear noise reduction was observed as the microjet pressure increased. The results showed that turbulent mixing noise reduction, as monitored by the overall sound pressure level at a shallow angle, correlated with the ratio of the microjet to the primary jet driving pressures (normalized by the ratio of the corresponding diameters). Semlitsch et al. [17] used implicit large eddy simulation to analyze the relationship between screech tone frequency and fluidic injection pressure of the microjets.

Rego et al. [18] utilized a Lattice–Boltzmann Method (LBM) to carry out numerical simulations of a flat plate placed in an irrotational hydrodynamic field of a jet with a round nozzle. Three cases were investigated with input flow characteristics determined from NASA wind tunnel experiments (see Brown [19]). Heated and cooled jets were analyzed at different Mach numbers, and the far-field noise was computed using the Ffowcs-Williams and Hawkings equation. Far-field spectral results showed a large noise increase for low to mid frequencies due to upstream hydrodynamic waves in the jet that propagate along the plate, 'scrubbing' over the surface, until it reaches the trailing edge of the plate. (This is referred to as the 'gust solution' in Afsar et al. [20]). As these waves interact with the trailing edge and the downstream turbulent mixing layer, they are scattered in the far-field as noise. A noticeable change in the far-field noise was observed when the trailing edge location was changed relative to the fixed nozzle. As the length of the plate was increased, the trailing edge was positioned in a region dominated by large-scale structures and a greater degree of low-frequency amplification was observed. When the plate was moved closer to the jet in the radial direction, the sound levels at lower frequencies increased.

Behrouzi and McGuirk [21] conducted experiments at the Loughborough University high-pressure nozzle test facility on a rectangular nozzle exhausting over a rectangular plate acting as an aft-deck. (A configuration similar to the one used in the current study.) The focus of the experiment was to study the effect of the rectangular plate on flow field development and to determine if any acoustic benefits could be obtained from the shielding effect that the aft-deck might produce. Schlieren imaging, pneumatic Pitot probes, and nonintrusive Laser Doppler Anemometry measurements were used to capture detailed information on the flow structures. The data revealed that the aft-deck created an asymmetry in the entrainment characteristics of the shear layers, inducing a net transverse pressure force on the plume by changing the inviscid shock cell structure. The presence of the aft-deck altered the net pressure force exerted on the flow which led to a transverse deflection of the jet. The aft-deck configuration produced a dramatic effect on the plume development, extending the potential core length slightly and reducing turbulence levels in the plume near field.

Bridges [22] further investigated this configuration by studying whether beveled edges on the rectangular nozzle would make a difference in the sound generated when compared to an aft-deck being added to the propulsion configuration (see also Zaman [23] and Zaman et al. [24]). Nozzles were fabricated with different aspect ratios of 2:1, 4:1, and 8:1. For each aspect ratio, three nozzle designs were tried: one basic nozzle with no extension and two beveled variants with bevel lengths of 1.3 and 2.7 inches. A rectangular nozzle with an aft-deck surface was also considered in which the rectangular nozzles had plates fitted to them such that when the plate surface was even with the inner lip of the nozzle, the surface and nozzle were effectively one piece. Five aft-deck lengths were tested: 1.3, 2.7, 4, 8, and 12 inches long. When the beveled nozzle was analyzed, it was found that the noise levels decreased in both azimuthal planes as the aspect ratio was increased, with much more dramatic reductions in the nozzle's major axis plane. These reductions in noise were lost when one lip of the nozzle was extended to make an aft deck. Low and high frequencies were examined with the aft deck extensions. For low frequencies, the sound levels were roughly the same on both sides of the plate. However, there were noticeable differences in the sound levels for higher frequencies on each side of the plate. This is due to the effect of the noise being shielded and reflected. The shielding effect seemed to increase as the length of the aft deck increased, but this was not the case for reflection. In addition, it is important to note that, for the largest plate lengths, the low-frequency amplification above the baseline jet noise was not dependent on the aspect ratio.

Seiner and Manning [25] considered the interaction between a supersonic jet and a flat plate from a rectangular nozzle. They showed that the distance between the nozzle exit and flat surface was an important parameter which can impact the screech noise. Ibrahim et al. [26] studied the effect of turbulence characteristics of jet flows on the radiated jet noise. Berland et al. [27] used compressible large eddy simulations to investigate the generation of screech tones from an under-expanded jet with a three-dimensional planar geometry.

In the present work, we aim to extend these results by examining the effect of sinusoidal surface deformations added to the trailing edge of the aft-deck plate. To this end, a suite of large eddy simulations have been performed, each targeting different geometrical configurations and flow conditions. Sound levels in the farfield were then computed to determine whether these deformations could potentially reduce the total noise emitted by the installed configuration. In these simulations, we consider a high aspect ratio rectangular jet. To simplify the nozzle geometry, we used a spanwise slice of width four times the height of the nozzle with periodic boundary conditions applied in the spanwise direction. Flow disturbances were imposed at the inflow boundaries to introduce three-dimensional flow structures into the jet flow. A high-order accurate solver was used to discretize the unsteady, compressible, conservative form of the filtered Navier–Stokes equations. The mean flow was determined using Reynolds Averaged Navier–Stokes (RANS) equations via a SST $k-\omega$ turbulence model (Menter [28]). The farfield noise radiation was evaluated using Ffowcs-Williams and a Hawkings acoustic analogy method.

We provide further details of the numerical formulation in Section 2, and present our results in Section 3.

2. Problem Formulation and Numerical Algorithms

2.1. Scalings

The governing equations considered here are expressed in terms of a generalized curvilinear coordinate system, subject to the following transformations:

$$\xi = \xi(x,y,z), \qquad \eta = \eta(x,y,z), \qquad \zeta = \zeta(x,y,z),$$

where ξ, η, and ζ are the body-aligned curvilinear coordinates corresponding to the streamwise, wall-normal, and spanwise directions, respectively, and x, y, and z are the (non-dimensional) global Cartesian coordinates of physical space. All dimensional spatial coordinates (x^*, y^*, z^*) are normalized by the reference length, D_j, the height of the nozzle (See Figure 1)

$$(x,y,z) = \frac{(x^*, y^*, z^*)}{D_j}, \tag{1}$$

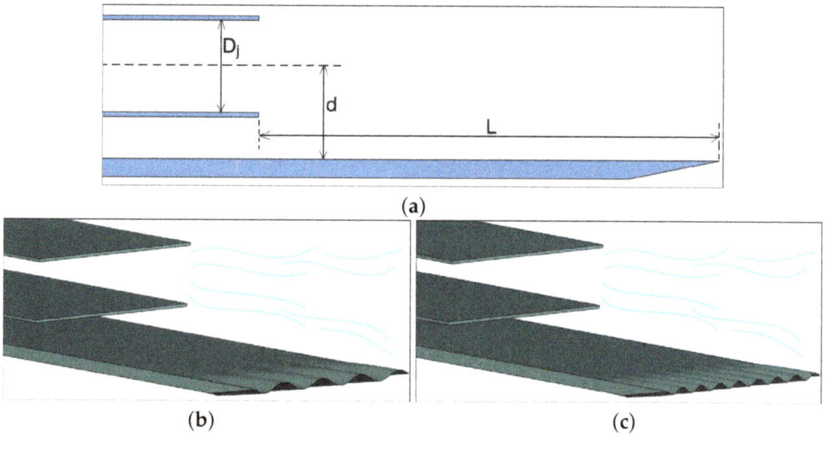

Figure 1. (**a**) The geometry of the nozzle and the plate; (**b**,**c**) two examples of trailing edge deformations of different wavenumbers.

The fluid velocity values are scaled by the jet velocity, V_j.

$$(u,v,w) = \frac{(u^*, v^*, w^*)}{V_j}, \tag{2}$$

Pressure values are scaled by the dynamic pressure, $\rho_\infty V_j^2$, and temperature values by the freestream temperature, T_∞. Reynolds number, Mach number, and Prandtl number are defined as:

$$Re = \frac{\rho_\infty V_j D_j}{\mu_\infty}, \qquad Ma = \frac{V_j}{a^*_\infty}, \qquad Pr = \frac{\mu_\infty C_p}{k_\infty} \tag{3}$$

where μ_∞, a_∞, and k_∞ are the freestream dynamic viscosity, speed of sound, and thermal conductivity, respectively. C_p is the specific heat at constant pressure. For all simulations, these values are initialized for air as an ideal gas.

2.2. Governing Equations

The filtered Navier–Stokes equations can be expressed in conservative form as:

$$\mathbf{Q}_t + \mathbf{F}_\xi + \mathbf{G}_\eta + \mathbf{H}_\zeta = \mathbf{S}. \tag{4}$$

Here, the subscripts denote partial derivatives with respect to time and the curvilinear coordinates. $\mathbf{Q}$ is the vector of conserved quantities:

$$\mathbf{Q} = \frac{1}{J}\{\rho,\ \rho\vec{u},\ E\}^T, \tag{5}$$

$\rho = \rho^*/\rho_\infty$ is the non-dimensional density of the fluid, $\vec{u} = (u, v, w)$ is the non-dimensional velocity vector in physical space, and E is the total energy. The flux vectors, $\mathbf{F}$, $\mathbf{G}$ and $\mathbf{H}$, are given by:

$$\mathbf{F} = \frac{1}{J}\left\{\begin{array}{c} \rho U \\ \rho\vec{u}U + \xi_{x_i}(p + \tau_{i1}) \\ EU + p\tilde{U} + \xi_{x_i}\Theta_i \end{array}\right\},$$
$$\mathbf{G} = \frac{1}{J}\left\{\begin{array}{c} \rho V \\ \rho\vec{u}V + \eta_{x_i}(p + \tau_{i2}) \\ EV + p\tilde{V} + \eta_{x_i}\Theta_i \end{array}\right\},$$
$$\mathbf{H} = \frac{1}{J}\left\{\begin{array}{c} \rho W \\ \rho\vec{u}W + \zeta_{x_i}(p + \tau_{i3}) \\ EW + p\tilde{W} + \zeta_{x_i}\Theta_i \end{array}\right\},$$

Here, the repeated indices indicate Einstein summation convention over $i \in 1, 2, 3$. U, V, W are the contravariant velocity components,

$$U = \xi_x u + \xi_y v + \xi_z w, \tag{6}$$
$$V = \eta_x u + \eta_y v + \eta_z w, \tag{7}$$
$$W = \zeta_x u + \zeta_y v + \zeta_z w, \tag{8}$$

τ is the shear stress tensor,

$$\tau_{ij} = \frac{\mu}{Re}\left[\left(\frac{\partial\xi_k}{\partial x_j}\frac{\partial u_i}{\partial\xi_k} + \frac{\partial\xi_k}{\partial x_i}\frac{\partial u_j}{\partial\xi_k}\right) - \frac{2}{3}\delta_{ij}\frac{\partial\xi_l}{\partial x_k}\frac{\partial u_k}{\partial\xi_l}\right], \tag{9}$$

and Θ is the heat flux,

$$\Theta_i = u_j\tau_{ij} + \frac{\mu}{(\gamma - 1)M_\infty^2 RePr}\frac{\partial\xi_l}{\partial x_i}\frac{\partial T}{\partial\xi_l} \tag{10}$$

Again, repeated indices (that are not present on the left-hand side) indicate summation. The Jacobian of the curvilinear transformation from the physical space to computational space is denoted by J, and $\mathbf{S}$ is a vector of prescribed source terms. Pressure, temperature, and density are related by the ideal gas equation of state:

$$p = \frac{\rho T}{\gamma M_\infty^2} \tag{11}$$

The dynamic viscosity and thermal conductivity k are related to temperature using the Sutherland's equations in dimensionless form:

$$\mu = T^{3/2}\frac{1 + C_1/T_\infty}{T + C_1/T_\infty}; \quad k = T^{3/2}\frac{1 + C_2/T_\infty}{T + C_2/T_\infty}, \tag{12}$$

For air at sea level, $C_1 = 110.4$ K, and $C_2 = 194$ K.

There are no explicit subgrid scale turbulence terms in Equation (4) (see Grinstein et al. [29]). Instead, the compressible Navier–Stokes equations are solved within the framework of an implicit large eddy simulation, where numerical filtering is applied to account for the missing sub-grid scale energy. The numerical solver uses high-order finite difference approximations for the spatial derivatives and explicit time marching. The time integration is performed using a second order Adams–Bashforth method [30]:

$$\mathbf{Q}^{n+1} = \mathbf{Q}^n + k\left[\sum_{\nu=0}^{K} \beta_\nu L(\mathbf{Q}^{n-\nu})\right] \tag{13}$$

Here, the constants β_ν are chosen to give either the maximum order of accuracy [30] or the lowest dispersion and dissipation. $L(\mathbf{Q})$ is the residual.

The spatial derivatives are discretized using the dispersion-relation-preserving schemes of Tam and Webb [31] or a high-resolution 9-point dispersion-relation-preserving optimized scheme of Bogey and Bailly [32]. To damp out the unwanted high wavenumber waves from the solution, high-order spatial filters, as developed by Kennedy and Carpenter [33], are used. No slip boundary conditions for velocity and adiabatic conditions for temperature are imposed at the solid surfaces. Sponge layers are imposed near the far-field boundaries in regions that are outside the flow domain of interest. The sponge layers combined with grid stretching act to damp-out unwanted waves returning from the farfield boundaries. For the LES simulations, the flow in proximity to the wall is modeled using the Werner–Wengler model (Werner and Wengle [34]), so some small turbulent flow structures may not captured adequately. At the inflow boundary, a constant mean flow in the core region is combined with a hyperbolic tangent function in the shear region, and disturbances in the form of a superposition of Fourier modes with random amplitudes, frequencies, and wavenumbers. These disturbances introduce three-dimensional flow variations into the jet flow.

The mean flows used to initialize the LES are obtained from RANS simulations, where a classical SST $k - \omega$ turbulence model (Menter [28]) is applied to account for missing fluctuations. This model was found to perform well for jet applications (see Mihaescu et al. [35])

In the present work, high-order, central-difference schemes are used to achieve increased resolution of the propagating disturbances. A shock-capturing technique suitable for simulations involving central differences in space is required to avoid unwanted oscillations that may propagate from discontinuities arising in supersonic flows. Shock capturing techniques are employed based on the general explicit filtering framework, a straightforward approach which introduces sufficient numerical viscosity in the area of the discontinuities, and negligible artificial viscosity in the rest of the domain. The technique of Bogey et al. [36] introduces selective filtering at each grid vertex to minimize numerical oscillations, and shock-capturing in the areas where discontinuities are present. This method has been proven to work efficiently for high-order accurate, nonlinear computations.

2.3. Ffowcs-Williams Hawkings Acoustic Analogy Method

The Ffowcs-Williams and Hawkings (FW-H) equation [37,38] is an inhomogeneous wave equation that can be derived by manipulating the continuity equation and the Navier–Stokes equations for a compressible fluid. The FW–H equation can be written as:

$$\begin{aligned} \frac{1}{a_0^2}\frac{\partial^2 p'}{\partial t^2} - \nabla^2 p' &= \frac{\partial^2}{\partial x_i \partial x_j}\{T_{ij}H(f)\} \\ &- \frac{\partial}{\partial x_i}\{[P_{ij}n_j + \rho u_i(u_n - v_n)]\delta(f)\} \\ &+ \frac{\partial}{\partial t}\{[\rho_0 v_n + \rho(u_n - v_n)]\delta(f)\} \end{aligned} \quad (14)$$

where u_i is the fluid velocity component in the x_i direction, u_n is the fluid velocity component normal to the surface at $f = 0$, v_i is the surface velocity components in the x_i direction, v_n is the surface velocity component normal to the surface, $\delta(f)$ is Dirac delta function, $H(f)$is Heaviside function, and $p' = p - p_0$ is the magnitude of the sound pressure disturbance at the far field.

The acoustic pressure in the farfield at a location, $\vec{\mathbf{x}}$, can be determined from:

$$p'(\vec{\mathbf{x}},t) = p'_T(\vec{\mathbf{x}},t) + p'_L(\vec{\mathbf{x}},t) + p'_Q(\vec{\mathbf{x}},t) \quad (15)$$

where

$$p'_T(\vec{\mathbf{x}},t) = \frac{1}{4\pi}\frac{\partial}{\partial t}\int_S \left[\frac{\rho_0 U_n}{r}\right]_{ret} dS \quad (16)$$

$$p'_L(\vec{\mathbf{x}},t) = \frac{1}{4\pi a_0}\frac{\partial}{\partial t}\int_S \left[\frac{L_r}{r}\right]_{ret} dS + \frac{1}{4\pi}\int_S \left[\frac{L_r}{r^2}\right]_{ret} dS \quad (17)$$

Given a point, $\vec{\mathbf{y}}$, in the near-field acoustic source on surface S, $r = |\vec{r}| = |\vec{\mathbf{x}} - \vec{\mathbf{y}}|$ is the distance from the surface source location to the farfield position $\vec{\mathbf{x}}$; U_n is the scalar product of $\vec{U} = \rho\vec{u}/\rho_0$ and the unit normal to the surface S at $\vec{\mathbf{y}}$; L_r is the scalar product of $L_i = P_{ij}n_i + \rho u_i u_n$ and the unit vector in the direction from $\vec{\mathbf{y}}$ to $\vec{\mathbf{x}}$; $P_{ij} = -\tau_{ij} + p\delta_{ij}$; τ_{ij} is the viscous stress; ρ is the instantaneous density; and ρ_0 and a_0 are the ambient density and sound speed, respectively.

In Equation (15), $p'_Q(\vec{\mathbf{x}},t)$ is the contribution from the quadrupole, but it is neglected here. The integrals in (16) and (17) are calculated at the emission time $\tau = t - r/a$, where a is the speed of sound and t is the reception time. The flow variables at τ must be interpolated from the flow data at multiple time iterations, t, due to its dependency on $\vec{\mathbf{y}}$, which varies over the surface S surrounding the jet.

For the numerical implementation of the FW-H method, a conical surface that includes the entire jet is considered. The primitive variables and viscous stresses are interpolated from the flow domain to the surface at every b time steps (the value of b depends on the desired time resolution) and written out to a large number of files for post-processing. Given a reception time t, the emission time τ is calculated for every small surface element ΔS on the conical surface. Within a specific surface element ΔS on the surface, an interpolation is necessary to obtain the flow data at the emission time τ, which is needed to calculate the integrals in (16) and (17). Next, we determine the time derivatives of U_n and L_r and evaluate the integrals in Equations (16) and (17). The acoustic pressure then is calculated from Equation (15).

3. Results and Discussion

Results are reported and discussed in this section for a high aspect ratio rectangular jet exhausting over a flat surface located underneath the jet flow. Because the aspect ratio of the nozzle exit is high, we simplify the problem by considering only a portion from the nozzle along the spanwise direction. This simplification is based on the assumption that the flow is statistically two-dimensional (an assumption also considered by [24]). Thus, the spanwise length of the flow domain is set to be four times the height of the nozzle, and periodic boundary conditions are imposed in the lateral direction.

The layout of the configuration is illustrated schematically in Figure 1a. The flat plate is oriented parallel to the jet axis and located beneath the jet exit. The vertical separation of the plate from the jet axis (d) and the distance from the jet exit to the trailing edge of the plate (L) were both varied in this study. These distances are specified below and in Table 1 in terms of multiples of the nozzle height (D_j). The vertical separation considered were 0.65, 0.75, and 1 times the nozzle height, while the horizontal distances to the trailing edge considered were 4, 5, and 6 times the nozzle height. A 3D view of the geometry of the nozzle and the plate with deformed trailing edge are shown in Figure 1. The deformations in the spanwise direction are prescribed as a sine function while the amplitude is modulated by a hyperbolic tangent function in the streamwise direction. Two combinations of wavenumbers and amplitudes are shown in Figure 1b,c.

The Reynolds number, based on the jet velocity and the height of the nozzle, is 500,000, and the acoustic Mach number is 0.8. The amplitude of velocity disturbances imposed at the nozzle exit are restricted to less than 3% of the jet velocity.

Table 1. Position of the plate with respect to the nozzle (see also Figure 1).

Case	Vertical Stand-Off Dist. (d)	Streamwise Dist. from the Nozzle Exit (L)
1	0.75 D_j	4 D_j
2	0.75 D_j	5 D_j
3	0.75 D_j	6 D_j
4	1.0 D_j	5 D_j
5	0.65 D_j	5 D_j

3.1. RANS Results

The mean flow and turbulent kinetic energy (TKE) used to initialize the LES are obtained from separate two-dimensional RANS simulations using a $\kappa - \omega$ turbulence model. The RANS results shown in Figures 2–6 are reported first to demonstrate the effect of the plate location on the jet mean flow and TKE. Plots of mean velocity magnitude (left) and turbulent kinetic energy (right) are shown in Figure 2 for two selected cases, corresponding to the largest and the smallest offset distance of the plate from the jet axis. These color plots indicate that there is a visible effect from the plate on both the mean flow and the turbulent kinetic energy. For the first configuration ($d = 1.0D_j$, $L = 5D_j$), which corresponds to the largest distance d from the center of the nozzle, the jet is significantly deviated downward by the plate (resembling the 'Coanda' effect). This deviation is also present in other cases, but they do not appear to be as significant as it is for this case. The smallest deviation from the jet axis occurs for the configuration ($d = 0.65D_j$, $L = 5D_j$), which corresponds to the smallest offset distance, d.

From the plots of TKE shown on the right side of Figure 2, there appears to be an attenuation of the TKE in the bottom shear layer of the jet as a result of the interaction with the plate boundary layer. This is potentially due to the viscous dissipation in the boundary layer that is developing on the surface of the plate. Further downstream of the plate trailing edge, the magnitude of the TKE appears to recover somewhat as the two shear layers show similar intensities. This is further revealed by the profiles of TKE across the jet shown in Figure 6.

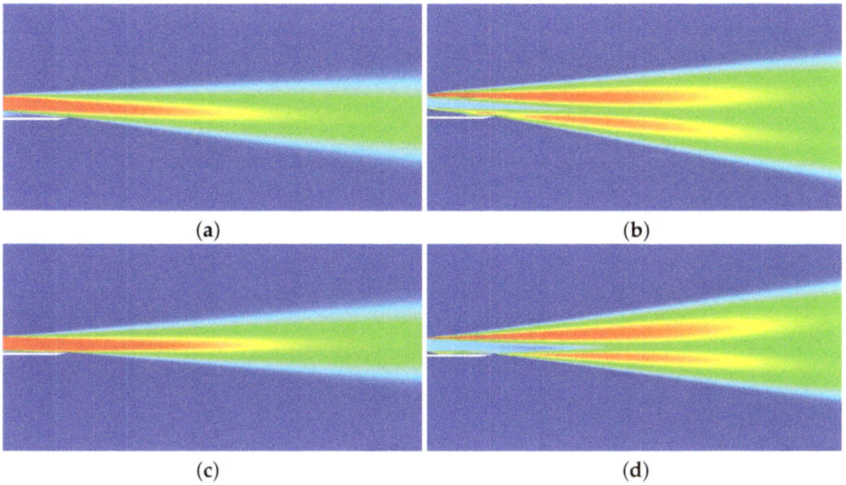

Figure 2. Mean velocity magnitude (**left**) and turbulent kinetic energy (**right**) distributions from RANS simulations of two cases. Top (**a**,**b**): $d = 1.0D_j$, $L = 5D_j$; Bottom (**c**,**d**): $d = 0.65D_j$, $L = 5D_j$.

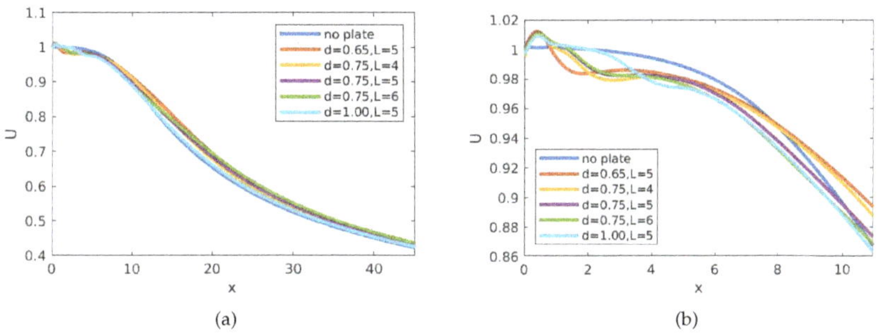

Figure 3. Mean streamwise velocity along the jet centerline from RANS: (**a**) x from 0 to 45; (**b**) details in the core region with x from 0 to 12.

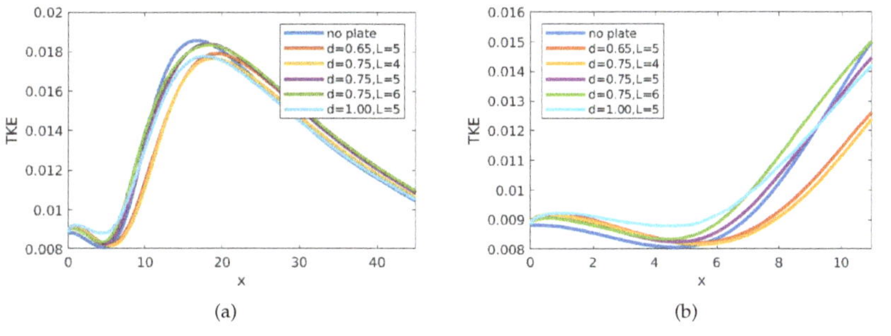

Figure 4. Turbulent kinetic energy along the jet centerline from RANS: (**a**) x from 0 to 45; (**b**) details in the core region with x from 0 to 12.

Center-line mean velocity distributions are plotted in Figure 3 for several tested configurations along with the results obtained from the baseline free jet (no plate). The presence of the plate seems to slow down the flow inside the potential core as seen in the zoomed-in plot shown in Figure 3b. While all curves seem to follow the same trend in the downstream region, there is a slight acceleration for the cases that include the plate.

This does seem to depend on the distance of the jet axis to the plate, d; the smaller this distance, the higher the acceleration. TKE distributions along the jet center line that are plotted in Figure 4 reveal a decrease of TKE in the downstream region for the cases that involve the plate. This decrease is more significant for the configurations with the smallest d and smallest L (yellow and red curves in Figure 4).

In Figures 5 and 6, we compare cross-flow vertical profiles of mean velocity magnitude and turbulent kinetic energy among the five configurations, at four axial locations, $x = 4D_j$;, $x = 10D_j$, $x = 25D_j$, and $x = 45D_j$; the results from the free jet case are also included. They all show that the largest deviation of the jet from the original position occurs for $d = 1.0D_j$ and $L = 5D_j$, corresponding to the largest distance of the plate from the jet (this was also observed in the contour plots of the mean velocity and TKE). Figure 6 shows that the reduction of the TKE in the lower shear layer is more significant for the configuration with the smallest distance from the trailing edge. Figures 5d and 6d indicate that the jet flow becomes more symmetric further downstream for all normalized profiles, except the case corresponding to $d = 1.0D_j$, $L = 5D_j$.

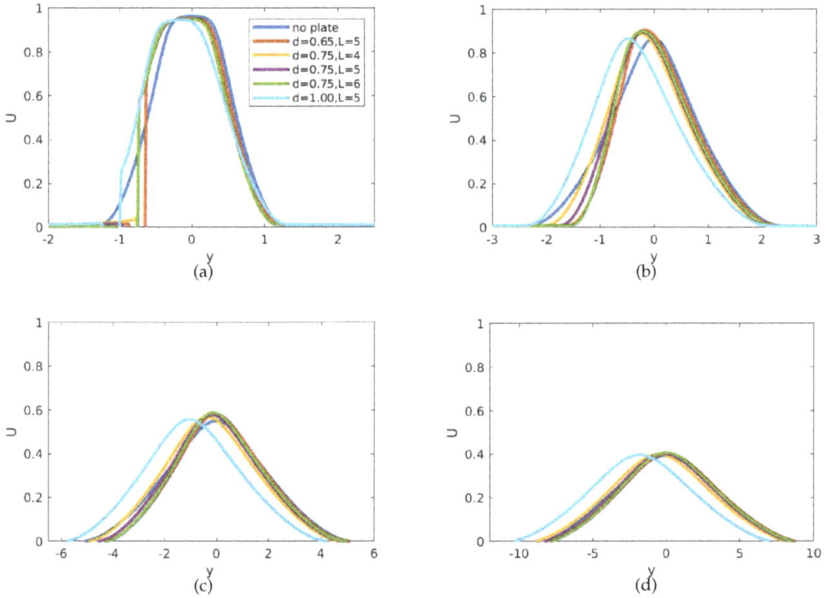

Figure 5. Streamwise velocity profiles along vertical direction, for different configurations: (**a**) $x = 4D_j$; (**b**) $x = 10D_j$; (**c**) $x = 25D_j$; (**d**) $x = 45D_j$.

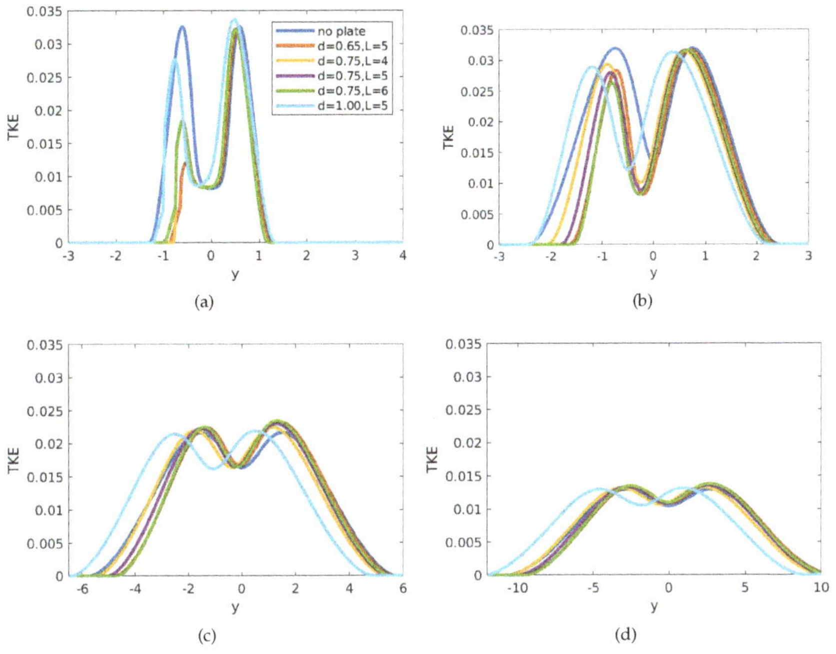

Figure 6. Turbulent kinetic energy profiles along vertical direction, for different configurations: (**a**) $x = 4D_j$; (**b**) $x = 10D_j$; (**c**) $x = 25D_j$; (**d**) $x = 45D_j$.

3.2. LES Results

A suite of LES runs were carried out corresponding to the geometrical configurations given in Table 1, and the trailing edge deformations characterized by the function

$$g(x,z) = 0.5A((1+\tanh(\sigma_x(x+x_0))) * \sin(2\pi(wz)z/L_z) \tag{18}$$

where $x_0 = -1.2$ is the streamwise location on the plate where the deformation begins to gradually increase from zero to the maximum amplitude at the trailing edge location; $\sigma_x = 2$ is a parameter controlling the 'thickness' of the tangent hyperbolic function; and L_z is the spanwise width of the plate. The amplitude, A, and wavenumber, k, are varied as given in Table 2. (See also Figure 1 for a visual representation of the trailing edge deformations.) As mentioned previously, the spanwise length of the flow domain is four times the height of the nozzle, and periodic boundary conditions are imposed in the lateral direction with the assumption that the aspect ratio of the rectangular nozzle is very high. This lateral dimension of $4D_j$ was selected to ensure that all relevant turbulent flow structures are captured accurately in the spanwise direction. The mesh consists of approximately 10 millions grid points, clustered mostly in proximity to the jet region. Grid stretching is used in the farfield along with an imposed sponge layer condition at the outflow boundary to ensure that the fluctuations are gradually dissipated while they leave the flow domain.

Figure 7 shows iso-surfaces of Q-criterion colored by the streamwise velocity component. ($Q = 1/2[|\mathbf{\Omega}|^2 - |\mathbf{S}|^2]$, where $\mathbf{S} = 1/2[\nabla\mathbf{v} + (\nabla\mathbf{v})^T]$ is the rate-of-strain tensor, and $\mathbf{\Omega} = 1/2[\nabla\mathbf{v} - (\nabla\mathbf{v})^T]$ is the vorticity tensor.) Acoustic waves radiating from the jet are plotted in gray contours. (The white lines are caused by gaps between blocks from the MPI decomposition). The attenuation of acoustic waves propagating from the jet underneath the plate can be observed. This is a result of the shielding provided by the plate causing the waves to not be as intense in this region.

Table 2. Amplitudes and wavenumbers of the trailing edge.

Treatment	Amplitude, A	Wavenumber, wz
1	$0.04\ D_j$	10
2	$0.06\ D_j$	8
3	$0.08\ D_j$	6
4	$0.12\ D_j$	5
5	$0.14\ D_j$	4
6	$0.16\ D_j$	3

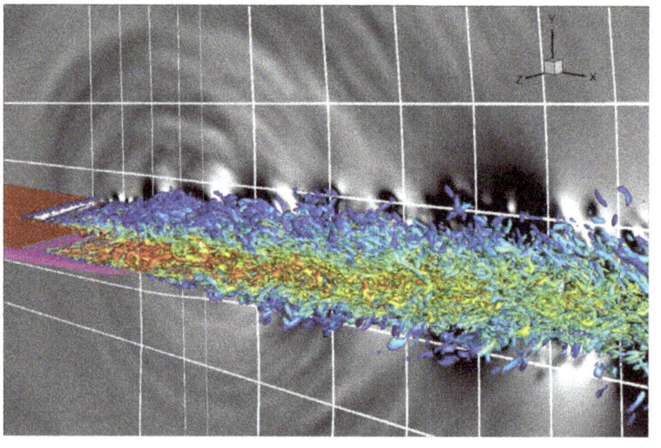

Figure 7. Iso-surfaces of Q-criterion colored by the streamwise velocity and pressure contours in gray.

The mean centerline velocity distributions are compared against RANS results in Figure 8. While there are some slight differences between the two in the potential core region, overall the agreement is good, showing similar decay in the downstream flow. (The LES profiles show some fluctuations from insufficient time spans used for the time averaging).

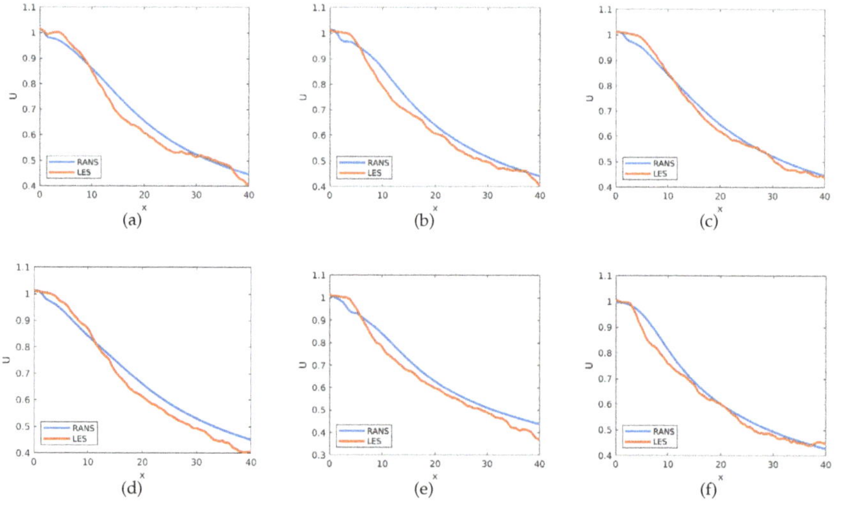

Figure 8. Mean streamwise velocity along the jet centerline: (**a**) $d = 0.65D_j$, $L = 5D_j$; (**b**) $d = 0.75D_j$, $L = 4D_j$; (**c**) $d = 0.75D_j$, $L = 5D_j$; (**d**) $d = 0.75D_j$, $L = 6D_j$; (**e**) $d = 1.0D_j$, $L = 5D_j$; (**f**) no plate.

Acoustic spectra have been calculated at three probe locations relatively close to the jet (in the nearfield). These probes are located at large angles with respect to the jet axis since it was observed that jet surface interaction noise dominates at large polar angles (see [39] for a study of AR8 nozzle and [20] where various rectangular nozzle jets were considered). The Ffowcs-Williams Hawkins surface is shown in Figure 9. The vertical side in the downstream was placed at 30 nozzle diameters from the nozzle exit plane. (Varying this distance did not affect the results significantly.) Two probes are located above the jet while the third is located underneath the jet (see Figure 9). The coordinates of the probes are listed in Table 3. A fourth probe was placed in the far field at 100 equivalent diameters and 50 deg angle with respect to the jet axis to compare the numerical results with experimental results collected by Bridges et al. [40] (see also Bridges [22]). Results were analyzed by using the sound pressure level and overall sound pressure level plots. The Strouhal number, St_e, was computed by the formula:

$$St_e = \frac{fD_j}{V_j} \tag{19}$$

where f is the frequency, D_j is nozzle height, and V_j is the jet velocity.

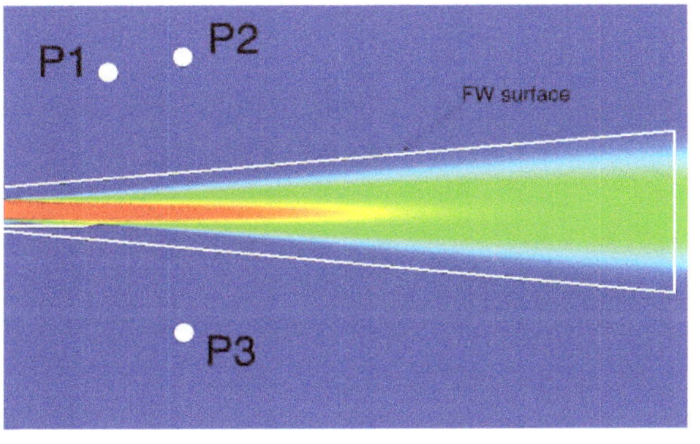

Figure 9. Locations of the three Probes, and the Ffowcs-Williams integration surface.

Figure 10 shows the comparison between our acoustic spectrum and the spectrum obtained from measurements (Bridges et al. [40]). Bridges et al. [40] conducted an extensive set of measurements to acquire acoustic data on single-flow convergent rectangular nozzles of aspect ratios 2:1, 4:1, and 8:1, at Mach numbers 0.7 and 0.9. Our comparison is for the largest aspect ratio 8:1 and Mach number 0.9. The agreement in Figure 10 is fairly good in the low frequency range of the spectrum. For the high frequency range the spectrum curve obtained from numerical simulations is slightly lower, which is most likely due to the mesh resolution of the LES not being sufficient to fully resolve the smallest flow scales that are responsible for the highest frequency noise.

Table 3. Probe locations.

Probe	X-Location	Y-Location
1	0	5
2	1	5.3
3	1	−3

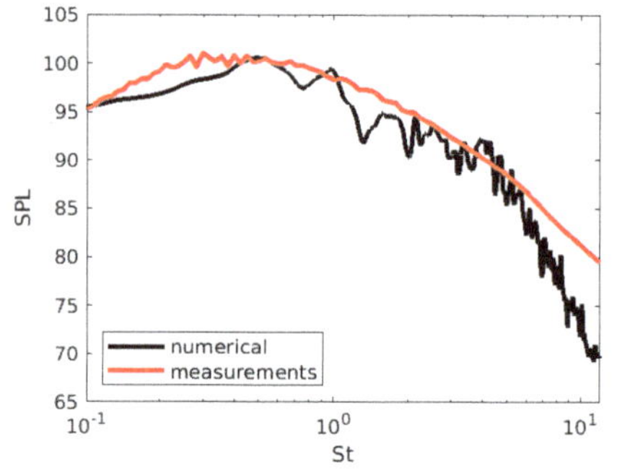

Figure 10. Comparison between numerical results and measurements (Bridges et al. [40]).

In general, the effect of trailing edge deformation on the radiated noise was small. We noticed both an increase and a decrease in noise, depending on the combination of amplitudes and wavenumbers characterizing the deformations. A visible reduction in noise was achieved for treatment 5 given in Table 2 ($w_z = 4$, $A = 0.14$). The acoustic spectra at the three probe points are plotted for this case in Figure 11a–c.

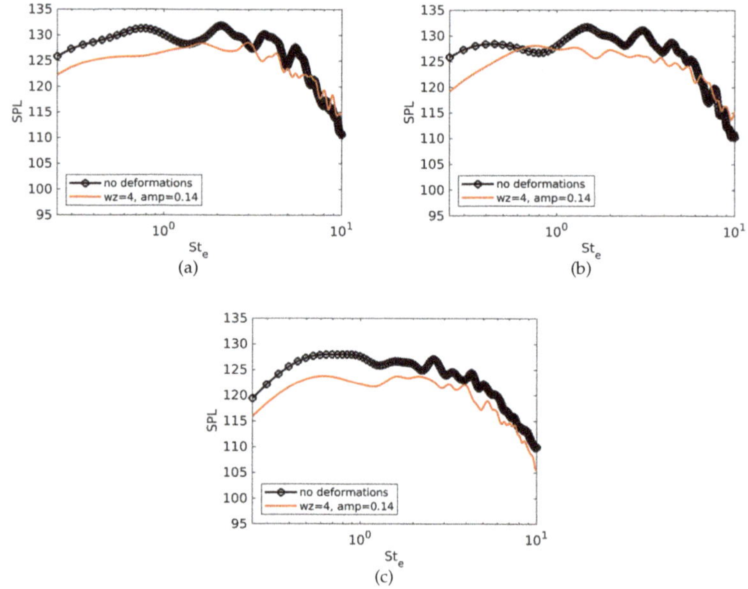

Figure 11. Typical acoustic spectra that show a reduction: (**a**) Probe 1; (**b**) Probe 2; (**c**) Probe 3.

The overall sound pressure levels (OASPL) observed at the probe points are plotted in Figures 12 and 13. In Figure 12, the OASPL's are plotted for the five configurations that do not involve trailing edge deformations. For the probes that are located above the jet, there is a significant increase in the radiated noise for the case corresponding to the smallest distance from the nozzle center ($d = 0.65D_j$, $L = 5D_j$). This is to be expected for this case since the end of the potential core is the closest to the trailing edge of the plate. For the probe located under the jet, there is an increase in the radiated noise for the case corresponding to the shortest plate length. This is an indication of a reduced shielding effect.

Figure 13 displays the OASPL results at all three probe locations for each of the tested configurations and prescribed trailing edge deformations. Each plot displays the results for all of the treatments given in Table 2 at one of the fixed plate configurations given in Table 1. When the data are broken down by each plate placement location, there were two apparent plate configurations that yielded a noticeable reduction in sound levels at the probe locations. We will look further into these configurations to determine if the trailing edge deformations provide an additional positive effect.

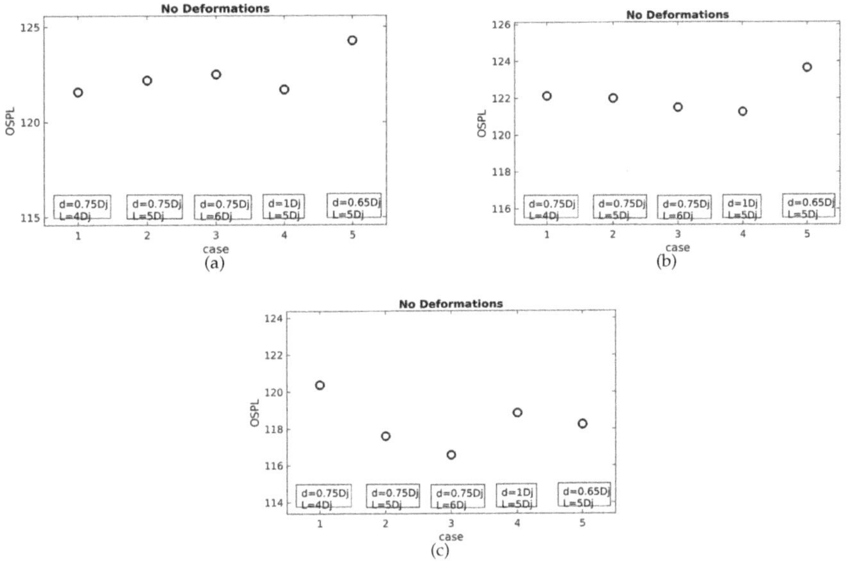

Figure 12. Level of reduction of the overall sound pressure level for all probes compared to no edge deformations: (**a**) Probe 1; (**b**) Probe 2; (**c**) Probe 3.

There appeared to be an overall reduction in noise for most of the treatments in case 1 ($d = 0.75D_j\ L = 4D_j$). The average overall sound pressure level for the no deformation case was 121.367 dB. For the $w_z = 10\ A = 0.04$ treatment, the average was 120.467 dB, and, for the $w_z = 6\ A = 0.08$ case, the average was 120.4 dB. The results for the Overall Sound Pressure Level data can be seen in Table 4. Although there was a reduction in noise for all of the deformations at this plate placement location, the two deformations mentioned above demonstrated the greatest reduction in noise.

Table 4. Overall Sound Pressure Level results for the probes corresponding to the plate placement of $d = 0.75\ D_j\ L = 4\ D_j$.

Probe	No Def.	wz = 10 A = 0.04	wz = 6 A = 0.08
1	121.5	120.7	120.8
2	122.1	121.0	121.6
3	120.5	119.7	118.8

For case 5 ($d = 0.65D_j\ L = 5D_j$), there was also a reduction in noise for all of the deformations applied to the trailing edge of the plate. The average overall sound pressure level for the probes for the no deformation case was 122.067 dB. For the $w_z = 10\ A = 0.04$ treatment, the average was 120.633 dB, and for the $w_z = 4\ A = 0.14$ treatment the average was 121.133 dB. The Overall Sound Pressure Level results for this plate placement location can be seen in Table 5.

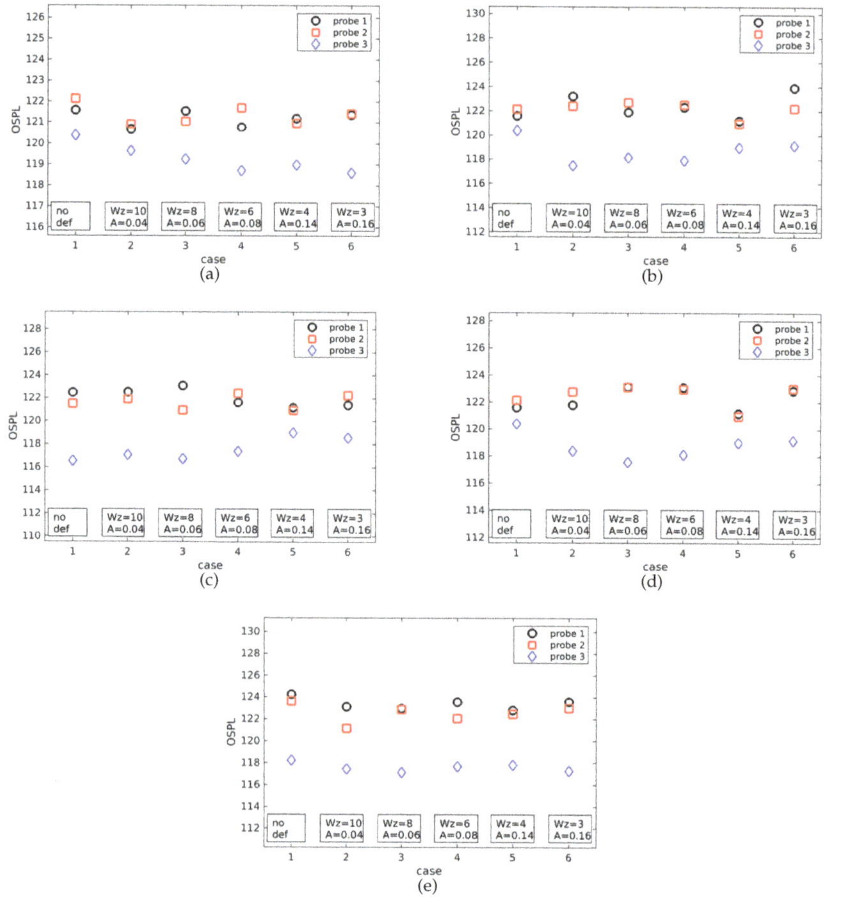

Figure 13. Overall sound pressure level for trailing edge deformations: (**a**) Case 1; (**b**) Case 2; (**c**) Case 3; (**d**) Case 4; (**e**) Case 5.

Table 5. Overall Sound Pressure Level results for the probes corresponding to the plate placement of $d = 0.65\ D_j\ L = 5\ D_j$.

Probe	No Def.	wz = 10 A = 0.04	wz = 4 A = 0.14
1	124.2	123.2	122.9
2	123.7	121.3	122.6
3	118.3	117.4	117.9

The OASPL results from these configurations yielded the best results for overall noise reduction among those tested. It would be worthwhile to consider other deformations that are close to these treatments as well as other nearby plate placements to determine if further noise reduction could be achieved.

4. Conclusions

In this work, jet flow over various surface deformations at the trailing edge of a plate installed under a high aspect ratio rectangular jet were simulated to determine the effect of the jet-surface interaction on the noise radiated to the farfield. RANS calculations and a suite of LES simulations were performed with a high-order accurate flow solver.

Numerous configurations and flow conditions were investigated, and the emitted noise from the jet-surface interactions calculated using Ffowcs-Williams and the Hawkings acoustic analogy method.

The RANS results, consisting of centerline velocity distributions, showed that the presence of the plate appeared to slow down the flow inside the potential core. For some of the tested cases, a slight acceleration of the flow further in the downstream was observed. The amount of acceleration appears to depend on the offset distance of the plate from the center of the jet, d. TKE distributions along the jet centerline revealed a decrease in TKE magnitude in the downstream region, which was expected. This decrease was more significant for the configuration with the smallest d and smallest L. Cross-flow profiles of mean velocity magnitude and turbulent kinetic energy at different axial locations showed a clear deviation of the jet from the axis of symmetry near the trailing edge. The TKE profiles indicated that the jet is slightly skewed around the trailing edge of the plate in a manner resembling the Coanda effect. Despite this deviation, the cross-flow profile becomes more symmetric further downstream. A reduction of the TKE in the lower shear was observed for all cases with the plate under the jet. This is most significant for the configurations with the smallest distance to the trailing edge.

For the LES analysis, various results consisting of iso-surface of Q-criterion, centerline velocity, acoustic spectra, and overall sound pressure levels were presented and discussed. Overall, good agreement was found between the centerline velocities from RANS and LES, with some small discrepancies in the potential core region. Acoustic spectra plotted for the configurations without trailing edge deformations confirmed a reduction of the noise under the plate from the shielding effect. For most cases involving the trailing edge deformations, there did not appear to be a significant change to overall sound levels. However, some cases did demonstrate a noticeable increase or decrease in emitted noise. In the best cases, the overall sound pressure levels were decreased from 1 to 5 dB.

The results presented in this work provide evidence that further refinement of the configurations described herein may yield additional noise reductions. Future research aimed at analyzing the interactions of the jet flow with different types of surface deformations (e.g., non-periodic and/or streamwise-oriented) may uncover a greater degree of noise reduction, beyond what our results have shown.

Author Contributions: C.N.H. ran simulations, post-processed and analyzed the results, and wrote the results and discussion section. A.S. analyzed the results, contributed to the discussion of the results, and wrote the Problem formulation and the numerical algorithm section. M.A. analyzed the results and wrote the Introduction section. E.C. contributed to the discussion of the results and proofread the manuscript. All authors have read and agreed to the published version of the manuscript.

Funding: The effort at Mississippi State University was supported by NASA EPScoR Project Number 80NSSC17M0039. All simulations were performed on Talon and Shadow HPC systems at the High-Performance Computing Collaboratory, Mississippi State University.

Institutional Review Board Statement: Not Applicable.

Informed Consent Statement: Not applicable.

Data Availability Statement: Not Applicable.

Conflicts of Interest: The authors declare no conflict of interest.

References

1. Brown, C.A.; Wernet, M.P. Jet surface interaction test: Flow measurements results. In Proceedings of the 20th AIAA/CEAS Aeroacoustics Conference, Atlanta, GA, USA, 16–20 June 2014; AIAA Paper 2014-3198.
2. Brown, C.A. Jet-Surface Interaction Test: Far-Field Noise Results. *J. Eng. Gas Turbines Power* **2013**, *135*, 071201. [CrossRef]
3. Podboy, G. *Jet-Surface Interaction Test: Phased Array Noise Source Localization Results*; ASME GT2012-69801; American Society of Mechanical Engineers: New York, NY, USA, 2012.
4. Bridges, J.; Wernet, M.P. *The NASA Subsonic Jet ParticleImage Velocimetry (PIV) Dataset*; NASA TM 2011-216807; NASA: Washington, DC, USA, 2011.

5. Brown, C.; Bridges, J. *Small Hot Jet Acoustic Rig Validation*; NASA TM 2006-214234; NASA: Washington, DC, USA, 2006.
6. Gutmark, E.J.; Grinstein, F.F. Flow control with noncircular jets. *Annu. Rev. Fluid Mech.* **1999**, *31*, 239–272. [CrossRef]
7. Gao, N.; Sun, H.; Ewing, D. Heat transfer to impinging round jets with triangular tabs. *Int. J. Heat Mass Transf.* **2003**, *46*, 2557–2569. [CrossRef]
8. El Hassan, M.; Meslem, A.; Abed-Meraim, K. Experimental investigation of the flow in the near-field of a cross-shaped orifice jet. *Phys. Fluids* **2011**, *23*, 045101. [CrossRef]
9. Violato, D.; Ianiro, A.; Cardone, G.; Scarano, F. Three-dimensional vortex dynamics and convective heat transfer in circular and chevron impinging jets. *Int. J. Heat Fluid Flow* **2012**, *37*, 22–36. [CrossRef]
10. Shakouchi, T.; Iryama, S. Flow characteristics of submerged free jet flow from petal-shaped nozzles. *J. Fluid Sci. Technol.* **2014**, *9*, 1–7. [CrossRef]
11. Malla, B.; Gutmark, J. Single Expansion Ramp Nozzles: Impact of Ramp Length on Flow and Acoustics. *AIAA J.* **2021**, *59*, 3152–3164. [CrossRef]
12. Mancinelli, M.; Di Marco, A.; Camussi, R. Multivariate and conditioned statistics of velocity and wall pressure fluctuations induced by a jet interacting with a flat plate. *J. Fluid Mech.* **2017**, *823*, 134–165. [CrossRef]
13. Proenca, A.; Lawrence, J.; Self, R. Investigation into the turbulence statistics of installed jets using hot-wire anemometry. *Exp. Fluids* **2020**, *61*, 220. [CrossRef]
14. Mohan, N.K.D.; Doty, M.J. Active Chevrons For Jet Noise Reduction. In Proceedings of the 24th International Congress On Sound In addition, Vibration, London, UK, 23–27 July 2017. [CrossRef]
15. Butler, G.; Calkins, F.T. Initial attempts to suppress jet noise using piezoelectric actuators. In Proceedings of the 9th AIAA/CEAS Aeroacoustics Conference, Hilton Head, SC, USA, 12–14 May 2003; No. 2003-3192. [CrossRef]
16. Zaman, K.B.M.Q. *Jet Noise Reduction by Microjets—A Parametric Study*; AIAA-2009-3129; NASA/TM-2010-215846; NASA Glenn Research Center: Cleveland, OH, USA, 2010.
17. Semlitsch, B.; Cuppoletti, D.; Gutmark, E.; Mihăescu, M. Transforming the Shock Pattern of Supersonic Jets Using Fluidic Injection. *AIAA J.* **2019**, *57*, 1851–1861.
18. Rego, L.; Avallone, F.; Ragni, D.; Casalino, D. Noise Amplification Effects due to Jet-Surface Interaction. In Proceedings of the AIAA Scitech 2019 Forum, San Diego, CA, USA, 7–11 January 2019; AIAA 2019-0001.
19. Brown, C. Jet-Surface Interaction Test: Far-Field Noise Results. In Proceedings of the ASME Turbo Expo 2012: Power for Land, Sea and Air, Copenhagen, Denmark, 11–15 June 2012; pp. 1–13.
20. Afsar, M.; Leib, S.; Bozak, R. Effect of de-correlating turbulence on the low frequency decay of jet-surface interaction noise in sub-sonic unheated air jets using a CFD-based approach. *J. Sound Vib.* **2017**, *386*, 177–207. [CrossRef]
21. Behrouzi, P.; McGuirk, J.J. Underexpanded Jet Development from a Rectangular Nozzle with Aft-deck. *AIAA J.* **2015**, *53*, 1287–1298.
22. Bridges, J.E. Noise from Aft Deck Exhaust Nozzle Differences in Experimental Embodiments. In Proceedings of the 52nd AIAA Aerospace Meeting, National Harbor, MD, USA, 13–17 January 2014; AIAA 2014-0876.
23. Zaman, K.B.M.Q.; Fagan, A.F.; Clem, M.M.; Brown, C.A. Resonant Interaction of a Rectangular Jet with a Flat-plate. In Proceedings of the 52nd AIAA Aerospace Meeting, National Harbor, MD, USA, 13–17 January 2014; AIAA 2014-0877. [CrossRef]
24. Zaman, K.B.M.Q.; Brown, C.A.; Bridges, J.E. *Interaction of a Rectangular Jet with a Flat-Plate Placed Parallel to the Flow*; AIAA Paper 2013-2184; NASA/TM-2013-217879 (E-18684); AIAA: Reston, VA, USA, 2013. [CrossRef]
25. Seiner, J.M.; Manning, J.C. Supersonic Jet Plume Interaction with a Flat Plate. In Proceedings of the International Power Lift Conference, Santa Clara, CA, USA, 17–19 March 1987.
26. Ibrahim, M.K.; Sawai, T.; Obase, K.; Mori, K.; Nakamura, Y. Experimental Investigation of Screech Tone Characteristics of Jet Interaction with a Flat Plate. *AIAA J.* **2009**, *47*, 2031–2038.
27. Berland, J.; Bogey, C.; Bailly, C. Numerical study of screech generation in a planar supersonic jet, Physics of Fluids. *Am. Inst. Phys.* **2007**, *19*, 075105.
28. Menter, F.R. Two-Equation Eddy-Viscosity Turbulence Models for Engineering Applications. *AIAA J.* **1994**, *32*, 1598–1605.
29. Grinstein, F.F.; Margolin, L.G.; Rider, W.J. *Implicit Large Eddy Simulation*; Cambridge University Press: Cambridge, UK, 2007.
30. Butcher, J.C. *Numerical Methods for Ordinary Differential Equations*; John Wiley: Hoboken, NJ, USA, 2003. [CrossRef]
31. Tam, C.K.W.; Webb, J.C. Dispersion-relation-preserving finite difference schemes for Computational Aeroacoustics. *J. Comput. Phys.* **1993**, *107*, 262–281.
32. Bogey, C. ; Bailly, C. A family of low dispersive and low dissipative explicit schemes for flow and noise computation. *J. Comput. Phys.* **2004**, *194*, 194–214. [CrossRef]
33. Kennedy, C.A.; Carpenter, M.H. *Comparison of Several Numerical Methods for Simulation of Compressible Shear Layers*; NASA Technical Report NASA-97-TP3484; NASA: Washington, DC, USA 1997.
34. Werner, H.; Wengle, H. Large-eddy simulation of turbulent flow over and around a cube in a plate channel. In *Turbulent Shear Flows 8*; Springer: Munich, Germany, 1993; Volume 8.
35. Mihaescu, M.; Semlitsch, B.; Fuchs, L.; Gutmark, E. Assessment of turbulence models for predicting coaxial jets relevant to turbofan engines. In Proceedings of the Conference on Modelling Fluid Flow (CMFF 12), The 15th International Conference on Fluid Flow Technologies, Budapest, Hungary, 4–7 September 2012.

36. Bogey, C.; Bailly, C.; Juve, D. A shock-capturing methodology based on adaptative spatial filtering for high-order nonlinear computations. *J. Comput. Phys.* **2002**, *228*, 1447–1465. [CrossRef]
37. Brentner, K.S.; Farassat, F. An Analytical Comparison of the Acoustic Analogy and Kirchhoff Formulations for Moving Surfaces. *AIAA J.* **1998**, *36*, 1379–1386. [CrossRef]
38. Ffowcs-Williams, J.E.; Hawkings, D.L. Sound Generation by Turbulence and Surfaces in Arbitrary Motion. *Philos. Trans. R. Soc. Lond.* **1969**, *A264*, 321–342.
39. Goldstein, M.E.; Leib, S.J.; Afsar, M.Z. Generalized rapid-distortion theory on transversely sheared mean flows with physically realizable upstream boundary conditions: application to trailing edge problem. *J. Fluid Mech.* **2017**, *824*, 477–512.
40. Bridges, J.E.; Brown, C.A.; Bozak, R.F. Experiments on exhaust noise of tightly integrated propulsion systems. In Proceedings of the 20th AIAA/CEAS Aeroacoustics Conference, Atlanta, GA, USA, 16–18 June 2014; AIAA Paper 2014-2904.

MDPI
St. Alban-Anlage 66
4052 Basel
Switzerland
Tel. +41 61 683 77 34
Fax +41 61 302 89 18
www.mdpi.com

Aerospace Editorial Office
E-mail: aerospace@mdpi.com
www.mdpi.com/journal/aerospace

www.ingramcontent.com/pod-product-compliance
Ingram Content Group UK Ltd.
Pitfield, Milton Keynes, MK11 3LW, UK
UKHW060123010826
14089UKWH00006B/118

9 783036 548982